LIFE HISTORIES OF NORTH AMERICAN WILD FOWL

by Arthur Cleveland Bent

IN TWO PARTS

PART II

Dover Publications, Inc., New York

Published in the United Kingdom by Constable and Company Limited, 10 Orange Street, London W. C. 2.

This new Dover edition, first published in 1962, is an unabridged and unaltered republication of the work first published by the United States Government Printing Office. Part I was originally published in 1923 as Smithsonian Institution United States National Museum *Bulletin 126;* Part II was originally published in 1925 as Smithsonian Institution United States National Museum *Bulletin 130.*

INTRODUCTION

The same general plan has been followed and the same sources of information have been utilized. Nearly all those who contributed material for former volumes have rendered similar service in this case. In addition to those whose contributions have been previously acknowledged our thanks are due to the following new contributors:

Photographs have been contributed by Charles Barrett, W. J. Erichsen, Audrey Gordon, W. E. Hastings, A. B. Klugh, G. M. McNeil, C. W. Michael, J. A. Munro, J. R. Pemberton, J. K. Potter, M. P. Skinner, and F. W. Walker.

Notes and data have been contributed by H. P. Attwater, A. C. Bagg, D. B. Burrows, K. Christofferson, H. B. Conover, M. S. Crosby, S. T. Danforth, A. D. Henderson, R. W. Jackson, J. W. Jacobs, W. DeW. Miller, Catharine A. Mitchell, J. A. Munro, J. R. Pemberton, F. J. Pierce, R. D. Camp, A. J. van Rossem, William Rowan, M. P. Skinner, T. C. Stephens, W. A. Strong, and C. L. Whittle.

THE AUTHOR.

Manufactured in the United States of America

Dover Publications, Inc.
180 Varick Street
New York 14, N. Y.

TABLE OF CONTENTS

LIFE HISTORIES OF NORTH AMERICAN WILD FOWL ORDER ANSERES (PART)

By Arthur Cleveland Bent
of Taunton, Massachusetts

Family ANATIDAE, Ducks, Geese, and Swans

GLAUCIONETTA CLANGULA AMERICANA (Bonaparte)

AMERICAN GOLDENEYE

HABITS

Spring.—With the breaking up of winter in Massachusetts, when the February sun has loosened the icy fetters of our rivers and the ice cakes are floating out of our harbors, the genial warmth of advancing spring arouses amorous instincts in the breasts of the warm-blooded goldeneyes. The plumage of the drakes has reached its highest stage of perfection; their heads fairly glisten with metallic green luster, in sharp contrast with their spotless white under parts; and their feet glow with brilliant orange hues. They must seem handsome indeed to their more somber companions of the opposite sex, as they chase each other about over the water, making the spray fly in ardent combat. They are strenuous, active suitors, and their courtships are well worth watching.

Courtship.—This interesting performance, the most spectacular courtship of any of the ducks, has been fully described in detail by Mr. William Brewster (1911). Rather than attempt to quote from such an exhaustive account, I would refer the reader to this excellent article, which is well illustrated and worthy of careful study. I prefer to quote Dr. Charles W. Townsend's (1910) account of it, which is more concise and yet quite complete; he writes:

One or more males swim restlessly back and forth and around a female. The feathers of the cheeks and crest of the male are so erected that the head looks large and round, the neck correspondingly small. As he swims along, the head is thrust out in front close to the water, occasionally dabbling at it. Suddenly he springs forward, elevating his breast, and at the same time he enters on the most typical and essential part of the performance. The neck is stretched straight up, and the bill, pointing to the zenith, is opened to emit a harsh, rasping double note, *zzee-at*, vibratory and searching in character. The head is then quickly snapped back until the occiput touches the rump, whence it is brought forward again with a jerk to the normal position. As the head is returned to its place the bird often springs forward kicking the water in a spurt out behind, and displaying like a flash of flame the orange-colored legs.

As these courtships begin on warm days in February and last through March, probably many pairs are mated before they migrate to their breeding grounds in April. Doctor Townsend writes me that he saw a pair copulating at Barnstable, Massachusetts, on March 28. Mr. Charles E. Alford (1921) writes:

Though the habit of lying more or less prone upon the water is common to most females of the Anatidae when they desire to pair, the duck goldeneye carries this performance beyond all normal bounds; her behavior on such occasions being, indeed, scarcely less amazing than that of the drake. With neck outstretched and her body quite limp and apparently lifeless, she allows herself to drift upon the surface exactly after the manner of a dead bird. When first I witnessed this maneuver I was completely deceived, for she remained thus drifting toward the shore, and with the male swimming round her for fully 15 minutes before actual pairing took place. This occurred on February 2, 1920, a beautiful springlike day, the whole of that month being unusually mild and sunny.

Nesting.—The American goldeneye, so far as I know, invariably places its nest in a cavity in a tree, preferably in a large natural cavity and often entirely open at the top. Considerable variation is shown in the selection of a suitable nesting site, which depends on the presence of hollow trees. Near Eskimo Point, on the south coast of the Labrador Peninsula, I found a nest on June 10, 1909, in a white birch stub on the bare crest of a gravel cliff over 100 feet above the beach. The stub, which stood in an entirely open place, was 6 feet in circumference and about 18 feet high, broken and open at the top down to about 12 feet from the ground. A female goldeneye flew out of the large cavity, in which were 15 handsome, green eggs on a soft bed of rotten chips and white down. The nest was about a foot below the front edge of the cavity. I have never seen another nest in such an open and exposed situation.

Mr. Brewster (1900) found this species breeding abundantly at Lake Umbagog, in Maine, in 1907, and made some valuable and interesting observations on its breeding habits. About the location of its nest, he says:

All the whistlers' nests which I have examined have been placed over water at heights varying from 6 or 8 to 50 or 60 feet and in cavities in the trunks of large hardwood trees such as elms, maples, and yellow or canoe birches. As the supply of such cavities is limited, even where dead or decaying trees abound, and as the birds have no means of enlarging or otherwise improving them; they are not fastidious in their choice, but readily make use of any opening which can be made to serve their purpose. Thus it happens that the nest is sometimes placed at the bottom of a hollow trunk, 6, 10, or even 15 feet below the hole at which the bird enters, at others on a level with and scarce a foot back from the entrance, which is usually rounded, and from 6 to 15 inches in diameter, but occasionally is so small and irregular that the whistler must have difficulty in forcing its bulky body through. I remember one nest to which the only access was by means of a vertical slit so narrow and jagged that it would barely admit my flattened hand.

In North Dakota, in 1901, we found goldeneyes nesting commonly in the timber belts around the shores of the lakes and along the streams in the Stump Lake region.

The goldeneyes choose for their nesting sites the numerous natural cavities which occur in many of the larger trees. They seem to show no preference as to the kind of tree and not much preference as to the size of the cavity, any cavity which is large enough to conceal them being satisfactory.

The occupied cavity can usually be easily recognized by the presence of one or two pieces of white down clinging to its edges; sometimes considerable of the down is also scattered about on the nearest branches. The first nest that we found, on May 30, was in an exceedingly small cavity in a dead branch of a small elm, about 10 feet from the ground. We heard a great scrambling and scratching going on inside, as the duck climbed up to the small opening, through which she wriggled out with some difficulty and flew away. I measured the opening carefully and found it only 3 inches wide by 4½ inches high; the cavity was about 3 feet deep and measured 6 inches by 7 inches at the bottom. The fresh eggs which it contained were lying on the bare chips at the bottom of the cavity, surrounded by a little white down.

On June 1 we explored a large tract of heavy timber on a promontory extending out into the lake for about half a mile, where we located five nests of the American goldeneye. The first nest was about 20 feet up in a cavity in the trunk of a large swamp oak and contained 4 eggs, apparently fresh. The second was in the trunk of a large elm and held only 1 egg, evidently a last year's egg. The third, which held 5 eggs, was in an open cavity in an elm stub about 12 feet from the ground. None of these eggs were taken and doubtless the sets were incomplete. While climbing to a Krider hawk's nest I noticed an elm stub nearby with a large open cavity in the top, which on closer investigation was found to contain a goldeneye's nest with 10 eggs buried in a mass of white down. The stub was about 10 feet high and the cavity about 2 feet deep; the bird was not on the nest, but the eggs proved to have been incubated about one week. A pair of western house wrens also had a nest in the dead branch above the cavity.

The fifth and last nest was found while walking along the shore, by seeing the goldeneye fly out over our heads from a small swamp oak on the edge of the woods. I could almost reach the large open cavity from the ground; the opening was well decorated with the tell-tale down, and at the bottom of the cavity, 2 feet deep, was a set of 14 eggs, in which incubation had begun, and one addled last year's egg, completely buried in a profusion of white down, so well matted together that it could be lifted from the eggs without falling apart, like a soft warm blanket.

In the Lake Winnipegosis region in Manitoba, where large hollow trees are scarce, we found the goldeneyes making the best of rather poor accommodations. We examined four nests all of which were in small, hollow burr oaks (*Quercus macrocarpa*) which were about the only trees in which suitable hollows could be found; the entrances to all of these nests were not over 5 feet from the ground; in some cases the trees were so badly split that the eggs were partially exposed to wind and rain and much of the down from the nests had been blown out onto surrounding trees and bushes; two such nests, found on June 2, with incomplete sets, were at the bottoms of large cavities, practically on a level with the ground in old stubs so badly cracked that the eggs were plainly visible. We were told that the "wood ducks," as they are called, would desert their nests if the eggs were handled, which proved to be true in the only two instances where we tried it.

According to Mr. John Macoun (1909) a nest was found by Mr. William Spreadborough "in a hollow cottonwood log on the ground," near Indian Head, Saskatchewan. He also quotes Mr. G. R. White as saying that the "nest is composed of grass, leaves, and moss and lined with feathers." I have never seen anything but rotten chips and down in a goldeneye's nest, and I doubt if any outside material is ever brought in. Probably the duck does not always take the trouble to clean out a cavity, but lays its eggs on whatever accumulation of rubbish happens to be there. The down is added as incubation advances until a thick warm blanket is provided to cover the eggs, when necessary, during the absence of the bird. I have a beautiful nest of this species in my collection, taken in 1901, with a thickly matted down quilt over the eggs which, though repeatedly handled, has retained its shape and consistency up to the present time.

According to Rev. F. C. R. Jourdain the goldeneye has been frequently induced to nest in nesting boxes in Germany. Mr. A. D. Henderson tells me that he has tried the experiment successfully near Belvedere, Alberta.

The down in the goldeneye's nest is large, light and fluffy; it is practically pure white in color. The breast feathers in it are pure white.

Eggs.—The goldeneye ordinarily lays from 8 to 12 eggs; 5 or 6 eggs sometimes complete the set; I have found as many as 15 and

Mr. Brewster has found 19. Mr. Brewster (1900) thinks that two females sometimes lay in the same nest, and says " several of the rounded, pure white, thick shelled eggs of the hooded merganser are somtimes included in a set of the green, thin-shelled eggs of the whistler."

The eggs of the goldeneye are handsome and easily distinguished from those of any other North American duck except its near relative, the Barrow goldeneye. In shape they vary from elliptical oval to elliptical ovate; a few specimens before me are almost ovate. The shell is thin, with a dull luster. The color is usually a clear, pale " malachite green," varying in the darker specimens to a more olivaceous or " pale chromium green "; various shades of color often occur in the same set. The measurements of 84 eggs, in various collections, average 59.7 by 43.4 millimeters; the eggs showing the four extremes measure **65** by 44, 59 by **45.5**, **48.8** by 43.5, and 59 by **41.2** millimeters.

Young.—Incubation is performed entirely by the female and lasts for a period of about 20 days. Only one brood is raised in a season. The young remain in the nest for a day of two, until they are strong enough to make the perilous descent to the ground or water. Many of the earlier writers have asserted that this, and other species of tree-nesting ducks, carry the young to the nearest water in their bills, but their observations seem to be based largely on hearsay or on insufficient evidence. Mr. Brewster's (1900) study of this species has given us positive evidence to the contrary. Although he personally missed the opportunity of seeing the performance, his trustworthy assistant, R. A. Gilbert, gave the following graphic account of what he saw, when the young were ready to leave the nest:

> At 6.45 the old duck appeared at the entrance to the nest, where she sat for five minutes moving her head continually and looking about in every direction included within her field of vision; then she sank back out of sight, reappearing at the end of a minute and looking about as before for another five minutes. At the end of this second period of observation she flew down to the water and swam round the stub three times, clucking and calling. On completing the third round she stopped directly under the hole and gave a single loud cluck or call, when the ducklings began scrambling up to the entrance and dropping down to the water in such quick succession as to fall on top of one another. They literally *poured* out of the nest much as shot would fall from one's hand. One or two hesitated or paused for an instant on reaching the mouth of the hole, but the greater number toppled out over the edge as soon as they appeared. All used their tiny wings freely, beating them continuously as they descended. They did not seem to strike the water with much force.
>
> While this was going on the old duck sat motionless on the water looking up at the nest. When the last duckling dropped at her side she at once swam off at the head of the brood, quickly disappearing in a flooded thicket a few rods away.

Dr. W. N. Macartney (1918) observed a similar performance near Dundee, Quebec; he writes:

On the afternoon of July 7 the old duck was seen at the foot of the tree, standing on the ground. She gave several low quacks or calls, and out of the hole in the tree overhead promptly tumbled about a baker's dozen of fledgling ducks. They were unable to fly, but were sufficiently grown to be able to ease their fall to the earth, and not unlike a flock of butterflies, they came down pell-mell, fluttering and tumbling, some of them heels over head, until they reached the ground, unharmed. The tree was nearly but not quite perpendicular, so they were unable to scramble down. The old bird gathered them in a bunch and piloted them along the fence for some 3 or 4 rods to the river. Down the rocky shore they went and into the water. The old duck then sank low in the water and the ducklings gathered over her back in a compact clump. She took them across the bay to a bed of rushes, some 10 rods distant, where they disappeared from sight.

Very little seems to be known about the food of the young, but probably they are fed largely on insects and soft animal food. Dr. Charles W. Townsend (1913) gives the following account of the behavior of a mother goldeneye and her young on a Labrador stream:

The old bird crouched low in the water, her golden eyes shining very prominently, and uttered hoarse rasping croaks. The young, whose eyes were gray-blue and inconspicuous, at once scattered, diving repeatedly, and disappeared in the bushes, while the mother kept prominently in view within 20 yards of the canoe leading us downstream. After repeatedly swimming and flying short distances ahead of the canoe for half a mile or so, croaking all the time, she disappeared around a bend and undoubtedly flew back to the young. Near at hand the young made no sound, but at a distance a loud beseeching peep was uttered.

Plumages.—The downy young goldeneye is quite distinctively colored and marked; it also has a carriage all its own, for it walks in a more upright position than other young ducks and it carries its head in a more loftly and perky attitude, which gives it a very smart appearance. The upper part of the head, down to a line running straight back from the commissure to the nape, is deep, rich, glossy "bone brown"; the throat and cheeks are pure white, the white spaces nearly meeting on the hind neck; the upper parts vary from pale "clove brown" on the upper back to deep "bone brown" on the rump; these colors shade off to "hair brown" on the sides and form a ring of the same around the neck; the posterior edges of the wings are white, and there is a white spot on each scapular region and one on each side of the rump; the belly is white. The colors become paler with age.

The first feathers appear on the flanks and scapulars and then in the tail while the bird is very small. According to Millais (1913):

Three nestlings hatched by Mr. Blaauw, at Gooilust, in Holland, on June 20, 1908, began to show feathers on the scapulars on July 18th. On August

8 they were completely feathered except for the flight feathers, which were just beginning to grow. At this date the irides were chocolate brown and the legs and toes yellowish. On August 25 the young birds were able to fly.

Early in the fall, as soon as the young birds have attained their full growth, the first winter plumage begins to develop. This plumage in the male is entirely different from the adult plumage and closely resembles that of the female. The young male may be distinguished from the female by its decidedly larger size; it also has less gray on the breast (which decreases toward spring), the back is darker gray, the head is darker and more or less mottled with dusky, and there is a more or less distinct suggestion of the white loral spot, which increases toward spring. This plumage is worn all through the first winter and spring, with slight and gradual changes toward maturity by a limited growth of new feathers; the head becomes darker and greener, the loral spot whiter, and the scapulars are changed. Individuals vary greatly in the time and extent of these changes. I have a young male in my collection, taken on May 27, which is still in the first winter plumage. In July the young male passes into the eclipse plumage, in which it can be distinguished from the adult by the wings, which are not molted until later. The change from the eclipse into the adult winter plumage is very slow in young birds, lasting well into the winter, and it is not until this molt is completed that old and young birds become indistinguishable.

The adult male assumes a semieclipse plumage late in July or in August, involving principally the head and neck, which becomes brown and mottled like that of the young male; the white loral spot partially disappears; the scapulars resemble those of the young male, and there are brownish feathers in the flanks. This is followed by a complete molt into the winter plumage, which is sometimes prolonged until late in the fall, but not so late as in the young bird.

The molts and plumages of the female are parallel with those of the male, but old and young birds are not so easily recognized. I believe that specimens showing the orange zone in the bill and the well-marked black band across the white space in the wing are old birds. The white neck of the adult female is acquired during the first spring.

Food.—While with us on the coast the goldeneye feeds largely on small mussels and other mollusks, which it obtains by diving in deep water or by dabbling in the shallows near the shore, it feeds to some extent also on the seeds of eel grass (*Zostera marina*). The stomach of a bird taken by Dr. John C. Phillips (1911) in a lake in Massachusetts "contained seeds of pondweed, water lily, bayberry, and burr reed, buds and roots of wild celery, and bits of water boatmen, and dragonfly nymphs."

On the Pacific coast Mr. W. L. Dawson (1909) found it feeding on mussels, crabs, marine worms, and on the remains of decayed salmon. On inland streams it may often be seen in the rapids chasing young trout fry or other small fish; tadpoles, fish spawn, and the larvae of insects are also eaten. Audubon discovered it hunting for cray-fish in the clay banks of our inland rivers. Throughout the interior, in fresh-water lakes and streams, it lives largely on vegetable food; it feeds on a great variety of aquatic plants, such as teal moss (*Limnobium*), flags (*Iris*), duckweed, pondweed, water plantain, and bladderwort, according to Doctor Yorke (1899).

Behavior.—The flight of the goldeneye is exceedingly swift and strong. About its breeding grounds among the lakes and streams of eastern Canada it is very active on the wing, circling high in the air about the lakes or flying up and down the streams above the tree tops, singly or in pairs, the female usually leading; it seems to show some curiosity or anxiety as to the intentions of the intruder, for it often repeats its flight again and again over the same course. The vibrant whistling of its wings in flight is audible at a long distance and has earned for it the popular name of "whistler" or "whistle-wing."

Mr. W. L. Dawson (1909) has thus graphically described it:

> Of all wing music, from the droning of the rufous hummer to the startling whirr of the ruffed grouse, I know of none so thrilling sweet as the whistling wing note of the goldeneye. A pair of the birds have been frightened from the water, and as they rise in rapid circles to gain a view of some distant goal they sow the air with vibrant whistling sounds. Owing to a difference in wing beats between male and female, the brief moment when the wings strike in unison with the effect of a single bird is followed by an ever-changing syncopation which challenges the waiting ear to tell if it does not hear a dozen birds instead of only two. Again, in the dim twilight of early morning, while the birds are moving from a remote and secure lodging place to feed in some favorite stretch of wild water, one guesses at their early industry from the sound of multitudinous wings above, contending with the cold ether.

When migrating, goldeneyes travel in small flocks usually high in the air. When rising from a pond they usually circle about for a few times, gradually climbing upward, and fly off at a considerable height; even on the seashore they are seldom seen flying for any distance close to the water. They can usually be recognized by their short necks, large heads, and stout bodies, as well as by the large amount of white in their plumage. This latter character has given them the name of "pied duck" or "pie bird" among the natives of the eastern Provinces.

The goldeneye is an expert diver; and although at times it uses both wings and feet under water, its method of diving, with wings pressed close to the sides, shows that it generally uses its feet alone. Dr. Arthur A. Allen writes to me that he has "seen goldeneyes using their wings, half spread, when feeding normally." When undisturbed it dives with great ease; the bill is pointed forward until it touches the water, when the bird slips out of sight without an effort, causing hardly a ripple. But when alarmed it plunges forward and downward with great vigor, cleaving the water as it does so.

Mr. F. S. Hersey timed a goldeneye diving and found that it dove with great regularity, remaining under for 21 seconds and on the surface for 13 seconds between dives. Although it usually feeds in rather shallow water, it can dive to great depths in search of shellfish if necessary; for this reason it is called "le plongeur" by the French residents of southern Labrador.

J. G. Millais (1913) narrates the following interesting incident, illustrating its power as a plunger:

No ducks are more bold in the "headers," they will take from the clouds when pursued by a raptorial bird. I was collecting birds one day in February, 1882, on Loch Leven, the Inverness-shire sea loch, when I heard the sound of goldeneye, accompanied by a peculiar hum of something passing through the air. On looking up I was just in time to see the interesting spectacle of a peregrine making a stoop at three goldeneyes. The ducks at this moment were high, I should say 80 yards in the air, and closed their wings as they heard or saw the peregrine coming, and dropped as if shot to the surface of the water. On striking the water there was no pause, they just passed out of sight, rising nearly 100 yards away, and flying low over the water. The peregrine, after its unsuccessful "stoop," did not pursue them. Like the long-tailed duck, but scarcely with the same skill in starting, the goldeneye has the power of opening its wings immediately on reaching the surface of the water, and commencing to fly. I have seen other ducks act in a similar manner when chased by peregrines, but none displayed such promptitude or fell from such a height as did these goldeneyes.

He says further:

In clear water it is easy to note the powerful strokes of the legs of these ducks, which seem to beat with great rapidity under water and much power. The stroke is more or less parallel to the wings; the head is held out straight in front. I have watched for hours the male goldeneye that lived for three years on the island below Perth bridge, and used to find his food at the bottom of the river in some 8 to 10 feet of water. In summer this water was as clear as crystal, and from the bridge above the observer could note every movement on the part of the bird. It always proceeded to a depth of 8 to 10 feet of water, and began to dive. On reaching the bottom, it at once commenced to turn the stones over with the bill, and from under these various water insects were found or caught as they attempted to escape. Sometimes it would find a small batch of young fresh-water mussels, and these it would devour very quickly one after the other, like a duck taking grain out of a pan. It never stayed under water more than a minute, even when finding food abundant in one spot, but came up, rested a moment or two on the surface, and dived again. All food was swallowed where it was found, and small pebbles and fairly large stones were pushed over in the search. Several times I saw the bird just move a flat stone. It would go all around it and try it from every point. If unsuccessful it would come to the surface and rest awhile, and then go down again for another effort. In a lake the goldeneye will dive in perpendicular position, but in flowing water it dives in a slant against the stream or tideway. Their bodies are very light, and bounce up to the surface like a cork immediately they cease to push downward with the feet. In still water the goldeneye often dives in circles to get to the bottom.

The goldeneye is not much given to vocal performances. The courtship note of the male has been described by Doctor Townsend (1910) as "a harsh, rasping, double note, *zzee-at*, vibratory and searching in character." Elon H. Eaton (1910) says that the male when startled or lost has a sharp *cur-r-rew*. Edward H. Forbush (1912) credits the female with "a single whistling peep." And Ora W. Knight (1908) has "heard the parents utter a low-pitched quack to call their young." M. P. Skinner says that "the quack of this duck seems harsher than the mallard's."

Game.—During the four years that I lived on the coast our most interesting winter sport was whistler shooting. Long before daylight we braved the winter's cold and pushed out our skiff to our blind among the ice cakes. We wore white nightgowns over our clothes, white caps and gloves, and sometimes had our gun barrels whitewashed, for the goldeneyes are very wary birds and it is necessary to remain motionless and invisible to be successful. The wooden decoys are placed, as soon as it is light enough to see, in some convenient open space, preferably off the mouth of some fresh-water creek. The blind is made of ice cakes or snow, high enough to conceal the gunners. With the coming of the daylight birds begin to move; large gray gulls are seen flapping slowly up the bay to feed on the mud flats; a flock of black ducks flies out from an open spring hole where it has been feeding all night. The winter sunrise is beautiful, as the rosy dawn creeps up from the cold, gray sea and sends a warm glow of color over floating ice and banks of snow. Our eyes are trained seaward to catch the first glimpses of incoming whistlers. At last a black speck is dimly discerned in the distance against a pink cloud; on it comes straight toward the blind, and we recognize it as an old cock whistler, the advance guard of the morning flight; he circles, sets his wings and scales down over the decoys; in our eagerness we betray ourselves by a sudden movement; he sees us and scrambles upward into the air to escape, but it is too late, the guns speak and the first kill is scored. Soon a small flock of five birds comes in, the shrill whistling of their wings sending a thrill of pleasure through our chilled veins; they scale down toward the decoys, but see the blind, wheel, and fly off without offering us a shot; they settle in the water away off among the floating ice and it is useless to stalk them. We have been too conspicuous to the keen eyes of the birds and must conceal ourselves better; so we pile up more ice around the blind and keep more quiet. Better luck follows in consequence, for the ducks decoy well, if their suspicions are not aroused, and during the next two hours we have good sport. By the time the early morning flight is over, an hour or two after sunrise, we have had enough of it and are glad to return home with a small bag of the keen-witted goldeneyes.

Winter.—To the residents of New England the hardy goldeneye, or "whistler" as it is more often called, is known chiefly as a winter resident or an early spring and late fall migrant, mainly along the seacoast now, though formerly, when less persecuted by gunners, it was often seen in inland ponds, where it is now seldom seen. It is an exceedingly wary and sagacious bird, soon learning to desert dangerous localities, but frequenting freely and regularly such places as the Back Bay basin in the city of Boston, where it is free from molestation.

On the coast, goldeneyes spend their days playing or feeding off the beaches, just beyond the breakers, swimming about among the ice cakes or flying into the tidal estuaries to feed. At night they usually fly off shore, where they can sleep in safety, bedded on the open ocean. They leave the marshes or ponds near the sea at, or within a few minutes of, sunset.

Goldeneyes linger to spend the winter as far north as they can find open water, in the interior as well as on the coast. In the swift rapids and open air holes of our large rivers they find congenial resorts as far north as Iowa, where they congregate in thousands. E. S. Currier (1902) says of the winter habits of the goldeneye on the Mississippi River:

The goldeneyes are very playful and, as spring approaches, noisy. The swift current is constantly forcing them toward the ice at the lower end of the pool, so that they are obliged to take wing and go to the other end of the air hole frequently. They rise on rapidly beating wings, the clear whistling ringing across the dark water and white ice fields, and scurrying upstream in irregular groups, drop in again with a noisy splash. This drifting down and flying back again seems to be enjoyed as much by the ducks as is coasting by the children.

Each group of arrivals is received with many bows and much flapping of wings by the ones on the water, and the penetrating cry of the drakes "*speer*" "*speer*" reaches to a great distance. It is a scene of great activity from daylight until darkness sets in, and makes winter less dreary to the birds of this locality.

The greatest movements take place about sundown when they all head for a favorite air hole (usually the largest) on whirring wings. Here they settle in with much bustle and confusion, playing and feeding until darkness sets in. They spend a great part of the night on thin, new ice at the edge of the open water. As a rule, unless migration is on or the ice is running, there is little movement during the night; but frequently you hear the noisy whistling of the wings of some belated or disturbed bird, soon followed by the distant splash as it strikes the water again.

As winter abates and the increasing warmth causes the ice to give way, followed by the great break-up as the ice goes out, the duck is at its best. The moving ice fields then keep them on the watch, and as the open water they are in narrows, they spring up and fly over the grinding, churning mass, drop into the next clear space upstream.

The instant they hit the water they go to playing, chasing each other, and diving to great distances. At times a part of the flock will rise and depart for some distant open water, soon followed by others, and then perhaps by the remainder, and that particular place will be deserted by them for the time being. Again they will congregate in an open space with the ice rumbling and roaring around them on all sides, seemingly loath to leave, but when another change takes place in the ice and a block sweeps toward them they are forced to leave.

Goldeneyes are often common on the larger northern lakes as long as they remain unfrozen; sometimes they are caught in the ice or perish through inability to find open water; but they are so hardy and such strong fliers that they do not suffer so much in this way as some other species.

M. P. Skinner has sent me the following notes on the winter habits of the goldeneye in the Yellowstone National Park:

The goldeneye is a winter visitor in the proportion of one male to three females. Usually these ducks frequent the larger lakes and streams, but I once found some in a pool of Pelican Creek under the lee of a high bank, and frequently on the reservoir near Mammoth where they dive for their food. Once I flushed a single bird from an irrigation ditch 6 feet wide. In winter, the only time they are at all common, they frequent the streams (Gardiner, Firehole, and Gibbon Rivers) kept open by hot water from hot springs and geysers. They are seldom seen on shore or standing on stones, although I have seen them on the edge of the ice along the Lamar River.

These birds are wilder than the more common Barrow goldeneye; they are here so short a time that they remain exceptions to the general rule that the wildfowl become extremely tame under the absolute protection afforded. Whenever they see me approaching they will swim together in a dense flock. They like swift water and are experts at shooting down the rapids. They are at times associated with buffleheads and mallard; sometimes this goldeneye and the Barrows are together, but more often the two species keep apart. Possibly rivalry of males extends to their cousins, but this is a weak explanation, for the Barrow males are often amicable among themselves when in small flocks containing both sexes.

The European goldeneye, which is supposed to be subspecifically distinct from our bird, may be added to our list on the strength of the capture of a female, supposed to be of the European race, on St. Paul Island, Alaska, on November 27, 1914, reported by Dr. G. Dallas Hanna (1916). He says of this specimen:

It is the same size as specimens from the Commander Islands and China; and while these are somewhat larger than birds from the Atlantic coast region of Europe, they are smaller than those from continental North America.

DISTRIBUTION

Breeding range.—The North American form breeds mainly north of the United States, entirely across the continent. South to Newfoundland (Humber and Sandy Rivers), northern New Brunswick (Northumberland County), central Maine (Washington to Oxford Counties), New Hampshire (Umbagog Lake and Jefferson region), northern Vermont (St. Johnsbury), northern New York (Adirondacks), northern Michigan (Neebish Island, Sault Ste. Marie), northern Minnesota (Lake County); northern North Dakota (Devils Lake), northwestern Montana (Flathead Lake and Glacier National Park), and the interior of British Columbia.

North to the limits of heavy timber in central Alaska (Yukon Valley), southern Mackenzie (Fort Rae, Great Slave Lake), the southwest coast of Hudson Bay (York Factory), and the northeast coast of Labrador (near Nain). Replaced in northern Europe and Asia by a closely allied race.

Winter range.—Cold coasts and large lakes south of frozen areas. On the Atlantic coast commonly from Maine to South Carolina; more rarely north to the Gulf of St. Lawrence and Newfoundland and south to northern Florida (Wakulla County). Rarely to the Gulf coasts of Mississippi, Louisiana, and Texas. On the Pacific coast from the Commander and Aleutian Islands to southern California (San Diego) and casually to central western Mexico (Mazatlan). On the Great Lakes (Michigan, Erie, and Ontario). Irregularly north in the interior to southern British Columbia (Okanogan Lake), northwestern Montana (Teton County), and the valleys of the Missouri and Mississippi Rivers, as far as Nebraska and Iowa; and south to Colorado (Beasley Lake and Barr Lake) and Arkansas (Big Lake) and occasionally to Arizona (Tucson) and Texas (Galveston and Corpus Christi).

Spring migration.—Northward and northwestward and away from the coasts. Early dates of arrival: Southern Maine, inland, March 27; Quebec, Montreal, March 19; Ontario, Ottawa, February 14; Minnesota, Heron Lake, March 14; Manitoba, southern, March 29; Alberta, Edmonton, April 6; Montana, Great Falls, March 9; Mackenzie, Fort Resolution, May 7, and Fort Simpson, April 28; Alaska, Nulato, May 3.

Average dates of arrival: Southern Maine, inland, April 5; Quebec, Montreal, April 4, and Lake Mistassini, May 3; Ontario, Ottawa, April 12; Minnesota, Heron Lake, March 25; northern North Dakota, April 20; southern Manitoba, April 21. Leaves the Massachusetts coast by May 1 or earlier.

Fall migration.—Southward and southeastward and toward the coasts.

Early dates of arrival: Massachusetts, October 8; Virginia, Alexandria, October 8. Average dates of arrival: Massachusetts, Woods Hole, November 15; Virginia, Alexandria, October 26; Iowa, Keokuk, November 24. Late dates of departure: Quebec, Montreal, November 7; Manitoba, Aweme, November 10.

Casual records.—Four records for Bermuda (April 10, 1854; December 29, 1874; February 5, 1875; and January 22, 1876). Two records for Pribilof Islands (May 6, 1917, and January 1, 1918). Accidental in Cuba and Barbados.

Egg dates.—North Dakota: Nineteen records, May 10 to June 11; 10 records, May 21 to June 1. Manitoba: Five records, June 2 to July 5. Labrador Peninsula: Three records, May 3 to June 30.

GLAUCIONETTA ISLANDICA (Gmelin)

BARROW GOLDENEYE

HABITS

This species has been well named, the Rocky Mountain goldeneye, for outside of the vicinity of the Continental Divide in the northern States and in southern Canada it is nowhere in this country an abundant species at any season. It is so rare throughout most of its American range that few ornithologists have ever seen it in life.

For this reason it is not strange that it was overlooked by some of the earlier writers and that until recently its distribution was so poorly understood. Wilson makes no mention of the species, and it was entirely overlooked by Audubon, who may have regarded it as a summer plumage of the common species. Even Coues (1874) refers to it as "the most northerly species of the genus, having apparently a circumpolar distribution, breeding only (?) in high latitudes, and penetrating but a limited distance south in interior"; the question mark is his and it is interesting to note that his doubt was removed by finding it breeding in the Rocky Mountains of Montana. It is now known to breed in these mountains as far south as Colorado, east in Canada to the north shore of the St. Lawrence, in Greenland, and in Iceland; but it is far from being circumpolar, for it occurs on the Old World continent only as a straggler; and it is not known to breed north of the Arctic Circle.

Courtship.—J. A. Munro (1918) has given us an interesting account of this species, in British Columbia, from which I quote as follows:

> The birds first begin to appear on Okanogan Lake early in March, but are not plentiful until the small mountain lakes are free of ice, early in April. The lakes selected for courtship, and later for the rearing of the young, are usually quite open and free of tules; hence the goldeneyes are always con-

spicuous and much easier to study than ducks that breed in the sloughs and hide their young in the thick vegetation. Generally by the 15th of April each little lake has its flock of courting goldeneyes, often 30 or 40 on a sheet of water of 50 acres extent or less. In these flocks adults and immatures are present in about equal numbers. The young of either sex do not breed until the second year, and do not assume their breeding dress until the second fall after they are hatched; that is, when they are over a year old.

The courtship display is witnessed in the flocks just prior to their splitting up into pairs. It is attended by much solemn bowing on the part of the drake, with a frequent backward kick, sufficiently strong to send a jet of water several feet into the air. His violet head is puffed out to the greatest possible extent, and altogether he is a handsome bird as, in a frenzy of sexual excitement, he swims up to the soberly attired duck. Sometimes the entire flock will commence to feed as if at a given signal, and again all the birds will simultaneously take wing and circle about the lake several times before once more splashing down to resume their courtship.

He also contributes the following note:

Two mated pairs in a small lake in the hills were under observation for two hours. The males acted as if extremely jealous of each other and on several occasions left their mates and engaged in spirited encounters. They rushed together over the surface with much splashing, and when about to meet rose upright and buffeted each other with their wings. A female, whose mate had been killed on an adjacent pond, flew into this lake and immediately one of the mated drakes left his mate and dropped in beside her, when he began bowing, but the strange female did not respond. The male then dove several times trying to rise beneath her, at which she flew some distance away and the drake then rejoined his mate.

M. P. Skinner contributes the following:

Early in the winter the ducks outnumber the drakes; but as spring approaches the proportion becomes more equalized. By February 1 the tendency of the flocks to pair off becomes noticeable; courting begins about the same time and lasts until June in some cases. Almost all the flocks are broken up by April 1. While the drakes do most of the "chasing" and "dancing," the females sometimes go through similar movements. The drake swims across the water with jerky motions, not necessarily toward the duck, occasionally an extra kick raises the breast above the surface and at the same time the bill is pointed up and opened and shut twice. Then the neck is stretched backward until the head rests on the lower back, then forward to the normal position, ending with a kick backwards that throws up a little spurt of water. The duck is frequently chased by the drake, with his head and neck stretched out horizontally in front and almost on the water surface.

Nesting.—Dr. T. M. Brewer (1879) published an interesting paper on this species which added greatly to our knowledge of it at that time. Edwin Carter, "who was probably the first to actually secure the nest and eggs of this species within the limits of the United States," sent to Doctor Brewer considerable information about the breeding habits of the Barrow goldeneye in Colorado. He says:

They nest in hollow trees, and it is surprising to see to what small cavi-

(1877) I have examined a great many trees, and every one that had a suitable opening either contained an occupant or indicated recent nesting by eggshells and other marks.

Maj. Allan Brooks (1903) found the Barrow goldeneye breeding quite commonly in La Hache Valley, in the Cariboo district of British Columbia. He says:

One set of eggs was taken from a hole in a dead Douglas fir, 50 feet from the ground, probably the deserted nest of a flying squirrel. The tree stood about 400 yards from the nearest water. The eggs (7) at this date (17th June) contained large embryos. I saw another nesting hole but was unable to reach it. The female brought 14 young ones out from this.

Mr Munro (1918) says of the nesting habits of this species:

By May 1 all breeding birds are mated and scattered over the country, seldom more than one or two pairs on a lake. The Barrow goldeneye shows a marked predilection for lakes that are strongly alkaline, even if they are poor in aquatic vegetation and in the midst of an open country with the nearest timber a half mile or more away. Such lakes are rich in small crustaceans, the chief food of this duck, and no doubt the lakes are occupied on account of the food provided, without reference to the availability of nesting sites.

An abandoned flicker's hole is usually selected for the nest, frequently in a dead yellow pine, for in this tree decay is rapid, and the hole soon becomes much enlarged. One can generally tell if the hole is occupied, by the fragments of down adhering to the rough bark at the entrance. The tree is often so much decayed that a single tug at the bark near the hole will remove the whole adjacent surface, exposing the gray-green eggs where they lie in the clinging soft down. It is rather hard to locate the nest when the tree selected by the bird is in heavy timber a half mile or more from the lake, but when the female is sitting it may be done by making an early morning trip to the lake, remaining under cover, and waiting for her to come to the lake to feed. She generally arrives between 9 and 11 and immediately joins the drake. After splashing and preening her feathers, she feeds most industriously for perhaps an hour and then flies directly back to the nest. I include here data for three nests taken in the Okanogan region.

Okanogan, British Columbia, May 12, 1916. A nest containing 11 fresh eggs was found in the hayloft of a deserted log barn, on the shore of a lake. The eggs were placed in a hollow scooped in the straw under a heavy beam which rested on the piled-up straw. The loft was well lighted through the spaces between the logs and by a large opening at one end. This situation is of course, most unusual, but it had apparently been used some years before the nest was found. I had seen broods of young on this lake in previous years, when I was not able to find the nest. The birds would generally alight on top of a chimney in an unused house close by before flying into the barn.

Farneys Lake, Okanogan, May 31, 1912. A nest with seven partly incubated eggs was placed in a large cavity in a yellow-pine stump, standing in 8 inches of water on the shore of the lake. The cavity containing the eggs was 18 inches above the water, and the eggs were in plain view of a person standing several feet away.

Rollings Lake, May 26, 1917. A nest containing seven fresh eggs was found in an old fir stub standing in 18 inches of water near the shore of the lake. The top of the stub had rotted out to a depth of 2 feet and the eggs were at the bottom of this cavity. Down could be seen protruding through a small hole in the stub, a few inches above the eggs.

John G. Millais (1913) says of the breeding habits of this species in Iceland:

Barrow's goldeneye arrive at their breeding places about the end of March, in flocks, and at once proceed to pair. I have been unable to discover any ornithologist who has seen the courtship display of this species, but I have little doubt that when we are able to procure specimens alive from Iceland, and keep them in good health, we shall find that it is much the same as that of the common goldeneye. The nest is usually placed in a hole in the bank of a stream flowing into a lake, in a hole in the lava rocks close to the water, or on some low island under bushes of dwarf willow, dwarf birch, amongst coarse grass or low scrub, such as *Empetrum nigrum* or *Azalea procumbens.* I found two nests just tucked in under large stones, and not 2 feet above the level of the stream. They are also said to nest in the turf walls of the sheep shelters.

He also quotes Riemschneider as saying:

The nest was always placed in more or less of a hollow, in natural hollows of the rocks, in covered-over cracks in the lava, or, as already mentioned, in the outer walls of peat shelters, erected for sheep, where a few blocks of peat have been taken out to form a nesting place, and even, and that not seldom, inside the shelter, in which case the food rack or a place like it would serve as a nesting place; as exit for flight the door of the shelter would in such a case be used. Such customs have given rise to the Iceland names for the species. In the natural hollows, holes in the rocks, fissures, etc., the nest is placed now in the foreground, now so far inside that you could not reach to it from the entrance opening, but were obliged to lift off the stones covering it for this purpose. Whilst as a rule the position of the nest is to be found approaching the level of the surface of the ground, I saw a nest in the Kalvastrond which was built in a hollow in the lava at more than twice a man's height. In the nest trough, which was formed to begin with in the food racks of the stalls, by pulling together dry grass stalks and other remnants of food round the nest, there was a very ample, delicate lining of whitish down, which had a very small admixture of fine, dry parts of plants. The eggs, 12 to 15 in number, and only exceptionally more, are distinguished from other ducks' eggs by their pure, blue-green color, are rather bulgy in shape, and have a smooth, not very shining shell.

From the above accounts it will be seen that the Barrow goldeneye prefers to nest in hollows; the absence of suitable trees in Iceland forces it to select other cavities; but in this country it seems to nest in situations similar to those chosen by the common goldeneye and, like that species, it lines the nest cavity with pure white down, scantily at first, but more profusely as incubation advances; probably no other material is brought into the cavity, but undoubtedly whatever material it finds there is not wholly removed. The down in the nest is indistinguishable from that of the common goldeneye.

Eggs.—The set consists of anywhere from 6 to 15 eggs, but probably the usual number is in the neighborhood of 10. The eggs are practically indistinguishable from those of the common golden-eye, though they may average slightly larger. The shape varies from elliptical ovate or elliptical oval to nearly oval. The color varies from "deep lichen green" to "pale olivine" or "pale glass green"; when freshly laid some of the darkest eggs may approach a pale shade of "malachite green." The measurement of 79 eggs, in various collections, average 61.3 by 44 millimeters; the eggs showing the four extremes measure **66.6** by 42, 61 by **47.2**, **57.3** by 42, and 61.9 by **41.4** millimeters.

Rev. F. C. R. Jourdain, who has studied the Barrow goldeneye in Iceland, writes me that incubation is performed by the female alone and that it is said to last for four weeks; he says that the female sits very closely and at times has to be removed by force, if one wants to see the eggs. The male remains close at hand during incubation.

Mr. Munro, however (1918), writes:

Immature males leave the country with the adult males in May, soon after the females have begun to brood their eggs. I have never seen an adult male at this season. Mr. Allan Brooks is of the opinion that the males go directly to the coast at this time.

Young.—He says of the young:

May 22 is the earliest date on which I have seen the young, and by August 1 they are full grown. At this time they are remarkably tame, allowing an approach to within a few yards and then, if alarmed, swimming to the middle of the lake, rather than taking wing. This fearlessness is characteristic until the shooting season opens in September, when they soon become wary. At this time, the birds rise from the water as one approaches, but almost invariably circle about the lake several times and then fly toward anyone standing on the shore, thus affording an easy shot. By the last week in October, when the common goldeneye, redhead, and scaups are returning from the north, the last of the Barrow goldeneyes have left.

Plumages.—The downy young of the Barrow goldeneye is very much like that of the common goldeneye. The upper half of the head, from below the eyes, and the hind neck are deep "bone brown" or "seal brown"; the upper parts are "bone brown," relieved by white on the edge of the wing and by scapular and rump spots of white; the lower half of the head and the under parts are white; there is a brownish gray band around the lower neck.

The plumage changes are similar to those of the common golden-eye. Of the development of the plumage in the young male, Mr. Millais (1913) says:

In November the new inner scapulars appear, and these at once give a character to the identification of the species. The black portion of the inner scapulars is much extended in Barrow's goldeneye, whereas in the common goldeneye it is confined to the margin of the feathers. At this date, too, the first white feathers come in between the bill and the eye. These increase in number throughout the winter, whilst numbers of pure white feathers come on the chest until the brown of immaturity disappears. Thus the advance of plumage continues to take place until March, when the young male has gained a considerable portion of its first spring dress, which is more or less similar to the adult male, except that the black and white scapulars are never fully attained, nor are the hind neck or flanks complete. The wings and tail still show the bird to be immature until the latter part of June or early July, when the usual complete molt takes place, the whole of the bird going into a partial eclipse similar to the adult male. In September the eclipse is shed, and all traces of immaturity have disappeared, so that in the following month, or, more correctly speaking, November, the bird is adult, at about 16 months.

The same writer describes the semieclipse plumage of the male, as follows:

At the beginning of July the adult male undergoes a fairly complete change to an eclipse plumage, although the white feathers in front of the eye are never completely lost. In this month the head and neck become a somewhat dirty gray brown, very light in the throat; the flanks, hind neck, and upper mantle, also portion of the lower neck and chest are brown with gray edgings; mantle, scapulars, brown, with light brown or gray edgings or tips; the whole bird now resembles a somewhat dirty-looking female, but its sex can easily be recognized by its superior size, small white feathers on the head, and by the wings, which always remain the same, which, with the tail and part of the back and tail coverts, are only molted once in the season. The adult male has scarcely assumed its eclipse dress before it again commences to molt into winter plumage, and in the case of all these ducks the process of change at this season may be said to be practically continuous.

The female undergoes the same sequence of plumage to maturity as the female goldeneye, attaining full maturity at an age of about 15 months. The females of these two species are very difficult to distinguish at any age. Mr. Millais (1913) says:

The characters of the female Barrow's goldeneye, apart from superior size, are the black back and tail, blackish head and longer crest, and general difference of a more intensified black and white. The yellow bill spot is also more extensive.

William Brewster (1909*a*) has made an exhaustive study of this subject, and I would refer the reader to his excellent paper on it. I would refer the reader also to a paper on this subject by H. F. Witherby (1913) and a still more exhaustive treatise by Maj. Allan Brooks (1920).

Food.—The food of the Barrow goldeneye seems to be the same as that of the common species. Dr. F. Henry Yorke (1899) records it as feeding on minnows and small fishes, slugs, snails, and mussels, frogs, and tadpoles, in the way of animal food; he has also found in its food considerable vegetable matter, such as teal moss, blue flag, duckweed, water plantain, pouchweed, water milfoil, water starwort, bladderwort, and pickerel weed. Mr. Munro (1918) says:

The feeding habits of the two species of goldeneye are identical. Both species are greatly attracted by the small crawfish lurking under large stones in shallow water. While hunting these shellfish, the ducks work rapidly along the shore, diving every few minutes, to probe under the edges of the large stones. They invariably try to submerge even if the water is not deep enough to cover their backs, and I have never seen them dipping as redheads and scaups frequently do. One can follow the goldeneye's movement as it encircles the large stones, by the commotion on the surface and by frequent glimpses of the duck's back. In shallow water, the birds remain below from 15 to 20 seconds, the crawfish being brought to the surface to be swallowed. By the end of winter the feathers on the forehead are generally worn off, through much rubbing against stones in this manner of foraging. When feeding in deep water, over the beds of *Potamogeton*, they stay in the same place until satisfied. In such places the small snails and crustacea that attach themselves to the stems of *Potamogeton* form their chief food, but little vegetable matter being taken beyond what is eaten with the shells. The small shellfish are swallowed while the birds are below the surface of the water, unlike the procedure followed with the larger crawfish. Their stay under water is of fairly uniform duration, ranging from 50 to 55 seconds. At the beginning of the dive the tail is raised and spread to its full extent.

Behavior.—He writes of its habits:

As far as I have been able to observe, there is no difference in the flight of the two species of goldeneye. Both have the same clumsy way of rising, and of flying close to the surface before attaining any speed; once under way they travel swiftly, and one's attention is held by the distinctive, musical whistle of their wings. Both the Barrow goldeneye and the American goldeneye are less gregarious than others of our ducks with the exception of the mergansers. I have never seen the Barrow goldeneye in large flocks except in the mating season. When feeding, two or three birds together are the rule, and five or six the maximum number noted.

Dr. D. G. Elliott (1898) writes of its behavior:

I have found it at times quite numerous on the St. Lawrence near Ogdensburg, and have killed a goodly number there over decoys, and some specimens, procured on these occasions, are now in the Museum of Natural History in New York. The two species were associated together on the river, and I never knew which one would come to the decoys, but I do not remember that both never came together, unless it might be the females, for, as I have said, it was difficult to distinguish them without an examination.

The birds would fly up and down the river, doubtless coming from, and going to, Lake Erie, stopping occasionally in the coves to feed, and floating down with the current for a considerable distance, when they would rise and fly upstream again. My decoys were always placed in some cove or bend of the stream where the current was least strong, for I noticed the birds rarely settled on the water where it was running swiftly. This duck decoys readily in such situations, and will come right in, and if permitted, settle among the wooden counterfeits. They sit lightly upon the water and rise at once without effort or much splashing. The flight is very rapid, and is accompanied with the same whistling of the wings so noticeable in the common goldeneye. In stormy weather this bird keeps close to the banks, seeking shelter from the winds. It dives as expertly as its relative, and frequently remains under water for a considerable time. The flesh of those killed upon the river was tender and of good flavor, fish evidently not having figured much as an article of their diet.

Maj. Allan Brooks writes that "the note is a hoarse croak. They have also a peculiar mewing cry, made only by the males in the mating season."

Mr. Millais (1913), who studied this species extensively in Iceland, writes of its behavior there as follows:

On the water the male of this species looks a larger, clumsier, and blacker bird than the common goldeneye. It seemed to me that it sits higher on the water, and was a bird that commanded instant attention. In summer the males, which, when the ducks have begun to sit, consort in small parties of two to six, or more, are exceptionally tame, and will permit an approach to within a few paces, if the observer moves slowly to the banks of the river where they are feeding or resting. In rising to fly they are somewhat clumsy, and run along the surface with considerable splashing, but they did not seem to me to make nearly so much noise in flight as the common species. The "singing" or "ringing" note is heard, but it is neither so loud nor so metallic. On June 27 the males were still in their breeding dress. A few seem to keep on the river near the nesting females, as if for form's sake, but the majority were resorting to the great lake of Myvatn, where the parties seemed to increase in size day by day. Females, with young, often floated past me while I was trout fishing, and once I had to draw in my line to prevent hooking a too confiding mother. Whilst watching males on feed, it struck me that they were less expert than the common goldeneye, and had more difficulty in getting under water. There was more noise and splash to get under, but once below the surface they seemed to be skilled performers of the highest order. I saw them more than once, from the high bank where my tent was pitched, feeding in exactly the same manner as the common species, turning over all the small stones, and probing beneath all large ones, and into holes. They stay less time under water in shallows than in the deep water of the lake, the time occupied being a half to one minute. On the river they reappeared again and again at the same spot, only pausing for a moment's rest and splash down again, whereas on the lake they would often keep moving forward in their dives, and take up a fresh position every time. They will stay and fish in very rough streams, edging into the current and out again as soon as they rise, but do not like such wild places as the harlequin.

In Iceland their enemies seem to be Richardson's skua, *Stercorarius parasiticus* L., which regularly attacks the females of all diving ducks and seizes their young, and the Iceland falcon, *Falco rusticolus islandicus*, which kills a few of the adults. There was hardly a morning or evening when I stayed at Myvatn, in June-July, 1889, that we did not see one or other of these two species harrying the ducks. Sitting in the tent to escape the awful plague of flies, a sudden roar of startled ducks would be heard, and on my going to investigate there was the falcon, with perhaps two young birds in attendance, bearing off some victims of its prowess. None of the ducks seemed to be unusually scared when the falcons passed by, as they often did, by day and night. They crouched on the water or rushed with their broods under the banks and hid as well as possible. It was only after the stoop and kill, when the bird of prey came on to their own level, that there was a general stampede of these ducks in the immediate vicinity of the murder.

Mr. Lucien M. Turner, in his unpublished notes, says:

These goldeneyes are common along the entire coast of Labrador and occur in scattered flocks of two or three to rarely more than a dozen in number. I have never found the nest or eggs of this species and am not positive that it breeds on the coast; although it occurs during the summer season, arriving by the 1st of June, and remains until early November. During the latter part of its stay it frequents the places of swift tide currents where it dives with wondrous celerity to procure its food from the bottom of soundings. Numbers of these birds were seen in a pocket some 2 miles from the mouth of the Koksoak River and on the left band. This place is locally known as Partridge's fishing place. I was camped there while delayed by stormy weather, and early each morning great numbers of these birds frequented the inner portion of the cove. Strange to say, these birds never made the same noise with their wings when they flew into the cove that they did when they flew out. The tempestuous weather and great distance prevented me from securing any specimens in that locality. This was late in September, and these birds congregated in large numbers, for some of the flocks certainly contained 200 individuals and were doubtless preparing to journey to the southward. None were seen after the last of September. I observe that these ducks are nocturnal in their habits and especially noisy toward the approach of day. They search for fresh feeding grounds from daylight to sunrise and appear to be very quiet, unless disturbed, during the midday hours.

Mr. Skinner says, in his notes:

I usually find these ducks by ones and twos and small groups, but once I found a flock of 85 swimming in a compact group off the shore of Yellowstone Lake. When in pairs, it is the female that takes alarm and flushes first. They take great delight in "shooting the rapids"; nothing in the Gardiner River, at least, being too rough for them. They drop down over a fall 3 feet high, and at the bottom go out of sight in the foam and spray, but nevertheless keep right on swimming along. Should they tire of this boisterous sport, they are quick to take advantage of any eddy, or rest behind a bowlder. As a rule these are the tamest of our ducks; on the reservoir and other roadside waters they are unalarmed even while the big autos go thundering past. If I approach a flock too closely, the Barrows swim away, or go coasting down the rapids, instead of flushing as the mallards do. But if they do flush, they go only a little way, come back, and drop down again into the water without hesitation or fear.

I do not believe the Barrows seek the society of other ducks, but common tastes bring them to many other ducks, mostly deep-water ducks, such as mallards, mergansers, buffleheads, ruddy ducks, bluebills, and American golden-eyes. Sometimes the last named and the Barrows are together, but more often the two species keep apart, possibly due to the rivalry of the males.

Flight is low and labored at first, as they rise against the wind, and they are often compelled to kick the water for the first 20 or more wing strokes. I have seen them start and later strike the water again where the average pitch of the rapids was 6 per cent or less. Rising against a strong head wind, they do much better, and then they may fly at a greater height than in calm weather, say of 30 or 40 feet. Once I was passing up a narrow stream with the wind behind me when I found a half dozen Barrows before me. They could not fly up into my face and the canyon was too high and narrow to fly out sideways, so the ducks were obliged to swim down past me and rise behind my back. Apparently they can not jump up as the mallard does so frequently. But once under way, flight is swift and powerful, giving rise to a distinct whistling sound.

These ducks bathe by standing almost erect on the water and rapidly flapping their wings so as to throw the water forward and over them; later by plunging under and shaking themselves at the same time. They usually dress their feathers while on the water, turning far over, first on one side, then on the other, to get at places ordinarily under the surface. Once I found a Barrow preening on a half submerged bowlder.

Winter.—The winter home of the Barrow goldeneye is not far south of its summer range, and its migration is not much more than a movement off its nesting grounds to more satisfactory feeding grounds. It seems to be fairly common on the Gulf of St. Lawrence in winter, where it frequents about the same resorts as the common goldeneye, but farther south on the Atlantic coast it is rare. It winters on some of the large lakes and rivers of the interior, as well as on the Pacific coast.

J. A. Munro (1918) says that "throughout the winter months it is found on the seacoast, in the many sheltered estuaries from Puget Sound to Hecate Strait and Dixon Inlet." He also writes to me "that a few birds winter on the Okanogan River below Okanogan Falls in a rapid stretch of water with strong bottom, where crawfish are very abundant. They are usually in company with the common goldeneye."

Mr. Skinner writes to me:

While the majority migrate on the freezing of the waters, a few remain along the Gardiner and Yellowstone Rivers all winter, and become even tamer than usual, entitling them to be rated as resident (in Montana) at all seasons. Barrows that had wintered along the Gardiner in 1920–21 began leaving about February 25, or about the time that waters elsewhere were beginning to open. First to go were the males, then the females and immatures; until only one was left on March 1, and that one went next day. But cold, freezing temperature brought them back, a female on the 7th and six drakes and five females on the 12th; then they decreased again. By the 24th they were abundant at the outlet of Yellowstone Lake at 1,000 feet higher altitude or just over 7,700 feet above sea level.

DISTRIBUTION

Breeding range.—In North America, a few breed on the Labrador Peninsula from the Gulf of St. Lawrence (Point des Monts) to northern Labrador (Davis Inlet). But the main breeding range is in the vicinity of the Rocky Mountains. East to western Alberta (Banff), northwestern Montana (Glacier National Park) and central northern Colorado (Boulder County). South to southwestern Colorado (Dolores County). West to southwestern Oregon (Crook County and Douglas County), central British Columbia (Okanogan and La Hache Valleys), to the coast of southern Alaska (Chilkat and Sitka), and south central Alaska (Lake Clark). North to northern Mackenzie (Fort Anderson rarely) and Great Slave Lake

(Fort Rae and Providence). Breeds abundantly in Iceland and in Greenland up to 69° or 70° N.

Winter range.—From the Gulf of St. Lawrence southward along the coast regularly to eastern Maine (Washington County), rarely to southern New England and as a straggler beyond. On the Pacific coast from southern Alaska (Wrangell and Portage Bay) to central California (San Francisco Bay). Rarely and irregularly in the interior, south to southern Colorado (La Plata River), and north to southern British Columbia (Okanogan Lake) and northern Montana (Great Falls).

Spring migration.—Dates of arrival: Quebec City, April 14 to 16; Mackenzie, Fort Anderson, June 14. Late dates: Ontario, Toronto, April 18, 1885; North Carolina, near Asheville, May 6, 1893.

Fall migration.—Dates of arrival: Quebec, Montreal, October 23; Connecticut, East Haven, November 14; Massachusetts, Wareham, November 27; District of Columbia, November 22, 1889; Wisconsin, Lake Koshkonong, November 14, 1896.

Casual records.—All records east of the Rocky Mountain region and south of New England must be regarded as casual. Most of these records are based on females, incorrectly identified. The records given above, under migrations, are believed to be authentic, as are also the following: Michigan, Ottawa County, 1907, and Detroit River, April 1, 1905.

Egg dates.—Iceland: Fourteen records, May 19 to June 30; seven records, June 2 to 17. British Columbia: Five records, May 12 to 31. Alberta: Two records, May 28 and 30.

CHARITONETTA ALBEOLA (Linnaeus)

BUFFLEHEAD

HABITS

The propriety of applying the name " spirit duck " to this sprightly little duck will be appreciated by anyone who has watched it in its natural surroundings, floating buoyantly, like a beautiful apparition, on the smooth surface of some pond or quiet stream, with its striking contrast of black and white in its body plumage and with the glistening metallic tints in its soft fluffy head, relieved by a broad splash of the purest white; it seems indeed a spirit of the waters, as it plunges, quickly beneath the surface and bursts out again in full flight, disappearing in the distance with a blur of whirring wings.

Spring.—Although a hardy species and generally regarded as a cold-weather bird, the bufflehead is rather slow in making its spring migration; it follows gradually the retreat of winter, but lingers on the way. Dr. F. Henry Yorke (1899) says:

The first issue of these birds appears in the interior above the frost line late in the spring, a short time before the bluewinged teals arrive; and with the ruddy ducks are the last of the divers to travel northward. They soon depart to the far north, where they are followed by the second and third issues, which scatter over the country before they also follow the advanced flight.

Courtship.—Although the bufflehead lingers in some of our ponds until quite late in the spring and during some seasons is fairly common, I have not been particularly successful in studying its courtship on account of its shyness. On bright, warm days during the latter part of April or early in May the courtship of this species may be studied with some hope of success, though long and patient watching through powerful glasses may be necessary. The males are quite quarrelsome at this season and fight viciously among themselves for the possession of the females. The male is certainly a handsome creature as he swims in and out among the somber females, his bill pointing upward, his neck extended, and his beautiful head puffed out to twice its natural size and glistening in the sunlight. Standing erect he struts about, as if supported by his feet and tail, with his bill drawn in upon his swelling bosom, a picture of pride and vanity, which is doubtless appreciated by his would-be mate. Suddenly he dives beneath her and on coming up immediately deserts her and flies over to another female to repeat the process. He seems fickle or flirtatious in thus dividing his attentions, but perhaps they have not been graciously received or he has been rebuffed. Sometimes he becomes coy and swims away until she shows interest enough to follow him. Eventually he finds the one best suited to him and the conjugal pact is sealed.

Dr. Charles W. Townsend (1916) describes the courtship of the bufflehead as follows:

A group of 35 or 40 of these birds, with sexes about equally divided, may have been actively feeding, swimming together in a compact flock all pointing the same way. They dive within a few seconds of each other and stay under water 14 to 20 seconds, and repeat the diving at frequent intervals. Suddenly a male swims vigorously at another with flapping wings, making the water boil, and soon each male is ardently courting. He spreads and cocks his tail, puffs out the feathers of his head and cheeks, extends his bill straight out in front close to the water and every now and then throws it back with a bob in a sort of reversed bow. All the time he swims rapidly, and, whereas in feeding the group were all swimming the same way in an orderly manner, the drakes are now nervously swimming back and forth and in and out through the crowd. Every now and then there is a commotion in the water as one or more drakes dive, with a splashing of water, only to come up again in pursuit or retreat. As the excitement grows a drake flaps his wings frequently and then jumps from the water and flies low with outstretched neck toward a duck who has listlessly strayed from the group. He alights beside her precipitately, sliding along on his tail, his breast and head elevated to their utmost extent and held erect. He bobs nervously. And so it goes.

Nesting.—The center of abundance of the bufflehead in the breeding season seems to be in the wooded regions of Canada lying west and north of the Great Plains, where it is well distributed and in some places quite common. Sidney S. S. Stansell (1909) says that in central Alberta it is "about as common as the mallard; nearly every small pond has its pair, and some of them two pairs, of this beautiful little duck. When two or more pairs occupy a single pond, the males are usually very pugnacious, often quarreling and trying to drive each other off the pond for hours at a time." A set of 12 eggs in my collection collected by Mr. Stansell, near Carvel, Alberta, on June 11, 1912, was taken from an old flicker's nesting hole, 20 feet from the ground; as the eggs were nearly fresh, there was no lining in the nest except the chips left by the previous occupant.

Herbert Massey has sent me some data regarding a set of 10 eggs in his collection, taken by Mr. W. H. Bingaman, at Island Lake, Saskatchewan, 50 miles west of Prince Albert, on May 28, 1905. The nest was in an enlarged flicker excavation 15 feet from the ground and 2 feet from the broken-off top of a dead poplar tree; the eggs lay 15 inches below the very irregular opening, among rotten wood dust, flicker feathers, and light-colored down of the duck; the tree was 10 yards from the shore of the lake; the female was sitting and was secured. The collector says:

I found this species nowhere abundant in Saskatchewan, and two sets are the complete result of my season's work among this species. I took one other set at Montreal Lake, 95 miles northeast of Island Lake, and the nesting site was almost a duplicate of this one; it also contained 10 eggs.

Maj. Allan Brooks (1903) thus describes the breeding habits of the bufflehead in the Cariboo district of British Columbia:

Almost every lake has one or more pairs of these charming little ducks. Unlike Barrow's goldeneye, the nests were always in trees close to or but a short distance away from water. These nests were invariably the deserted nesting holes of flickers, and in most cases had been used several years in succession by the ducks. The holes were in aspen trees, from 5 to 20 feet from the ground, and the entrance was not more than 3¼ inches in diameter. The number of eggs ranged from 2 to 9, 8 being the average; in color they resemble old ivory, without any tinge of green. I have several times seen the eggs of this duck described as "dusky green," but these have evidently been the eggs of some species of teal. The female bufflehead is a very close sitter, never leaving the nest until the hole was sawed out, and in most cases I had to lift the bird and throw her up in the air, when she would make a bee line for the nearest lake, where her mate would be slowly swimming up and down unconscious of the violation of his home. In many cases the eggs had fine cracks, evidently made by the compression of the bird's body when entering the small aperture.

J. A. Munro found a nest—

in an old pileated woodpecker's hole near the top of a yellow pine stub, without bark or branches and broken off 40 feet from the ground. It stood among young Murray pines and poplars 20 yards from the shore of the lake. Down adhering to the entrance of the hole identified the nest as belonging to the bufflehead. The nest had been used the previous year by buffleheads and during the following winter by flying squirrels. This was indicated by a quantity of old bufflehead down, with fragments of eggshell adhering, lying at the bottom of the tree. To this down the flying squirrels had added a quantity of moss. Apparently the female bufflehead had removed the mixture of moss and down before commencing to lay.

Where trees are scarce, as in certain parts of Saskatchewan, the bufflehead is said to lay its eggs in a hole in a bank, after the manner of the belted kingfisher, using for this purpose the deserted burrow of a gopher near some small lake. Such cases must be exceptional, however. The down in the nest is small, light, and flimsy; it is "pallid purplish gray" in color, with small white centers. The breast feathers in it are pure white.

Eggs.—The bufflehead lays from 6 to 14 eggs, but 10 or 12 seems to be the usual number. M. P. Skinner writes to me that he "encountered a female on Yellowstone River with 16 well-grown young, and, as I could not find another parent, I have always assigned this extraordinary brood to the one mother."

The shape is bluntly ovate, elliptical ovate, or nearly oval. The shell is smooth and slightly glossy. The color varies from "ivory yellow" to "marguerite yellow" or "pale olive buff." The measurements of 86 eggs in various collections, average 48.5 by 34.7 millimeters; the eggs showing the four extremes measure **55** by 37, 53.5 by **38**, and **40** by **26** millimeters.

Plumages.—As might be expected, the downy young bufflehead closely resembles the young goldeneye in color pattern. The upper parts, including the upper half of the head from below the lores and eyes, the hind neck, the back and the rump, are deep rich "bone brown," with a lighter gloss on the forehead and mantle; the inner edge of the wing is pure white; there is a large white spot on each side of the scapular region and on each side of the rump; and an indistinct whitish spot on each flank. The under parts, including the chin, throat, cheeks, breast, and belly are pure white, shading off gradually into the darker color on the sides of the body and with an indistinct brownish collar around the lower neck.

In the juvenal plumage the sexes are much alike and resemble the adult female, except that the colors are duller and browner and the white cheek patches smaller than in the adult. The young male soon begins to differentiate from the young female, by increasing faster in size and by the development of the head, with a more conspicuous white patch. The progress toward maturity is very slow, and even in May the young male has only partially assumed the adult plumage; the tail and much of the body plumage has been renewed, the wings are still immature, and the head has acquired large white patches, but only a few of the purple feathers of the adult. A complete summer molt occurs in July and August, after which the adult plumage is gradually assumed and is completed in November and December. The young male thus becomes adult at an age of 17 or 18 months. The young female makes practically the same progress toward maturity.

I have never seen the eclipse plumage of the bufflehead, but according to Mr. Millais (1913) both old and young males assume "a fairly complete eclipse, resembling a similar stage of plumage in the goldeneye."

Food.—The bufflehead obtains its food by diving, usually feeding in small companies so that one or more remain on the surface to watch for approaching dangers while the others are below; sometimes only one remains above, but it is only rarely that all go below at once; should the sentinel become alarmed it communicates in some way with the others which come to the surface and all swim or fly away to a safe distance.

Neltje Blanchan (1898) describes its feeding habits very neatly, as follows:

> A bufflehead overtakes and eats little fish under water or equally nimble insects on the surface, probes the muddy bottom of the lake for small shellfish, nibbles the seawrack and other vegetable growth of the salt-water inlets, all the while toughening its flesh by constant exercise and making it rank by a fishy diet, until none but the hungriest of sportsmen care to bag it.

Audubon (1840) says:

> Their food is much varied according to situation. On the seacoast, or in estuaries, they dive after shrimps, small fry, and bivalve shells; and in fresh water they feed on small crayfish, leeches, and snails, and even grasses.

Ora W. Knight (1908) says that in the inland regions of Maine "they feed on chubs, shiners, small trout fry, and other small fish. Along the coast their diet is very similar." Other writers include in their food various small mollusks, crustaceans, beetles, locusts, grasshoppers, and other insects.

Dr. F. Henry Yorke (1899) lists the following genera of plants among the food of the bufflehead: *Limnobium*, *Myriophyllum*, *Callitriche*, *Utricularia*, and *Pontederia*. Vegetable matter seems to form only a small part of the food of this species and is eaten mainly during the summer.

Behavior.—The flight of the bufflehead is exceedingly swift and direct, generally at no very great elevation above the water, and is performed with steady and very rapid beats of its strong little wings. It rises neatly and quickly from the surface of the water and sometimes from below it, bursting into the air at full speed. When alighting on the water it strikes with a splash and slides along the surface. It generally travels in small irregular flocks made up largely of females and young males, with two or three old drakes.

It is one of the best of divers, disappearing with the suddenness of a grebe, with the plumage of its head compressed and its wings closely pressed to its sides. It can often succeed in diving at the flash of a gun and thus escape being shot. Under water it can swim with closed wings swiftly enough to catch the small fish on which it feeds so largely; but I believe that it often uses its wings under water for extra speed. It can also dive to considerable depths to secure its food from the bottom. Charles E. Alford (1920) says that it seldom or never dives to a greater depth than 2 fathoms. He timed a large number of dives and found that the period of immersion varied from 15 to 23 seconds, usually it was about 20, and the interval between dives varied from 4 to 8 seconds.

The following incident, related by Mr. Samuel Hubbard, jr. (1893), shows that its diving powers are sometimes taxed to the limit:

A broad, sandy bay made in from the harbor, the upper end of which terminated in a shallow slough about 18 inches deep. I waded across and was proceeding toward the beach, when my attention was attracted by a small bufflehead duck (*Charitonetta albeola*) commonly called butterball. He was swimming around in the slough and obtaining his food in the way common to his kind, by diving and picking up that which came his way. With an admiring glance at his beautiful plumage, I was about to pass on, when one of those pirates of the air, a duck hawk (*Falco peregrinus anatum*) came in sight. Without hesitating an instant, he made straight for my little friend and swooped at him. His long talons came down with a clutch, but they closed on nothing, for the duck was under the water. Undaunted the hawk hovered overhead, and as the water was clear and shallow, he could follow every movement of his prey. Again the duck came up; the hawk swooped to seize him, each move being repeated in quick succession and each dive becoming shorter and shorter. It was evident that the poor little hunted creature was getting desperate, for the next move he made was to come out of the water flying. The hawk promptly gave chase. There was some clever dodging in the air, but the duck, frightened and tired, soon saw that his swift pursuer was getting the best of it, so he closed his wings tight against his body and dropped like a stone into the water and plunged out of sight. Now comes the beginning of the end. While he was under water he either saw the hawk hovering over him or else he became bewildered, for he came again out of the water flying. Like lightning the hawk struck; there was a muffled "squawk," and the tragedy was ended.

Dr. J. G. Cooper (1860) writes:

I once saw a male that I had just wounded dive in clear water, and, seizing hold, by its bill, of a root growing under water, remain voluntarily submerged for almost five minutes, until he supposed all danger past, when, again ascending to the surface, he paddled off with great rapidity.

I cannot remember that I have ever heard its note, but Dr. D. G. Elliot (1898) says that "it utters at times a single guttural note, which sounds like a small edition of the hoarse roll of the canvas-back and other large diving ducks."

L. R. Dice (1920) says of its notes:

As a rule they are silent; only on a few occasions were any calls heard. Once while driving a pair in front of a blind to take pictures, the male and female became separated. Then the male gave a squeaky call, which the female answered with a hoarse *quack, quack*, and the male immediately flew to her side. At another time a female alighted in an eddy of the river and gave a low call, *quk, quk, quk, quk, quk, quk, quk*, slowly, and the male in a few minutes appeared and alighted beside her.

Fall.—In the fall this species is one of the later migrants, coming along with the hardier winter ducks. It is not of much account as a game bird; its body is small and its flesh is not particularly desirable, as it feeds so largely on animal food. It is, however, often very fat, from which it has derived the name of "butterball." It is apparently not regularly hunted or sought for by gunners, but is often shot while hunting other species.

Winter.—W. L. Dawson (1909) says of the habits of this species on the coast of Washington:

Buffleheads are among our most abundant ducks in fall and winter throughout the State. They are found alike in swift rivers and on placid mill ponds. Brackish pools and tide channels, tide flats, and tossing billows, all are alike to these happy and hardy little souls. Perhaps the greatest number, however, are found upon the bays and shallower waters of Puget Sound. They associate chiefly in little flocks of from half a dozen to 50 individuals, and they venture inshore, as often as they dare, to feed on the rising tide. When they reach us in October they are fat as butter (whence, of course, "butterball"), but they have gained their flesh on the cleaner feeding grounds of the northern interior. On a fare of fish and marine worms, which they obtain in salt water almost entirely by diving, their flesh soon becomes rank and unprofitable.

M. P. Skinner has sent me the following notes on the habits of buffleheads in Yellowstone Park:

As a rule these ducks are on the larger waters such as Yellowstone Lake and Yellowstone River, resorting to smaller lakes and ponds at very infrequent intervals. In stormy times, they are driven to quieter waters, but even then prefer to find a calm spot near shore of Yellowstone Lake or a back water on the river. When on streams, they do not care for the swifter water. They are fond of sitting on sand bars, gravel bars, mud points, and on the beaches about Yellowstone Lake. Many of these birds are to be seen all winter in openings in the ice on the lake, and on the river where kept ice-free by the current, along the Firehole River kept open by hot geyser water, and on the Gardiner River below the mouth of warm Boiling River. They are social and keep together in small, compact flocks. Similar food habits bring them in close contact with some ducks and the limited open water in winter with others. In these ways, they are often with mergansers, Barrow goldeneye, American goldeneye, canvasbacks, redheads, bluebills, coot, grebes, mallard, green-winged teal, baldpate, shovellers, ruddy ducks, geese, and swans. On the sandy beaches, they are often near spotted sandpipers, or pelican, if not actually with them.

From the above it will be inferred that the bufflehead winters as far north as it can find open water in the interior. On the coasts it is found as far north as New England and British Columbia. It seems to prefer to be on or near the frost line and does not go much south of the United States in winter.

Dr. Leonard C. Sanford (1903) writes:

The butterball is common on both coasts, and is fond of shallow, sandy bays, frequenting the tide rips and mouths of rivers, remaining through the coldest weather. A few years ago this bird was common all along the New England shore. Large numbers wintered on the sound between New Haven and Stratford, where the coast is shallow and sandy, early in the morning leaving the outer flats and feeding up the rivers. It was a simple matter to shoot them on their flight, as they came over the bars, low down and usually in the same course. Recently the butterball seem to have largely disappeared from the New England coast, though still common on bays farther south.

DISTRIBUTION

Breeding range.—Mainly in the interior of Canada. East to northern Ontario (probably), said to breed in New Brunswick and recorded once as breeding in southeastern Maine (Washington County). Has been recorded as breeding formerly, and probably only casually, south to southeastern Wisconsin (Pewaukee Lake), northern Iowa (Clear Lake, etc.), and Wyoming (Meeteetse Creek); but it evidently does not breed now anywhere south of the Canadian border except in northern Montana (Milk River, Flathead Lake, and Meagher County). West to central British Columbia (Sumas and southern Okanogan). There is a recent breeding record for California (Eagle Lake). North to west central Alaska (Kuskokwim River and the Yukon Valley), northern Mackenzie (nearly to the mouth of that river), Great Slave Lake (Forts Rae and Resolution) and the southwestern coasts of Hudson Bay and James Bay.

Winter range.—Mainly in the United States, entirely across the continent. South casually to Cuba; commonly to South Carolina, northern Florida (Leon County), the Gulf coasts of Louisiana and Texas; and less commonly or rarely to central Mexico and Lower California (San Quintin). North to the Aleutian and Commander Islands, the Alaska Peninsula, southern British Columbia (Okanogan Lake), northwestern Montana (Tetlow County), the Great Lakes (Michigan, Huron, and Ontario) and the coast of Maine.

Spring migration.—Northward and inland. Early dates of arrival: Pennsylvania, Renovo, February 29; Massachusetts, March 11; Ontario, Ottawa, March 26; Illinois, Shawneetown, February

27; Iowa, southern, March 1; Minnesota, Heron Lake, March 6; Alberta, Alix, April 24; Alaska, Cross Sound, April 13, and Craig, May 9; Pribilof Islands, May 19. Average dates of arrival: Pennsylvania, Renovo, March 18; Massachusetts, March 11; New Brunswick and Nova Scotia, March 22; Indiana, central, March 2; Illinois, northern, March 21; Michigan, southern, March 31; Ontario, southern, April 7, and Ottawa, April 24; Nebraska, Omaha, March 15; Iowa, southern, March 22; Minnesota, Heron Lake, March 26; South Dakota, central, April 8; Manitoba, southern, April 25; Saskatchewan, Osler, May 2; Mackenzie, Fort Simpson, May 11. Late dates of departure: North Carolina, Smith's Island, April 15; Massachusetts, Taunton, May 2; California, Los Angeles County, April 22.

Fall migration.—Gradual southward movement, mainly inland. Dates of arrival: Ontario, Ottawa, October 26; Nova Scotia, Sable Island, November 7; Massachusetts, October 8; Rhode Island, October 13; Pennsylvania, November 10. Late dates of departure: Alaska, Fort Reliance, October 7; Quebec, Montreal, November 1; Ontario, Ottawa, November 8.

Casual records.—Accidental in southern Greenland (Godhaven, 1827, and Frederikshaab, 1891). Two records for Bermuda (November, 1875, and December, 1845). Accidental in Cuba, Porto Rico, and Hawaiian Islands (Maui).

Egg dates.—British Columbia: Six records, May 15 to June 4. Alberta, Saskatchewan, and Manitoba: Five records, May 31 to June 11. Alaska: Two records, June 6 and 12.

CLANGULA HYEMALIS (Linnaeus)

OLDSQUAW

HABITS

Spring.—Oldsquaws, or long-tailed ducks, as I should prefer to have them called, are lively, restless, happy-go-lucky little ducks, known to most of us as hardy and cheery visitors to our winter seacoasts, associated in our minds with cold, gray skies, snow squalls, and turbulent wintry waves. Though happy and gay enough during the winter, the height of their merriment is seen in the spring or when the first signs of the breaking up of winter announce the coming of the nuptial season and arouse the sexual ardor of these warm-hearted little ducks. Early in the spring they become more restless than ever, as they gather in merry flocks in the bays and harbors of the New England coasts; the males, in various stages of budding nuptial plumage and fired with the enthusiasm of returning passion, gather in little groups about some favored female in fantastic postures, rushing, flying, quarreling, and filling the air with their musical love notes. If noisy at other times, they are still more so now, vieing with each other to make themselves seen and heard; it is a lively scene, full of the springtime spirit of joy, love, and life.

The increasing warmth of the April sun and the stimulus of the courtship activities start the restless birds on their spring migration by various routes to their summer homes on Arctic shores. While cruising along the north shore of the Gulf of St. Lawrence on May 23, 1909, we saw what was probably the last of the spring migration on the south coast of Labrador; between the Moisie River and Seven Islands we saw numerous large flocks of from 50 to 200 birds each, perhaps 1,000 or 1,500 birds in all. They were noisy and very active, on the water and flying about high in the air, and many seemed to be in summer plumage or changing into it. They were evidently preparing for their overland flight to Hudson Bay; we saw none farther east along the coast, and, from what we could learn from the natives, we inferred that very few migrate around the eastern coast of Labrador and that the bulk of the flight passes overland northwestward to Hudson Bay. Oldsquaws no longer breed on the south coast of Labrador, where Audubon found them, and probably very few still breed on the northeast coast; I saw none during the summer of 1913 even as far north as Cape Mugford, but I obtained a skin of a male in full breeding plumage from an Eskimo at Okak and saw a set of eggs in Rev. W. W. Perrett's collection, taken at Ramah. Lucien M. Turner's unpublished notes state that "they arrive at the mouth of the Koksoak River as soon as the ice breaks up; this being a variable date, of course influences the time from the 20th of May to the 10th of June. Their first appearance is usually in the smaller fresh-water ponds and lakes from which the ice earlier disappears, long before the sea ice in the coves and bays begins to move out." Probably these birds reach this portion of Ungava by the Hudson Bay route rather than by an outside route and through Hudson Straits, which are badly icebound at this season.

There is an extensive northward migration through the interior. E. A. Preble (1908) writes:

In the spring of 1904 I first saw this species at Fort Simpson May 10, from which date it was common. The birds, usually in small flocks, floated down with the current among the ice floes, occasionally rising and winging their way swiftly upstream to regain lost ground. The males played about on the water, chasing each other and uttering their loud, clear notes, which soon became associated in the mind with the long, cool evenings of the Arctic spring, with the sun hanging low in the northwestern horizon. When they are lightly swimming about, the long tails are elevated at an angle of about 45°, and with their striking color pattern the birds present a very jaunty appearance. They are usually rather tame, sometimes rising and coming to meet the canoe, and actually becoming less wild if shot at. When slightly wounded they are among the most expert of divers and are difficult to secure. The males played together considerably before the females arrived, but after that important event their gymnastic and vocal performances knew no bounds.

Dr. E. W. Nelson (1887) says that this is the first of the ducks to reach high northern latitudes in Alaska.

The seal hunters find them in the open spaces in the ice off St. Michaels from the 1st to the 20th of April, and the first open water in shore is sure to attract them. After their arrival it is no uncommon occurrence for the temperature to fall to 25° or 30° below zero, and for furious storms of wind and snow to rage for days, so the first comers must be hardy and vigorous to withstand the exposure.

W. Elmer Ekblaw has sent me the following attractive account of the arrival of the oldsquaws in northern Greenland:

The distinctive resonant call of the oldsquaw announces the arrival of real spring to the far Arctic shores. The earlier herald, the snow bunting, comes while yet the land is covered with snow, while still the ice lies solid and unbroken throughout the wind-swept fjords, and while yet the midnight sun is new and even the noonday is chill; the oldsquaw comes when the snow is gone from the valleys and the slopes, and the first saxifrage and willow have burst into blossom, when great dark leads and pools of open water break the white expanse of fjord ice, and when the sun at midsummer height is warm at midnight as at noon. When the challenging clarion of the oldsquaw rings out over the great north, spring has come.

The first few oldsquaws come winging noisily in along the open leads the first week in June. The males predominate in the first flocks, but by mid-June, when the immigration is at its height, the females appear to be as numerous as the males. Until the inland ponds and lakes are open, the oldsquaws frequent the leads and open pools in the fjord ice; they are most numerous along the shore where the tidal crack opens up the ice, and where the warmer fresh water coming down the slopes and hills melts away the ice foot from the shore. In this along-shore water they apparently find more food, or food more to their liking—generally crustacea and small fish.

From mid-June to the 1st of July the ice on the inland lakes melts away rapidly. Just as soon as the belts of open water show along the banks, the oldsquaws begin to leave the sea and enter the fresh-water lakes. In flight the female always leads, distinguishable by her plumper, dull-colored body and shorter tail; in swimming the male usually leads. Every lakelet has its pair or two of oldsquaws. Some of the pairs seems to be mated when they arrive in the northland, but many mate after their arrival. In the last two weeks of June the local mating season is at its height. Because the males are the more numerous, the rivalry for the females is very keen, and the fighting continuous. During this time, the lake-dotted plains and valleys in the flats about North Star Bay resound with the clamor and din of mating oldsquaws, and the birds may be seen flying swiftly from pool to pool, from point to point. The Eskimo consider them the swiftest flying birds of the northland.

Courtship.—The season of courtship is much prolonged or very variable with this species. This is one of the few ducks that have a spring molt and nuptial plumage; the time and extent of the molt varies greatly in different individuals; and the flush of sexual ardor is probably contemporary with the change of plumage. Consequently some individuals molt and mate before they start on the spring migration and others not until after they have reached their breeding grounds. Many males apparently do not acquire a full nuptial plumage during the whole summer and probably do not mate that season; I have seen males in Alaska in midsummer in practically full winter plumage and in various stages between that and full nuptial plumage; the full development of the latter seems to be rather rare.

John G. Millais (1913) writes:

As previously stated, the actual courtship of the male is generally aroused and brought about by the sexual desire of the female, and amongst ducks the females are very irregular as to the time of their coming into season. Thus only one or perhaps two females in a large flock may be well advanced in their summer plumage and their breeding instincts, and these are the special objects of desire of all the males. I have noticed a bunch of 8 or 10 females swimming apart and not a male going near them, whilst 10 or 15 males will crowd round some particular female and lavish upon her all their arts of charm. The most common attitude of the male in courtship is to erect the tail, stiffen the neck to its fullest extent, and then lower it toward the female with a sudden bow, the bill being held outward and upward. As the head curves down, the call is emitted. Sometimes the head is held out along the water before the female, who herself often adopts this attitude, or makes a "guttering" note of appreciation with head held in close to the body. Another common attitude of the male is to throw the head right back till it almost touches the scapulars, the bill pointing to the heavens. As the bird throws the head forward again the call is emitted. Many males will closely crowd round a female, all going through the same performance. It is not long before a fight starts amongst the males, so that the lady of the tourney is in the midst of a struggling clamorous mass of squabbling knights, each endeavoring to show his qualifications to love by his extravagant gestures or strength. To add to the confusion, any male long-tails in the neighborhood are sure to hear the noise and come flying in all haste to take part in the jousts. Even males still in full winter plumage will come and be almost, if not quite, as active as the rest. They advance with all haste, swaying from side to side, their sharp-pointed wings being only arrested when almost above the contest. Then they close the wings in mid-air and dash into the fray with all their ardour. So impetuous and gallant are males of this species that they will chase each other for long distances, falling often in the sea and sending the spray flying; down they go under the water and emerge almost together on the surface to continue the chase in mid-air. I have twice seen a male when flying seize another by the nape and both come tumbling head over heels into the sea in mad confusion.

In Mr. Hersey's notes, made at the mouth of the Yukon, I find the following account of the later courtship, observed on June 19, 1914, which shows that the birds are not all paired when they arrive on their breeding grounds.

To-day I watched the courtship of a pair of this species. A male and two females were swimming about in a small pond. As the male began calling another female joined the party. The male, however, paid all his attentions to one of the females and did not notice either of the others. As this favored bird swam slowly about her admirer followed, his head drawn in close to his shoulders and the bill pointing downward, the tip not more than an inch or two from the surface of the water. When within 6 or 7 feet of the female he would raise his head till it pointed straight upward and give a succession of deep notes not unlike the baying of hounds heard at a distance. These notes were usually in series of four or five, and with each the head was thrown still farther back. The long tail was carried straight out horizontally as a general thing, or depressed slightly, but at times was elevated to an angle of about 45°. After calling, the bird dropped his head to its former position close to the water. All this time the female kept up a low quacking. After several of these sallies she would face her suitor, extend the neck and head flat upon the water and swim toward him, turning when within a foot or two, and pass him whereupon he turned and the performance began all over again. After about an hour of this the female took flight closely followed by the male, and after circling the pond several times both birds returned to the water. The other two females had retired to the other end of the pond where they had been quietly feeding, but the male now chased both of these birds out of the pond and then returned to the remaining bird. I have several times seen a female flying closely pursued by two males, all three twisting and turning so that it was difficult for the eye to follow them, but the female always kept in the lead.

Mr. Ekblaw, in his notes, writes:

On July 1, 1914, near the little Eskimo village, Umanak, on North Star Bay, I was able to study the mating antics of the oldsquaw at close range. The day was ideally calm, clear, and mild, and the birds were unusually stirred by the "cosmic urge." Just across the steep ridge southeast of the house lies the broad, terraced, flood plain of a creek which now is a mere remnant of a stream unquestionably much larger in the past. The lowest terrace of this plain is one of such imperfect drainage that ponds and swales are numerous. About the shallow ponds and wet swales grasses and sedges grow in abundance. The ponds teem with tiny animal life. Here the oldsquaws breed and nest in numbers. It is one of their favorite haunts. I was concealed among the rocks of a ledge some 50 yards from a rather large, comparatively deep pond, where the ice was melted along the edges. In the open water, on the edge of the ice, and along the grass-covered banks, seven pairs of oldsquaws were distributed, and two males were struggling strenuously for an unmated female. The paired birds were swimming contentedly about the pool, busily preening their feathers on the ice, or sleeping cosily on the banks; the unmated female and the two males were strenuously sweeping the water or chasing over the pond in swift zigzag flight. Whenever one of the males attempted to mate with the female, the other invariably attacked, much to the evident displeasure of the female, who would then take quick wing, noisily protesting, and pursued by both males. Settled in the pool the males fought fiercely, splashing and churning the water. Neither seemed able to vanquish the other, and when I left my hiding place they were still struggling.

Nesting.—Audubon's (1840) historic account of finding the oldsquaw breeding on the southern coast of Labrador is now ancient history, but it is worth quoting as a record of conditions which no longer exist. He writes:

In the course of one of my rambles along the borders of a large freshwater lake, near Bras-d'Or, in Labrador, on the 28th of July, 1833, I was delighted by the sight of several young broods of this species of duck, all carefully attended to by their anxious and watchful mothers. Not a male bird was on the lake, which was fully 2 miles from the sea, and I concluded that in this species, as in many others, the males abandon the females after incubation has commenced. I watched their motions a good while, searching at the same time for nests, one of which I was not long in discovering. Although it was quite destitute of anything bearing the appearance of life, it still contained the down which the mother had plucked from herself for the purpose of keeping her eggs warm. It was placed under an alder bush, among rank weeds, not more than 8 or 9 feet from the edge of the water, and was formed of rather coarse grass, with an upper layer of finer weeds, which were neatly arranged, while the down filled the bottom of the cavity, now apparently flattened by the long sitting of the bird. The number of young broods in sight induced me to search for more nests, and in about an hour I discovered six more, in one of which I was delighted to find two rotten eggs.

The following extracts from Mr. L. M. Turner's notes will illustrate the nesting habits of this species in Ungava, where it probably still breeds regularly:

To the freshwater ponds, around whose margins high grasses and sedges grow, the oldsquaws resort to build their nests. The nest is composed of grass stalks and weeds to a depth of 2 or 3 inches, in which the first egg is deposited. This is covered with down plucked from the bird, and to it a greater quantity is added as the number of eggs increases. The eggs, in the clutch, vary from 5 to 17; 9 to 13 being the usual number. The distance of the nest from salt water varies greatly, for I have seen a nest, on a small, low island, not more than a yard from the edge of the water, and again I have found one that was more than half a mile, where a large lake on the level of the higher land was connected by a swampy tract with the head of a long and deep but narrow gulch through which a small stream coursed.

A few pairs breed on the Pribilof Islands. I saw a pair on the village pond on St. Paul Island, but did not have time to hunt for its nest. Mr. William Palmer (1899) found a nest and nine eggs about 40 feet from this pond on June 12.

It was placed on a little hillock on the killing ground. When flushed, about 10 feet off, the bird flew directly to its mate in the pond. Leaving the eggs, I returned soon, to find that she had been back, had covered them completely with down and dry, short grass, and returned to the pond.

The main breeding grounds of the oldsquaw, in North America, extend from the mouth of the Yukon all around the coast of northern Alaska and all along the Arctic coast of the continent and the northern islands to Greenland. Throughout the whole of this region it is one of the most abundant ducks. The nests are widely scattered over the tundra, but are more often found near the shores of the small ponds or on little islands in them; the nest is usually well concealed in thick grass or under small bushes, but it is often found in open situations. The female is a close sitter, and when she leaves the nest she covers it so skillfully with the dark sooty brown down, grass, and rubbish that it matches its surroundings and is about invisible. It is well that she does so for she has many enemies, wandering natives, roving dogs and foxes, jaegers, gulls, and other nest robbers that are always on the lookout for eggs. Mr. Hersey found a nest near St. Michael on July 5, 1915, containing six eggs; it was in an open situation in short grass 20 feet from the shore of a small pond, and while going for his camera, only a short distance away, some short-billed and glaucous gulls found the nest and destroyed the eggs.

Mr. Ekblaw, in his notes, says:

> The nests are built in small-cup-like hollows, sometimes in the grass near the pools, but more frequently among the rocks at considerable distances from any water. The sites are selected with a view to concealment. The nest is well lined with mottled brown down from the female's breast. When the eggs are covered with the down in the absence of the bird, or when she is brooding them, the nest is well-nigh impossible of detection. The female is not readily frightened from her nest; when she is driven off she simulates injury and distress to draw away the intruder, in the manner common to so many birds. The foxes take toll of the oldsquaw eggs as of the eggs of all the birds.

The down in the oldsquaw's nest is very small; it varies in color from "bister" to "sepia," and has small but conspicuous whitish centers.

Eggs.—The oldsquaw is said to lay as many as 17 eggs, but the set usually consists of from 5 to 7. Only one brood is raised in a season. In shape the eggs are ovate, elliptical ovate, elongate ovate, or even cylindrical ovate; they are often more pointed than ducks' eggs usually are. The surface is smooth, but not very glossy. The color varies from "deep olive buff" to "olive buff" or from "water green" to "yellowish glaucous." The measurements of 139 eggs, in the United States National Museum collection, average 53 by 37 millimeters; the eggs showing the four extremes measure **60** by 39, 58 by **40, 48** by 37.5, and 51.5 by **35.5** millimeters.

Young.—The period of incubation is said to be about three and a half weeks. It is performed by the female alone, but the male does not wholly desert her, remaining in the nearest pond and encouraging her with demonstrations of affection. About the time that the young are ready to hatch he flies away and joins others of his kind on the seacoast. As soon as the eggs are hatched and the young are strong enough to walk, the mother leads them over the perilous journey to the nearest water. Mr. Turner watched a mother bird conducting her brood of 13 young for more than half a mile from a swampy tract through a long and deep but narrow gulch, through which a small stream flowed, down to a cove into which it opened. His notes state:

> The old bird was much disturbed when I came upon her and she pretended to be wounded. She fluttered and waddled about in a frantic manner, but while chasing her I saw the young and could then have easily taken the old

bird in my hand, as she actually fluttered at my feet, so intent was she to withdraw my attention from her young. I retired, and with peculiar call she gathered the young ones and began her march. I followed them to the salt water where the mother seemed frantic with joy, as she flopped around like a tame duck at the approach of rain. The young were not more than two days old and had awaited until they had sufficient strength to undertake the long journey. They took to the water as though they had been accustomed to it for weeks. I must confess that I felt pleased that I did not molest them for I have seldom seen anything that afforded me greater satisfaction than to witness the pleasure evinced by the old bird when she had her young on the bosom of the sea where she felt so secure.

There are so many enemies to be contended with at this critical period that it is a wonder that any of these ducks ever succeed in raising a brood. It is only by good luck in many cases that the nest is not discovered and robbed; and only eternal vigilance and a constant struggle on the part of the devoted mother serves to protect the little ones from their enemies. In the instance just related several occupied fox dens were within a short distance of the nest, yet the eggs were hatched and the young were conducted away in safety.

The following incident is related by Mr. Millais (1913):

I watched a newly hatched brood of long-tailed ducks one day for a long time and noticed that they took very little food for themselves. They caught a few flies, but most of their food was obtained by the mother diving incessantly and bringing up substances from the bottom and placing it before her brood. When she appeared they all kept up a gentle "peeping" sound and kept close together in a bunch, seldom running to catch flies as other young ducks do. After watching these birds for some time I wandered up the river to the Lake of Myvatn to look at a scoter's nest, and on returning witnessed the attack of two Richardson's skuas, a black and a white bellied one, on the same brood of long-tailed duck. The method of attack was exactly the same as I have seen employed by carrion crows in Hyde Park. One skua swooped down and distracted the mother's attention to one side by hovering over the water. The anxious parent opened her bill and gave a series of grating calls. As the marauder came to the level of the water the long-tailed duck with raised crest made a fierce rush of a few yards at it and in this short space of time the second skua swooped down, picked up a nestling, and swallowed it alive, head first. The frantic mother then darted in the other direction when the skua that had first attacked nimbly picked up a duckling and swallowed it whilst mounting into the air. These skuas, which are plentiful at Myvatn, must commit considerable havoc amongst the very young ducks and doubtless constitute their chief enemies. Mr. Manniche, whom I had the pleasure of meeting in Denmark in 1911, tells me that the glaucous gull is equally mischievous in destroying the young of long-tails and king eiders in East Greenland and probably Buffon's skua is another successful pirate.

The young are taught to dive at an early age, but at first they are not very successful. Mr. Hersey's notes state that on July 5, 1915, a female and eight recently hatched young were "seen on a small pond. The female did not fly but swam around the young calling softly to them. At an apparent signal from the mother all would dive, but the young were unable to stay below the surface more than four or five seconds. The parent would then come to the surface and again try to coax them to follow her below. She did not attempt to lead them to the shore." When the young are fully grown and able to fly they all leave the small ponds and sheltered places where they have been living and move off to the shore. This often occurs quite late in the season, for small downy young are often to be found up to the middle of August or later; the hatching date is very variable, as the eggs are so often disturbed and a second or third laying made necessary. The lateness of the fall migration gives a long breeding season and plently of time to make several attempts at raising a brood.

Mr. Ekblaw's notes state that:

While the females are brooding the eggs the males and the sterile females fly and swim about the ponds in promiscuous flocks. These sterile females are restless and active, quite different from the mated females during the mating season. In the mating season the female of a pair usually rests quietly on the bank of the pond, apparently heavy with eggs, while her mate swims about near her. The sterile females are noisy and uneasy.

These sterile females and the males leave the land and the environs of the shore about the 20th of July and seek the outer skerries and open water. The nesting females take their little ducklings down to the salt water as soon as they can toddle along, and from then until they hie themselves away to the southland they spend all their time along the rocky shores in the pleasant coastal bays and fjords or about the icebergs and floes. Very frequently two mother birds join their flocks, and then, when swimming about in the open sea, one mother leads the flock while the other brings up the rear.

The young birds grow fast and quickly develop strength for swimming far and fast. They can dive as well as they can swim. They soon lose their first down and take on a juvenal dark-brown downy covering, into which the feathers gradually come. It is not until they are fully grown, about the last week in August or first week in September, that they are able to fly, and then, fat and plump and strong, they start southward by easy stages, developing wing power as they go. Long before the elders begin leaving, the oldsquaws are gone from the coast, and then winter soon sets in.

Plumages.—The downy young oldsquaw is very dark-colored above, very deep, rich "clove brown," becoming almost black on the crown and rump, and paler "clove brown" in a band across the chest. This dark color covers more than half of the head, including the crown, hind neck, and cheeks; it is relieved, however, by a large spot below the eye and a smaller one above it of whitish, also an indistinct loral spot and postocular streak of the same. The throat is white and the sides of the neck and auricular region are grayish white. The belly is white. Both the dark and the light brown areas become duller and grayer with age.

The plumage appears first on the under parts; the breast and belly are fully feathered first, then the flanks and scapulars; the plumage covers the back, head and neck before the wings are grown, in September. All this takes place while the old birds are molting their wings and are flightless. Both old and young birds are able to fly by October and are then ready to start on their migrations.

Mr. Millais (1913) describes the juvenal plumage, as follows:

In first plumage, in September, the young male has the crown dark brown; the back of the neck is grayish-brown till it meets the mantle, which, with the wings, back, and tail, are black, with a dark-gray suffusion. A dark band of grayish-black also crosses the upper part of the chest, and these feathers, as well as the gray and spotted ones on the sides of the chest, are edged with light sandy-brown; the scapulars blackish-brown, edged with light sandy-brown; flanks gray, tinged with sandy-brown; thighs gray; breast, belly, and vent white. In many specimens only the center tail feathers are black, the rest being brown, edged with white, whilst some have a few sandy edged feathers on the upper tail coverts. Round the eye and lores whitish-gray; cheeks, throat, and chin brown-gray. In many specimens the secondaries are brown, and the breast spotted with brownish-gray. In this month the young male is no larger than the female, but by the end of October it has grown to nearly the full size of the adult male. By the end of this month, and during November, new feathers are rapidly coming in, and the immature feathers of the head are being replaced by others resembling those of the adult.

From this time on during the winter and spring there is a slow but steady progress toward maturity of plumage by a practically continuous molt. The crown, throat, and neck become gradually whiter, the brownish-black cheek patches develop and brownish-black feathers appear in the chest; the gray face begins to show, the back becomes blacker and the grayish-white scapulars and flank feathers appear. By the end of March the young male begins to look like the winter adult, but the colors are not so pure or so intense. In April the molt into the first nuptial plumage begins, in which the young bird can be distinguished from the adult by the faded and worn wings, the imperfect and mottled appearance of the breast, and the absence of the long tail feathers. During August and September a complete molt occurs, at which the wings and tail are renewed, the long tail feathers are acquired and the adult winter plumage is assumed; by the end of November, at the latest, this plumage is complete and the young bird may be said to be adult at seventeen months of age.

The seasonal changes of plumage in the adult male oldsquaw are unique, striking, and very interesting; it is one of the few ducks in the world to assume a distinctly nuptial plumage. The molt into this plumage begins in April and in the oldest and most vigorous birds it is completed in May; but in the younger birds and less vigorous individuals, the molt is prolonged into the summer and is often incomplete. Some birds acquire the full nuptial dress (in which the face is "smoky gray," with a white space around and behind the eye, the feathers of the upper back and scapulars are broadly edged with "sayal brown" and the rest of the head, neck, breast, back and wings are deep, rich "seal brown," nearly black on the upper surfaces) before they migrate north in the spring; I have seen birds in the full perfection of this plumage as early as May 10; but most birds reach their breeding grounds in various stages of transition, showing more or less white or gray feathers in the dark areas. There is no real eclipse plumage in the oldsquaw, but Mr. Millais (1913) says:

During August the male long-tailed duck completely changes the wing, tail, back, and black portion of the mantle and black breast band, these parts being replaced by the new winter plumage. The head, neck, and upper mantle, showing worn and faded plumage feathers, remain until shed at the end of September. The elongated scapulars are shed and not renewed until late September, but in late July a considerable number of new blackish and brown feathers come into the upper scapulars and mix with the old worn summer plumage feathers, whilst a number of new dark-gray feathers, similar to those worn by the scaup, tufted duck, and goldeneye male in eclipse, come unto the flanks and remain until shed again in early October. The reason of this, I take it, is that since Nature abhors sudden changes of color from dark to light, whilst the landscape is still under the warm colors of summer and autumn, the male long-tail only renews those parts of its plumage to the full winter dress which are directly in harmony with its surroundings, adding, however, temporary feathers, as it were, to carry it over the three temperate months when it hides in the shadows of banks or rocky inlets. Thus all the dark parts of its plumage are renewed once, and once only, and the light parts which would be noticeable are delayed by a temporary makeshift until such a time as concealment is no longer necessary.

Of the plumage changes of the female he says:

In first plumage the immature female closely resembles the young male, except that the color is somewhat paler. They are also easily recognized by October by the incoming feathers of the respective sexes. In the case of the young female the advance of plumage is somewhat slower than that of the male, and she only obtains a few of the winter dress feathers. In April the greater part of the adult summer plumage is assumed, but there is little or no change on the back, breast, and lower parts, whilst the wings and tail are not shed at all until the principal molt in July and August. The young female then gradually assumes her winter dress, which is complete in October. She is thus adult at 16 months, and will pair and breed the following summer.

Young females may be recognized during their first year by their imperfect and more or less mottled head pattern and by the broad, gray edgings of the upper back and scapulars. Adult females in winter have the feathers of the back, scapulars, and wing coverts broadly edged with "tawny" or "cinnamon brown," and the upper chest and flanks are suffused with lighter shades of the same. The partial prenuptial molt occurs in April, leaving the wings worn and faded; the head becomes largely brownish black, with whitish spaces before and behind the eye, and the upper parts are nearly uniform brownish black. In August and September a complete molt produces the winter plumage.

Food.—Most of the oldsquaw's food is obtained by diving in water of moderate depths to the beds of mussels (*Mytelis edulis*) and other bivalve and univalve mollusks, but many of these are picked up in shallow water, as the rising tide covers the ledges; as the mussels open their shells to procure food they are picked out by the ducks. Much food is also picked up along the beaches, such as shrimps, sand fleas, small mollusks, crustaceans, beetles, and marine insects, together with some seaweeds and a quantity of sand. On their breeding grounds the food consists largely of the roots, leaves, buds, and seeds of various aquatic plants; the young live largely on insects, larvae, and soft animal food which abounds in the tundra pools. Dr. F. Henry Yorke (1899) includes in their vegetable food teal moss (*Limnobium*), blue flag (*Iris*), duckweed, water plantain, pond-weed, and pickerel weed.

George H. Mackay (1892), writing of the habits of this bird on the Massachusetts coast, says:

> Oldsquaws do not seem to be at all particular in regard to their food, eating quite a variety, among which are the following: A little shellfish, very small, resembling a diminutive quahog (*Venus mercenaria*), but not one; sand fleas; short razor shells (*Siliqua costata*); fresh-water clams; small white perch; small catfish; penny shells (*Astarte castanea*); red whale bait (brit); shrimps; mussels; small blue-claw crabs; and pond grass. It was during the early part of the severe winter of 1888 that many oldsquaws sought the land. Alighting on the uplands adjacent to the north shore of the island, they came in flocks of a hundred or less, in order that they might obtain and eat the dried fine top grass (*Anthoxanthum odoratum*) which grows wild there; when engaged in plucking it their movements while on the ground were far from awkward, in fact rather graceful, as they ran quickly about gathering the grass, some of which was still in their mouths when shot.

William B. Haynes (1901) observes:

> Most authorities agree that the oldsquaw is unedible when killed on the Great Lakes, but here (Ohio) they vary their diet with worms and are far better eating than scaup or goldeneye. I have found the common angleworm and a large green worm resembling a cutworm in their throats.

Edwin D. Hull (1914) says that in Jackson Park, Chicago, in winter:

> The plants, rocks, and piers constitute a very favorable habitat for immense swarms of silvery minnows (*Notropis atherinoides*), which seem to be almost if not entirely the sole source of food for the old squaw in this locality. The stomach of an adult female found floating in a lagoon April 1, 1912, contained approximately 140 of these minnows, all entire, besides many fragments of the same fish, but no other food. The fish averaged about 2 inches in length.

Behavior.—When migrating, old squaws fly high in the air in irregular flocks or in Indian file, but at other times they fly close to the water or a few feet above it, but almost never in a straight line; they twist and turn suddenly, showing the breast and belly alternately like shore birds, swinging in broad circles most unexpectedly. Their flight is so swift and so erratic that it is very difficult to shoot them, but they are often very tame or stupid and are quite as likely to swing in toward a gunner's boat as away from it; then in turning they often bunch together so closely that a tempting shot is offered; I have seen as many as nine dropped out of such a bunch at one shot. I have seen them, when shot at, dive out of the air into the water, swim for a long distance under water, and then come out of the crest of a wave flying at full speed, as if they had never broken their flight. They can rise readily off the surface of even smooth water, and when alighting on it often drop in abruptly with an awkward splash. If there is a strong wind blowing they are more inclined to circle into the wind, glide down gently against it on set wings and alight with a sliding splash. Old squaws can generally be recognized at a long distance by their peculiar method of flight and by their striking color pattern, the white head and neck and the short, sharp-pointed, black wings being very conspicuous.

Toward spring they are particularly restless and active on the wing and often indulge in aerial evolutions, such as Mr. Mackay (1892) describes, as follows:

> These ducks have a habit of towering both in the spring and in the autumn, usually in the afternoon, collecting in mild weather in large flocks if undisturbed, and going up in circles so high as to be scarcely discernible, often coming down with a rush and great velocity, a portion of the flock scattering and coming down in a zigzag course similar to the scoters when whistled down. The noise of their wings can be heard for a great distance under such conditions. In one such instance, at Ipswich Bay, Mass., a flock of several hundred went up twice within an hour.

The old squaw swims low in the water, but makes rapid progress even in rough water; it rides easily over the ordinary waves, but dives under the crest of a breaker with good judgment and precision. It is one of the most expert of the diving ducks and will often dodge under at the flash of a gun; in diving the wings are partially opened as if they were to be used under water; probably they generally are so used, but not always. It can dive to great depths if necessary. Prof. W. B. Barrows (1912) says:

> Several observers mention the fact that it is often caught in the gill nets set in deep water for lake trout and whitefish (in the Great Lakes.) One fisherman at St. Joseph told me most positively that he had seen it caught repeatedly in net set at a depth of 30 fathoms (180 feet).

Dr. A. W. Butler (1897) and Mr. E. H. Eaton (1910) both make similar statements, and the latter says that "at Dunkirk, N. Y., between five and seven thousand have been taken at one haul"; this seems almost incredible.

When feeding in flocks their diving tactics are interesting to watch. L. M. Turner (1886) writes:

When searching for food they string out in a long line and swim abreast. At a signal one at the extreme end goes down, the rest follow in regular time, never all at once, and rarely more than two or three at a time. The last one goes down in his turn with the regularity of clockwork. As they dive they seem to go over so far as to throw the long tail feathers until they touch water on the other side. They remain under water a long time and usually come up near each other.

His notes state that they "remain under water for periods varying from 40 to 92 seconds." This last figure seems as if it might be an error, for the following observations by Mr. Seton Gordon (1920) seem to indicate great regularity in the diving periods of this species; he says:

On one occasion, December 16, I timed a drake during six dives, as follows: 37, 37, 37, 30, 37, 37 seconds. As will be seen, his periods of submersion were extremely regular. On December 18 I watched for some time a pair diving energetically. The drake kept under longer than the duck, half a dozen of his dives being as follows: 37, 42, 36, 35, 33, 32 seconds, and those of the duck, 33, 37, 35, 33, 33, 32 seconds. On emerging, the duck seemed to shoot up more buoyantly than the drake. In the afternoon I timed the drake for four dives, as follows: 42, 40, 42, 45 seconds. The periods during which the birds were above water between the dives I timed as follows: 10, 8, 6, 8, 7, 11 seconds. On December 21 I timed a pair diving and emerging almost simultaneously, as follows: 34, 32, 37, 38, 40, 43, 36 seconds. Before the two longest of these dives, the birds swam for some time on the surface of the water.

If there is any one thing for which the old squaw is justly notorious it is for its voice. It certainly is a noisy and garrulous species at all seasons, for which it has received various appropriate names, such as old squaw, old injun, old wife, noisy duck, hound, etc. The names south-southerly, cockawee, quandy, coal and candle light, as well as a variety of Indian and Eskimo names have been applied to it as suggesting its well-known notes; all of these are more or less crude imitations of its notes, which are difficult to describe satisfactorily, but when once heard are afterwards easily recognized, for they are loud, clearly uttered, and very distinctive. Mr. Francis H. Allen has given me the best description of it as "*ow-owdle-ow* and *ow-ow-owdle-ow* with a Philadelphia twang; that is, with a short *a* sound in the *ow*. The last syllable is higher pitched than the rest and is emphasized."

Rev. J. H. Langille (1884) describes it very well in the following words:

To my ear it does not recall the common name "south-southerly" given it on the Atlantic coast, but is well expressed by an epithet given it by the Germans about Niagara River, who call it the "ow-owly." *Ow-ow-ly, ow-owly, ow-owly*, frequently repeated in succession, the first two notes considerably mouthed, and the last syllable in a high, shrill, clarion tone, may suggest the queer notes to anyone whose ear is familiar with them. Not infrequently the last syllable is left out of the ditty, the bird seeming somewhat in a hurry, or the note becomes a mere nasal, *ah, ah, ah,* rapidly uttered.

Doctor Nelson (1887), referring to the notes heard on their breeding grounds, writes:

During all the spring season until the young begin to hatch, the males have a rich musical note, imperfectly represented by the syllables *A-léedle-a, a-léedle-a,* frequently repeated in deep, reedlike tones. Amid the general hoarse chorus of waterfowl at this season, the notes of the old squaw are so harmonious that the fur traders of the Upper Yukon have christened it the "organ duck," a well-merited name. I have frequently stopped and listened with deep pleasure to these harmonious tones, while traversing the broad marshes in the dim twilight at midnight, and while passing a lonely month on the dreary banks of the Yukon delta I lay in my blankets many hours at night and listened to these rhythmical sounds, which with a few exceptions were the only ones to break the silence. These notes are somewhat less common during the day.

Mr. Ekblaw, in the notes, describes them as follows:

The call is a loud, ringing *ong, ong-onk* that carries far and clear. The call is given with a quick hard recoil of the head with the emission of each syllable, as if requiring considerable force. The vibrant resonance of the call is undoubtedly due to the peculiar development of the voice box. At the base of the trachea, just at the junction of the bronchial tubes, is a coiled enlargement resembling a mellophone, with a tightly stretched membrane along one side, probably the mechanism by which the volume and quality of the call are produced.

Oldsquaws fly in flocks by themselves, but in their winter haunts they are associated more or less with red-breasted mergansers, scoters, eiders, and goldeneyes, frequenting similar feeding grounds. They are said to show a decided antipathy to cormorants and to leave their feeding grounds when these birds visit them. Mr. Millais (1913) says:

The principal enemies of the species are the cowardly white-tailed eagles, who kill numbers of half-grown young and wounded birds, the Greenland and Iceland falcons, the three long-tailed skuas, and the great blackback and glaucous gulls. Arctic foxes and polar bears also account for a good many before they can fly.

Fall.—By the time that the young birds are strong on the wing, during the early part of the fall, both old and young leave their breeding grounds in the north and begin to gather in large flocks on the Arctic coast. John Murdoch (1885) says, of the beginning of the migration at Point Barrow:

Through July and August they vary in abundance, some days being very plenty, while for two or three days at a time none at all are to be seen. At this season they fly up and down not far from the shore and light in the sea. Toward the end of August they are apt to form large "beds" near the station, and this habit continues in September whenever there is sufficient open water.

Many come from the east in September and cross the isthmus at Pergniak and continue on down the coast to the southwest. We noticed them going southwest past Point Franklin, August 31, 1883, in very large flocks. After October 1 they grow scarcer, but some are always to be seen as late as there is any open water.

In northern Labrador similar movements take place; "from the last of August to the middle of October immense numbers of these birds assemble in Hudson Straits," according to Mr. Turner's notes; he says that they disappear from Fort Chimo about the middle of November.

On the New England coast the first cold storm of late October brings a few scattering flocks of oldsquaws, but it is not until late in the fall or when early winter conditions are almost here that the heavy flight comes along; driven like snowflakes ahead of a howling norther, flock after flock of these hardy little sea fowl sweep and whirl over the cold gray waves; or high in the air they twist and turn, twinkling like black and white stars against the leaden sky. The shore birds' whistles are no longer heard; they have passed on to warmer climes; but like the cries of a distant pack of hounds the merry notes of the oldsquaws cheer the gunner's heart as he sits in his anchored boat behind a string of wooden decoys waiting for a shot at passing "coots." The main flight of scoters has passed, an occasional V-shaped flock of geese passes overhead, honking its warning of approaching winter, and soon long lines of brant will be looked for winging their apparently slow and heavy flight close to the water. As Walter H. Rich (1907) puts it:

Winter is close at hand. There is a sting in the wind, a nip in the air, and the fingers are numb and blue as they hold the gun barrels. But out on the water, careless of wind or wave, rides a flock of "squaws" making always a merry clatter. Ever and anon some of their number rise against the breeze to dart off at lightning speed, apparently in the mere enjoyment of flight, for, circling a half a mile about, they plump down again among their comrades, all the time noisily calling to each other. We might almost say they are the only song birds among the ducks, for really their notes are very pleasant to hear and quite musical in comparison with the usual vocal production of the family.

Game.—Oldsquaws are not held in much esteem as game birds; their flesh is rank, fishy, and tough; but there are gunners that will eat them. Many are shot, however, every fall by gunners who are out after scoters; later in the season large flocks of oldsquaws frequently pass over or along the line of boats anchored in their path. They decoy well to the wooden blocks used for this kind of shooting, and are often quite tame or full of curiosity. They often offer tempting shots, and their flight is so swift and erratic that it requires considerable skill and practice to hit a single bird; when they are flying before the wind one must hold well ahead of them. They are so tough that only a small portion of the birds shot down are killed, and it is almost useless to pursue the wounded ones, as they are more than a match for the gunner in rough water.

Winter.—Oldsquaws are common in winter as far north as southern Greenland and the Diomede Islands in Bering Straits, where they can find open water among the ice, but they are more abundant below the regions of frozen seas. Even on the New England coast they sometimes encounter ice conditions too severe for them. Mr. Mackay (1892) writes:

Although, as their Latin name expresses, they are particularly a cold-weather bird, it is a matter of interest that ducks with such Arctic proclivities should find the effects of the climate so rigorous at times on the New England coast that they are unable to sustain life and are in consequence obliged to succumb. Yet such is the case. It was during the winter of 1888, when, standing on the high land of Nantucket Island and looking seaward in any direction, nothing but ice was visible; for a month the harbor was closed and there was sleighing on it. There was no open water in sight except an occasional crack in the ice caused by the change of tides; most of the sea fowl had left this locality during the early stage of the severely cold weather. Many oldsquaws remained, however, until they were incapacitated through lack of food and consequent loss of strength from doing so. As a result it was a common occurrence to find them lying around dead or dying on the shore. Those that were alive were so weak they could not fly, and on examination proved to be nothing literally but skin and bone, others apparently had starved to death.

Referring to their habits here, he says:

Off the south side of Nantucket Island the oldsquaws collect in countless myriads. On February 19, 1891, I saw a flock of oldsquaws estimated to contain 2,000 birds off the south shore of Nantucket about 5 miles from the island, and I know of no better place to observe them in numbers. They arrive about the third to the last week in October, according to the weather, and remain until the latter part of November; most of them then move farther south. The height of their abundance is the first half of November. They congregate on "Old Man's Rip" and on "Miacomet Rip," shoal ground 2 to 3 miles from the south shore of the island, the water there being 3 to 4 fathoms deep. Here they live in security, with an abundance of food, during the day. About 3 o'clock p. m. they commence to leave this place for the Sound (the movement continuing until after dark) where they regularly roost, flying around that part of the island which affords them at the time the greatest shelter from the wind, returning on the following morning to their feeding ground by which ever route is the most favorable. An examination of the stomachs of some of those oldsquaws which I shot in the early morning coming from the Sound, showed them to be empty. I think occasionally on clear calm nights they remain on their feeding grounds, and do not go into the Sound to roost. They apparently prefer to feed in water not more than 3 to 4 fathoms deep, or shallower, unless compelled in order to obtain food. I have noticed north of Cape Cod during the winter months that some oldsquaws will feed and remain just back of the line of breakers on the beaches, and also around the rocks, but generally they are in small and detached groups of but few individuals.

Many oldsquaws spend the winter in the Great Lakes and in other large bodies of water in the interior, but it is decidedly a maritime species by preference. For a study of the habits of this species on Lake Michigan in winter, I would refer the reader to an excellent paper on this subject by Edwin D. Hull (1914) based on observations for three seasons at Chicago. I can not afford the space to quote from it as freely as it deserves. Severe winter conditions sometimes drive a few birds as far as the southern borders of the United States. Messrs. Beyer, Allison, and Kopman (1907) record the capture of one in Louisiana on February 13, 1899. "At the time of the capture of this specimen a severe blizzard was sweeping the South. Zero temperatures were reported at points near the Louisiana coast."

DISTRIBUTION

Breeding range.—Arctic coasts of both hemispheres. On the Labrador Peninsula south, in Audubon's time, to the southeastern coast of the peninsula (Bras d'Or), but now probably not much south of northern Labrador (Okak and Nain). Ungava Bay and the lower Koksoak River (Fort Chimo) and down the eastern shore of Hudson Bay, perhaps as far as Cape Jones. On Southampton Island and other lands north of Hudson Bay, and on the west coast of the bay at least as far south as Cape Fullerton and perhaps as far as Churchill. Along the entire Arctic coasts and barren grounds of Canada and Alaska. South on the Alaskan and Siberian coasts of Bering Sea to the Aleutian and Commander Islands and on all the islands in that sea. Along the Arctic coasts and barren grounds of Asia and Europe, south in Scandinavia to about 60° north. On the Faroe Islands, on Iceland, and on both coasts of Greenland. North on practically all Arctic lands as far as they have been explored up to 82° north.

Winter range.—In North America south on the Atlantic coast abundantly to southern New England, commonly to Chesapeake Bay and North Carolina, more rarely to South Carolina, Florida (Brevard and Leon Counties) and occasionally to the Gulf coast of Louisiana; north, when open water is to be found, to the Gulf of St. Lawrence and sometimes southern Greenland. On the Pacific coast south regularly to Washington, less commonly to California, as far south as San Diego; north to the Aleutian Islands, and sometimes to the Diomede Islands. In the interior it winters abundantly on the Great Lakes and more rarely or irregularly on other large bodies of water west and south to Nebraska (Omaha), Colorado (Barr Lake), and Texas (Lake Surprise). In southern Europe south to about 40° north, on the Black and Caspian Seas; and in Asia, south to Lake Baikal, China, and Japan.

Spring migration.—Mainly coastwise, but also overland to northern coasts. Dates of arrival: New Brunswick, Grand Manan, March 9; Ontario, Ottawa, April 2; Mackenzie, Fort Simpson, May 10; northern Greenland, Etah, May 20; Boothia Felix, latitude 70°, June 12; Winter Harbor, latitude 75°, June 22; Cape Sabine, latitude 78°, June 1; Fort Conger, latitude 81°, June 6. Dates of arrival in Alaska: Chilcat, March 11; Admiralty Island, April 17; St. Michael, April 1; Cape Prince of Wales, April 22; Kowak River, May 22; Humphrey Point, May 20; Point Barrow, May 15.

Late dates of departure: Rhode Island, May 4; Massachusetts, May 22; Maine, May 21; Gulf of St. Lawrence, May 23; Pennsylvania, Erie, May 18; Alberta, Fort McMurray, May 15; southern Alaska, May 19.

Fall migration.—Reversal of spring routes. Early dates of arrival: Great Bear Lake, August 28; Pennsylvania, Erie, September 13; Massachusetts, September 30; New York, Long Island, October 8 (average October 16). Late dates of departure: Northern Greenland, latitude 82°, September 16; Alaska, Point Barrow, December 9, and St. Michael, October 20.

Egg dates.—Arctic Canada: Fifty-three records, June 7 to July 18; 27 records, June 19 to July 4. Alaska: Sixteen records, May 22 to July 28; eight records, June 16 to July 9. Labrador: Three records, June 16, 17, and 27.

HISTRIONICUS HISTRIONICUS HISTRIONICUS (Linnaeus)

ATLANTIC HARLEQUIN DUCK

HABITS

The harlequin duck is a rare bird on the Atlantic coast of North America, where its chief summer home is in Labrador and Ungava. Comparatively little is known about it even there, as very little thorough ornithological work has been done in that largely unexplored region. But in western North America the species is widely distributed and in some sections of Alaska, notably the Aleutian Islands, it is very abundant. W. Sprague Brooks (1915) has recently separated the western bird, as a distinct subspecies, under the name *pacificus*. As this seems to be a well-marked form with a distinctly separate range, I have compiled a separate life history for it. Except for the descriptions of the eggs and plumages, which are the same for both forms, the following remarks refer mainly to the Atlantic form.

Spring.—Mr. Lucien M. Turner's notes state that—

They arrive at Fort Chimo by the 25th of May and then frequent the smaller fresh-water ponds and lakes. They retire to the seashore by the 5th of June, or even earlier if the ice has cleared from the beach. The outlying islets are favorite places in the earlier days after their arrival; but when the water is mostly clear of ice they prefer the rugged shores of the larger islands and shores of the mainland where the reefs and jagged, sunken rocks are to be found; these birds are rarely to be seen along shingly beaches unless they may be merely passing from one point to another.

He says of their behavior:

The males are extremely pugnacious and quickly resent the approach of another male toward their mates. They flop through the water with surprising speed toward the intruder with open mouth, uttering a hissing sound, and seize the offender by the body and quickly pluck out a beakful of feathers if the pursued bird does not dive or flutter away.

Nesting.—Audubon (1840) claims to have found them breeding on islands in the Bay of Fundy; he writes:

There they place their nests under the bushes or amid the grass, at the distance of 20 to 30 yards from the water. Farther north, in Newfoundland and Labrador, for example, they remove from the sea, and betake themselves to small lakes a mile or so in the interior, on the margins of which they form their nests beneath the bushes next to the water. The nest is composed of dry plants of various kinds, arranged in a circular manner to the height of 2 or 3 inches, and lined with finer grasses. The eggs are five or six, rarely more, measure $2\frac{1}{16}$ by $1\frac{9}{16}$ inches, and are of a plain greenish-yellow color. After the eggs are laid, the female plucks the down from the lower parts of her body and places it beneath and around them, in the same manner as the eider duck and other species of this tribe.

Dr. C. Hart Merriam (1883) contributes the following:

While in Newfoundland last winter I learned that these birds, which are here called "lords and ladies," are common summer residents on the island, breeding along the little-frequented watercourses of the interior. I was also informed, by many different people, that their nests were built in hollow trees, like the wood duck's with us. Mr. James P. Howley, geologist of Newfoundland, has favored me with the following response to a letter addressed to him on this subject: "I received your note inquiring about the harlequin duck, but delayed answering it till the arrival of one of our Indians. It is quite true the birds nest in hollow stumps of trees, usually on islets in the lakes or tarns of the interior. They usually frequent the larger lakes and rivers far from the seacoast, but are also found scattered all over the country."

Most of the eggs of the harlequin duck in collections came from Iceland, where the species breeds abundantly and where many nests have been found. John G. Millais (1913) gives the following attractive quotation from Reimschneider, illustrating the behavior and nesting habits of this species in Iceland:

This is the finest of all the species here. Their movements both on land and water are quick, skillful, and graceful; they run swiftly on dry land, and their gait reminds one very little of the waddling of other ducks, but in walking the small head with its beautiful beak is stretched rather forward, and the long tail pointing downward, with the proportionately slender body and the peculiar coloring, all give this bird a rather foreign appearance, though certainly not an unlovely one. The plumage of this small duck charmed me particularly when I saw it swimming upstream with unparalleled swiftness through the frothing foam of the Laxa, winding about through the eddies of the strongest breakers, and making use of the quieter places in the most skillful way. I then always had in mind the other much less common Icelandic name *Brindufa* (breaker dove). I have never seen the harlequin duck make an even temporary stay on the lake, but they always keep to the swiftly flowing rivers of the neighborhood; e. g., on the Laxa, where I visited a small breeding colony near the Helluvad farm. When I came to this place on June 24 I was several times obliged, in order to reach the nests, to ride through the water of the river to a series of small heath-overgrown rock islands upon which the ducks breed. Here I found, in addition to several nests of the *Fuligula marila*, four nests of the *F. histrionica;* it is certain that there were still more nests to be found close to. I put the number of pairs nesting at this place at from 10 to 12. The first nest, standing under a thick clump of heath, had a sort of bank of dry heath around the shallow hollow of the site of the nest. This hollowed-out basin contained the first half-finished lining of gray down mingled with fine dry grass. In the nest lay five eggs, which I took away, and which proved not to have been sat on at all. This nest had been hitherto untouched by human beings, but not so the others which I saw, and which had already lost some of their eggs. The next nest showed exactly the same construction, and in this the down lining was still altogether wanting. This one contained only two eggs. While the two first nests we have just described were some paces from the edge of the island, the next, unprotected by heath growth, was placed on a small piece of rock jutting out over the river. The basin contained a complete lining of gray down mixed with grass, and the loose edge of this was carefully pulled down over three eggs which were in the nest. The duck flew away from the fourth nest which I visited as soon as I was quite close to it, and this one again was placed more in the middle of the island under a clump of heath, and was very plentifully lined with down with an unusually small admixture of parts of plants; it contained three eggs.

The down in the harlequin duck's nest is "olive brown" or "drab," with rather large, but not very conspicuous, whitish centers. Small whitish breast feathers, with a pale brownish central spot and pale brownish tip, are usually found in the down.

Eggs.—The harlequin duck lays from 5 to 10 eggs, usually about 6. The shape varies from bluntly ovate to elongate ovate, and some eggs are quite pointed. The shell is smooth and slightly glossy. The color varies from "light buff" or "cream color" to paler tints of the same. The measurements of 90 eggs, in various collections, average 57.5 by 41.5 millimeters; the eggs showing the four extremes measure **61** by 42.5, 59 by **44**, **52** by 39, and 56.2 by **37.5** millimeters.

Young.—Incubation is performed entirely by the female, who also assumes full care of the young. Audubon (1840) writes:

The male leaves her to perform the arduous but, no doubt to her, pleasant task of hatching and rearing the brood, and, joining his idle companions, returns to the seashore, where he molts in July and August. The little ones leave the nest a few hours after they burst the shell, and follow their mother to the water, where she leads them about with the greatest care and anxiety. When about a week old she walks with them to the sea, where they continue, in the same manner as the eiders. When discovered in one of these small inland lakes, the mother emits a lisping note of admonition, on which she and the young dive at once, and the latter make for the shores, where they conceal themselves, while the former rises at a good distance, and immediately taking to wing, leaves the place for awhile. On searching along the shores for the young, we observed that, on being approached, they ran to the water and dived toward the opposite side, continuing their endeavors thus to escape, until so fatigued that we caught four out of six. When at sea, they are as difficult to be caught as the young eiders.

Mr. Millais (1913) says that the period of incubation is said to be three and one-half weeks.

It is presumed that the young are at first fed by the old bird direct from the bill, as newly hatched young always hold their bills upward to the beak of the foster parent, and will not at first pick up food for themselves. At first the food is principally the larvae of *Ephemerae*. The down period of the young is said by Faber to be about 40 days.

Mr. O. J. Murie has sent me the following interesting notes on the behavior of young harlequin ducks:

The harlequins acquire their love for rough water early, for the young are brought up among the rapids of northern rivers. Several broods of these ducklings were found on the Swampy Bay River, in northern Ungava. I saw the first family one day when we had paddled across the swift current above a rapid, to hunt for a portage. As we floated into a sheltered eddy near shore, a band of ducklings swam quietly out past our canoe. They appeared singularly unconcerned and unafraid. At first I did not recognize them as harlequins and they all looked the same size to me. But one of the Indians declared one of them was the mother. They swam around the base of a huge bowlder and headed deliberately into the swift water. In astonishment I watched them go bouncing down the rapid, around the bend out of sight.

A few days later I witnessed a still better exhibition. We stopped to camp at the head of a rapid which culminated in an abrupt fall of 20 or 30 feet. Here we found some more harlequins. I got two young and the mother between me and the fall and attempted to corner them for a photograph. There was but a narrow lane of comparatively quiet water near shore. As I neared the little group the mother flew upstream, and the little ones spattered up over the water, actually entering the edge of the swift current in order to get by me. Upon repeating the performance several times, I had an opportunity to perceive their wonderful knowledge of currents and their skill in navigating them. Finally, when pressing them close for a near approach, they again entered the swift water. At the same time the mother came flying low and passed downstream. This time the youngsters were evidently caught, for the current carried them out of sight over the falls. With a feeling of remorse I looked below. I had not intended to be the means of their destruction. At first I could distinguish nothing among the ripples and the foam-flecked current below. Then I saw them floating along, rising to shake the water from their down, then quietly preening themselves. Although they had clearly endeavored to avoid the falls, they were none the worse for the accident when it did happen.

Plumages.—The downy young is "bister" or "Prout's brown" above, including the top of the head down to the level of the eyes, the lores, hind neck, back and rump; the under parts, including the cheeks up to the eyes, are pure white; there is a white spot above and in front of the eye and an indistinct whitish streak on each scapular region; the front of the wing is margined with white. The juvenal plumage comes in first on the flanks and scapulars; the former are, at first, "olive brown" with white tips and the latter are "warm sepia."

In the juvenal plumage in the fall the young male resembles the adult female, but can be recognized by the looser texture of the plumage, the worn tail, the gray instead of brown flanks, and by having less white on the breast. J. G. Millais (1913) says:

Toward the end of November the young male begins to assume the adult dress rapidly; the tail and tail coverts are replaced by adult feathers; a tinge of burnt sienna appears on the long flank feathers; the wing coverts, the scapulars, mantle, and the whole of the adult feathers on head and neck come in, so that by the end of January a young male in my possession is almost like an adult, except for the smaller black and white bars on the sides of the chest, a brown rump and bill, mottled and immature under parts, and immature wings. The change then proceeds very slowly. From specimens in Mr. Schioler's collection it is clear that the male harlequin follows the same course of plumage as the long-tailed duck and the goldeneye. A greater or lesser part of the immature under parts are shed between the months of March and June, and the last signs of immaturity in the shape of the wings are not shed until late July or August, when the young male goes into an eclipse similar to the adult male. By September the new wings are obtained and the portions that were assumed as eclipse are being shed, so that it is not until November—that is, at 14 months—that the young male stands in full dress. It will breed in the following spring.

This does not agree with what specimens I have seen, which indicate that the young male makes but little progress toward maturity during his first year. Among the large flocks of immature males which we saw in the Aleutians in June very few birds showed anything approaching the adult plumage, and most of them could hardly be distinguished from females, except at very short range. Most of the specimens seen and collected were in worn immature plumage, dull brown above, with lighter edgings, wholly mottled below, and with varying amounts of the slaty feathers and the white markings of the adult on the head. Perhaps these were especially backward birds, and others, which we took to be adults, were normal or advanced young birds; but the latter were certainly in a very small minority.

The young female requires about the same time to reach maturity—about 16 months. Mr. Millais (1913) says:

There seems to be less difference between the young and adult female harlequin than almost any of the diving ducks. Yet the immature female, prior to February, when the new tail is assumed, can always be recognized by the worn ends and lighter colors of the tail and under parts. The under parts are not nearly so broadly speckled as the adult, and there is a greater area of white. The flanks are grayer, and have a sandy tinge. Also the white spaces about the eye are always more heavily edged with slaty-brown.

I should add to this that in the young female the head is usually duller brown and the feathers of the back show more light edgings.

Of the eclipse plumage Mr. Millais (1913) says:

The whole plumage of the adult male in eclipse is a uniform dark slate gray, the head and neck being somewhat darker, as well as the rump, under and upper tail coverts, which are almost black; the single white ear covert spot is retained, and the white space in front of the eye is dull white, both these parts being edged with black; long scapulars, lower neck, upper and lower flanks, sooty brown; about the end of August the wings and tail are shed (as usual only once). Like all the diving ducks, the male harlequin is practically in a state of molt from July 1 until it reaches the full winter plumage early in October.

Food.—Most of the harlequin duck's food is obtained by diving, but much of it is picked up along the shores or about the rocky ledges. On the inland streams where it breeds it consists largely of water insects and their larvae, among which the caddis fly is prominent; it also includes fish spawn, small fishes, small frogs, tadpoles, small fresh-water crustaceans and mollusks, and some aquatic plants. On the seacoast it feeds on similar kinds of marine animal life which it picks up on the kelp-covered rocks at low tide or obtains by diving in the surf along the shore or over the ledges; it apparently does not often dive for its food in deep water. The common black mussel (*Mytelus edulis*) is one of its main food supplies; these mollusks grow in immense beds on shallow ledges and are easily obtained; occasionally a large mussel has been known to trap the duck and cause its death by drowning. Small crustaceans, such as sand fleas and small gasteropods, are also picked up.

R. P. Whitfield (1894) gives the following account of the contents of a bird's stomach, taken on Long Island:

In December, 1893, Mr. William Dutcher brought to me the stomach contents of a harlequin duck (*Histrionicus histrionicus*) shot at Montauk Point, Long Island, about the 3d of the month. An examination of the material showed what an industrious collector the bird must have been, for it had in its crop remains of no less than three individuals of the small mud crab of our coast, *Panopeus depressa* Smith, one carapace being almost entire; besides remains of some other forms of Crustaceans. Of the little shell *Columbella lunata* (*Astyris lunata* of the Fish Commission Reports), there were no less than 39 individuals represented, besides several small Littorinas. This shell is seldom more than one-sixth of an inch long, and is usually quite rare on our shores. It could only have been obtained in such numbers by a sort of sifting of the bottom mud of the bays by the duck, and indicates how carefully the process had been carried on in order to obtain so small an article of food.

Behavior.—Mr. Millais (1913) describes the flight of this species as follows:

The beautiful markings of the male of this species are only noticeable when the observer is close at hand, so that they are not the easiest duck to identify except when in flight. The flight, at first somewhat laborious, is very rapid. The short, pointed wings are beaten swiftly, and the bird constantly swings from side to side, even more frequently than the long-tailed duck. The elevation is moderately high, performed at an altitude similar to the goldeneye, but when passing up or down stream it zigzags and turns, to accommodate its line to every bend of the stream, however slight. The harlequin never thinks of cutting off corners, and it would seem that it imagines its life depends on keeping exactly over the water, however much it bends or twists. I have seen harlequins fly religiously above a bend in a stream that formed almost a complete circle in its course, and yet the birds did not cut across it to shorten their route.

I have watched harlequin ducks in flight many times and have shot quite a few of them, but I never noticed any swinging from side to side, as referred to above, and several writers have referred to their flight as straight. They usually fly close to the water and often in such compact flocks that a large number may be killed at a single shot. They also swim in close formation, sometimes with their bodies almost touching.

Walter H. Rich (1907) says:

If a shot is fired at a flock on the wing they will sometimes plunge from the air into the water and after swimming below the surface again take wing, coming up a hundred yards away—seeming, the instant they reappear, to dash from the depths into the air at full speed, leaving the gunner inexperienced in their ways, and who perhaps had thought that by some miraculous chance he had killed the entire flock, to find that he doesn't care for that kind of duck after all. I passed through just such an experience once, and remember yet how disgusted and surprised I was when after steaming up to where the whole flock should have been dead—no duck—and what may have been their ghosts rising from their watery graves 60 yards away.

Harlequin ducks are fond of feeding in rough water along rocky shores or in the surf just off the beaches, where they ride the waves lightly and dive through the breakers easily and skillfully. They dive so quickly that they often escape at the flash of a gun. In diving the wings are usually half opened as if they intended to use the wings in flight under water, which they probably do.

The peculiar whistling note of this duck has been likened to the cry of a mouse, whence it has been called the "sea-mouse" on the coast of Maine. Mr. Bretherton (1896) describes it as "a shrill whistle descending in cadence from a high to a lower note, commencing with two long notes and running off in a long trill." Mr. Millais (1913) writes:

When first arriving at the breeding grounds in flocks in early May they are very restless, constantly flying to and fro, whilst the females utter their usual call of "*Ek-ek-ek-ek*," to which the males respond with a low or hoarse "*Hu*" or "*Heh-heh*." These calls they also frequently make in winter, and I have heard single females uttering their cry constantly when flying, as if they had lost their companions and were seeking them. When they are paired both sexes utter a different note, "*Gi-ak*," and this note is used at all times when the pair meet, until the males leave the females at the end of June.

Mr. Aretas A. Saunders writes me:

I heard these birds call several times. The call note is usually uttered when on the wing. It sounded to me like "*oy-oy-oy-oy*" rapidly repeated, usually seven or eight times. I never heard the note from any but the males, and it was usually uttered when in pursuit of one of the females.

Winter.—The winter home of the harlequin duck is on the seacoast. On the Atlantic coast they are not common south of Maine and not abundant even there. They are often seen about the rocky bays of the eastern Provinces in winter, but more often they frequent the outlying rocky islands and ledges. In spite of the brilliant coloring of the males they are surprisingly inconspicuous among the kelp-covered rocks and the wet, shiny seaweeds of varied hues. On the Atlantic coast they are widely known as "lords and ladies," and by the French inhabitants of Quebec they are called "canards des roches" or "rock ducks." They usually flock by themselves in small flocks, but are frequently associated with oldsquaws.

DISTRIBUTION

Breeding range.—Iceland, southern Greenland (north on the east coast to Scoresby Sound and on the west coast to Upernavik), the Labrador Peninsula (Nain, Lance au Loup, Fort Chimo, etc.), and Newfoundland (Hawks Bay, etc.). Birds said to breed in the Ural Mountains and the Yaroslav Government may be of this subspecies, but the breeding birds of eastern Siberia are probably referable to *pacificus*.

Winter range.—The Atlantic coast of North America, south regularly to the Bay of Fundy and the coast of Maine, more rarely to Long Island Sound and casually farther south. Resident in Iceland.

Spring migration.—Atlantic coast birds retire northward in February and some reach Greenland in March. Arrive at Fort Chimo, Ungava, May 25. Seen in the Gulf of St. Lawrence as late as May 29.

Fall migration.—Early dates of arrival: Maine, October 19; Massachusetts, November 1; Rhode Island, November 28.

Casual records.—Rare or casual on Lake Ontario (Toronto. October 20, 1894, and December 4, 1920). Accidental as far south as South Carolina (Mount Pleasant, January 14–16, 1918) and Florida

(Pensacola, March 20, 1886). Rare or casual in Scandinavia, Russia, Germany, Switzerland, Italy, and Great Britain.

Egg dates.—Iceland: Twenty-three records, May 20 to July 9; twelve records, June 6 to 30. Labrador: Two records, June 3 and 10. Greenland: One record, June 24.

HISTRIONICUS HISTRIONICUS PACIFICUS Brooks

PACIFIC HARLEQUIN DUCK

HABITS

I had always supposed that the harlequin duck was a comparatively rare and somewhat solitary species until I visited the Aleutian Islands in the summer of 1911; here we found this subspecies to be one of the commonest and most widely distributed of the ducks; we saw them in large or small flocks about all of the islands wherever they could find the rocky shores that they love to frequent. I saw more harlequin ducks here in one day than I have ever seen elsewhere in my whole life. Most of the birds were in large flocks, some of them in immense flocks, but they were also frequently seen in pairs, feeding about the kelp-covered rocks at low tide, among which they were surprisingly inconspicuous and were easily approached. Even the large flocks were not wild or shy, and we had no difficulty in shooting all we wanted. The large flocks were made up almost wholly of females and immature males, but they were usually led by two or three adult males. The presence of mated pairs and some small flocks of adult males led us to suppose that they were breeding there, perhaps back in the interior in the rocky canyons of the mountain streams, but we found no signs of nests around the shores. Similar gatherings of harlequin ducks are found all summer about the Pribilof Islands and all along the southern coasts of Alaska and British Columbia, as far south as Puget Sound. Nearly all, if not all, of these birds are probably immature birds which are not yet ready to breed, or unmated or barren birds, mainly the former. Some may be birds which have bred early, have lost their broods or their mates, and have returned to join their fellows in these summer-flocking resorts, which are practically the same as the winter resorts. The migrations of this species do not amount to much more than a brief withdrawal into the interior during the nesting season.

Courtship.—The best account that I have seen of the courtship of this species is by B. J. Bretherton (1896), as follows:

The writer has often watched the males in spring, calling, and the actions of these birds may justly be said to resemble the crowing of a rooster. In giving forth their call the head is thrown far back with the bill pointed directly upward and widely open; then with a jerk the head is thrown forward and downward as the cry is uttered, and at the same time the wings are slightly expanded and drooped. Afterwards they will rise in the water and flap their wings.

Charles W. Michael (1922), who has had exceptional opportunities to study the behavior of harlequin ducks at short range, describes another courtship performance, as follows:

When the birds appeared in front of camp on the morning of April 12 they were acting strangely. Apparently they were making love. They were bobbing and bowing to one another, swirling around, touching their bills together, and uttering little chatty sounds. One of the moves on the female's part was to slowly submerge her body until just her head and neck appeared above the surface of the water—a bold invitation on her part for attention. In spite of the wanton actions of the female, the love-making failed to reach the climax.

Nesting.—I have never found the nest of the harlequin duck, and I infer that few others have succeeded in doing so in North America, for surprisingly little is to be found in print about the nesting habits of this species. None of the well-known Alaskan explorers speak of finding nests, except Turner (1886), who says:

The nest and eggs were not procured, and the only nest I ever saw was near Iliuliuk village, on Unalaska Island. Two immense blocks of rock had become detached from the cliff above, and when they fell their edges formed a hollow place beneath. In under this I discovered a deserted nest, which the native who was with me asserted was that of a bird of this species. The form was similar to that of the nest of *C. hyemalis*, and in fact so closely resembled it that I persisted in it being of this bird until the native asked me if I did not know that the oldsquaw did not build in such places.

Major Bendire wrote to Dr. D. G. Elliot (1898):

The harlequin duck undoubtedly nests both in our mountain ranges in the interior—Rockies and Sierra Nevadas—as well as on many of the treeless islands of the Alaskan Peninsula and the Kurile Islands, and I have not the least doubt that it breeds both in hollow trees, where such are available, and either on the ground or in holes made by puffins where it can find such, not far from water.

Dr. E. W. Nelson (1887) writes of the breeding haunts of this species:

Among the host of waterfowl which flock to the distant breeding grounds of Alaska in spring, this elegantly marked bird is the most graceful and handsomely colored. As if conscious of its beauty, the harlequin duck leaves the commonplace haunts sought by the crowd of less noble fowls, and along the courses of the clear mountain streams, flowing in a series of rapids into the larger rivers, they consort with the water ouzel, Swainson's thrush, and such other shy spirits as delight in the wildest nooks, even in the remote wilderness of the far north. Dark lichen-covered rocks, affording temporary shelter to the broad-finned northern grayling or the richly colored salmon trout as they dart from rapid to rapid, steep banks overhung by willows and alders, with an occasional spruce, forming a black silhouette against the sky, and a stillness broken only by the voices of the wind and water, unite to render the summer home of these birds, along the Yukon, spots devoted to nature alone, whose solitude is rarely broken, and then only by the soft footsteps of the savage in pursuit of game.

Mr. D. E. Brown has sent me the following note:

On May 7, 1924, a fisherman flushed a female western harlequin duck from a set of seven eggs. This nest was near Port Angeles, Clallam County, Washington, and was on a rocky point of a swift running mountain stream.

Eggs.—Eggs of the Pacific harlequin duck are scarce in collections, and I have no measurements available for comparison, but they probably do not differ essentially in color, shape, or size from those of the Atlantic bird.

Plumages.—The sequence of molts and plumages of this western subspecies are apparently the same as those of its eastern relative.

Behavior.—Aretas A. Saunders writes me from Montana that:

While fishing they sit in midstream, facing the current, often where it is swiftest, paddling just enough to keep themselves stationary. Whenever they see a fish, they dive for it, and usually appear again, a considerable distance downstream with the fish. They dive down into the middle of swift rapids, in places where one would expect them to be dashed in pieces against the rocks, yet they always emerge again, unharmed. Whenever the birds go downstream they usually swim down, and from what I have observed, do this largely under water. As soon as they come to the surface they generally turn and face the current. I have never seen them swim upstream, even where the water is not swift, and believe that when they wish to go upstream they nearly always rise and fly. One afternoon I watched a male bird fishing at the edge of a large pool where the water was not swift. He took up a position to watch at the edge of the pool, standing with his feet and under parts in the water but his head and breast out. From this position he dove after fish whenever he saw them, but I could not make out that he was always successful in catching the fish.

Mr. Michael (1922) says:

Harlequins are expert swimmers and divers. They dive and swim under water with all the ease of a grebe, besides possessing the ability of the water ouzel to walk about on the river bed against the swift currents. When feeding, so far as we were able to observe, they show no preference as to depth of water. When working upstream along the shore they wade in the shallow water, prying among the stones. Where the water is deeper they tip up in the manner of mallard ducks, and where the water is still deeper they dive. They dive in water a foot deep and they dive in water 6 feet deep, always going down where there is a gravelly bottom. Most often they stay under water not more than 15 seconds. Often they stay down 20 seconds, and occasionally they remain under the water as long as 25 seconds. To leave the surface of the water they use their wiry tails as a spring to make the plunge and as they go down both wings and feet are used as a medium of propulsion. When once on the gravelly bottom the wings are closed, the head is held low, and the progress is made against the current, as they walk along poking amongst the stones. When coming to the surface they float up like bubbles, without movement of wings or feet. Their bills break the water and their bodies pop suddenly onto the surface where they rest a moment. While poising on the surface between plunges their bodies float high. When earnestly feeding, seldom more than 10 seconds elapse between plunges. The birds seldom dive simultaneously. The female usually acts first.

At times the harlequins choose the swiftest riffles, and when feeding there their method is the same as when in the less joyous waters. They apparently dive from any position with equal ease, but always as they go down they turn upstream, and even in the swiftest currents they come up in about the same spot at which they went down. When feeding in these racing waters they merely hesitate on the surface, and four or five dives are made in rapid succession. Such work as this is strenuous, but the birds are quite at home in the swiftest currents, and when tired from their exertions they swing into an eddy behind some snag or bowlder and rest as they bob about on the surface.

M. P. Skinner writes to me that they have been observed coasting down on the Yellowstone River almost to the brink of the Lower Falls, 308 feet high, and then, when it seemed as if they would surely go over, they would fly upstream again and repeat the performance.

Game.—As a game bird the harlequin duck is of little importance. It is a comparatively rare bird, or entirely unknown, in most of the regions frequented by gunners; and even where it is fairly common its haunts are rather inaccessible. Moreover, it lives so largely on animal food that its flesh is not particularly palatable. Among the natives of the Aleutian Islands and other parts of Alaska, however, large numbers are killed for food. Mr. Bretherton (1896) describes the method of hunting employed by the natives of Kodiak Island, as follows:

When first the writer went to Kodiak he tried hunting with a boat, relying on wing shooting to get his birds, but without much success; and seeing that the natives always got more birds, he changed his plan and took to the natives' method, as follows: When a band of ducks was seen feeding, a landing was made and the beach approached from the land, the hunter being careful not to be seen. By watching the flock it would be seen that they all dived about the same time, and the time they remained down was about the same length each time. When the last duck dives, the hunter runs toward them, dropping in the grass or behind a rock about the time he calculates the first duck should be coming up again. In this manner he can approach close to the flock, that nearly always feed in the shallow water along the shore. When the last run is made, the hunter, if an old hand, stands on the edge of the water, the gun at "ready," and a couple of extra shells in the hollow of his right hand, the flock all being down. The first duck that comes up gets it, and the second one gets the second barrel, and in this way, by sharp practice, it is often possible to bag six or seven out of one flock.

DISTRIBUTION

Breeding range.—Western North America and northeastern Asia. East in northwestern Canada probably to the Mackenzie Valley and Great Slave Lake, but nowhere else east of the Rocky Mountain region. South in the Rocky Mountain region to Montana (Glacier National Park, Chief Mountain Lake, etc.), Wyoming (Shoshone River), and Colorado (Blue River near Breckenridge). West to central California (west slope of Sierra Nevada Mountains) and Washington (Cascade and Selkirk Mountains) and the mountain regions of British Columbia, and Alaska (Sitka region, Sanakh Island, etc.). Westward throughout the Aleutian, Commander, and Kurile Islands. Probably on St. Matthew and St. Lawrence Islands. West in Siberia to Lake Baikal and the Lena River and east to Kamchatka and northeastern Siberia (Providence Bay, Marcova, etc.). North in summer and probably breeding to the Arctic coasts of Alaska (Barter Island) and Canada (Mackenzie Bay).

Winter range.—Mainly on the seacoasts, but also on inland waters, not far from the southern parts of its breeding range. Winters sparingly in its Rocky Mountain breeding range; other interior records are regarded as casuals. On the Pacific coast south to central California (Monterey Bay) and north to the Aleutian and Pribilof Islands. On the Asiatic side from the Commander Islands south to Japan.

Spring migration.—First arrivals reached Fort Simpson, Mackenzie, on May 25, 1904. Usually arrives at the mouth of the Yukon, Alaska, about June 1. A late date for Pierce County, Washington is June 5, 1915.

Fall migration.—Early dates of arrival: Washington, Kitsap County, September 10; California, San Louis Obispo County, October 8.

Casual records.—Rare or accidental in the interior as far south as Nebraska (Omaha, September 16, 1893 and 19, 1895) and Missouri (St. Louis, October 29, and Montgomery County, March 21, 1897).

Egg dates.—Alaska: Four records, June 13 to July 1. Mackenzie Bay: One record, June 20. Montana: One record, June 10. Washington: One record, May 7.

CAMPTORHYNCHUS LABRADORIUS (Gmelin)

LABRADOR DUCK

HABITS

What little there is known about the life history of this extinct species has already been published and repeatedly quoted by various writers. Probably nothing more of importance will ever be learned about its former abundance or its habits. It is doubtful if any more specimens will ever be brought to light. Therefore, in writing this obituary notice, it is necessary only to compile what has already been written in order to make its life history as nearly complete as possible.

Nesting.—It is supposed to have bred, formerly, from the south coast of Labrador northward, but there is very little positive evidence on which to substantiate even this indefinite statement and much less evidence on which to elaborate it. We might infer from what Coues (1861) says that the Labrador duck bred farther north and passed through Labrador on its migrations; he says:

I was informed that though it was rarely seen in summer, it is not an uncommon bird in Labrador during the fall.

William Dutcher (1894) undertook to obtain some further information regarding the occurrence of the species in Greenland, through Mr. Langdon Gibson, who accompanied the Peary expedition to that region in 1891, acting as ornithologist of the party. Although Mr. Gibson made numerous inquiries and showed pictures of the bird to various people along the coast, he could find no evidence to indicate that the Labrador duck had ever been seen there. A portion of his report is worth quoting in full:

In August, 1892 (the latter part, I believe), on our way home we touched at Godthaab, the largest town in Greenland. Here we were entertained by Herr Anderson, the Danish inspector of South Greenland, an accomplished naturalist, and at his house I had the pleasure of inspecting one of the finest collections of Arctic birds I had ever seen. I showed him my little pamphlet on the Labrador duck, and also presented it to him on my departure. He told me that his collection represented 20 years' work, and all the hunters in South Greenland (some 500 men) had instructions to bring to him any strange birds that they might get. In this way he has added to his collection from time to time many rare birds and eggs. In all this time he claims to have heard nothing of the Labrador duck, which I consider is substantial proof that within the last 20 years the Labrador duck has not visited Greenland. From Godthaab we came directly home to Philadelphia, and this ended my ineffectual attempts at learning something more definite regarding this species.

Audubon (1840) did not see a living specimen of this duck in Labrador, where it was supposed to be breeding commonly. It hardly seems likely that he could have overlooked it, if it had been there. Therefore, his brief account of its breeding habits must be considered unsatisfactory and unreliable. He says:

Although no birds of this species occurred to me when I was in Labrador, my son, John Woodhouse, and the young friends who accompanied him on the 28th of July, 1833, to Blanc Sablon, found, placed on the top of the low tangled fir bushes, several deserted nests, which from the report of the English clerk of the fishing establishment there, we learned to belong to the pied duck. They had much the appearance of those of the eider duck, being very large, formed externally of fir twigs, internally of dried grass, and lined with down. It would thus seem that the pied duck breeds earlier than most of its tribe.

Professor Newton (1896) writes:

This bird, the *Anas labradoria* of the older ornithologists, was nearly allied to the eider duck, and like that species used to breed on rocky islets, where it was safe from the depredations of foxes and other carnivorous quadrupeds. This safety was, however, unavailing when man began yearly to visit its breeding haunts, and, not content with plundering its nests, mercilessly to shoot the birds. Most of such islets are, of course, easily ransacked and depopulated. Having no asylum to turn to, for the shores of the mainland were infested by the four-footed enemies just mentioned, and (unlike some of its congeners) it had not a high northern range, its fate is easily understood.

Maj. W. Ross King (1866), who spent three years shooting in and about the Gulf of St. Lawrence, previous to 1866, says: "The pied duck or Labrador duck is common in the Gulf of St. Lawrence, and breeds on its northern shore, a short distance inland." The foregoing quotations, though meager and unsatisfactory, contain about all we know about the breeding habits of the Labrador duck.

Food.—Very little seems to have been recorded about its food and feeding habits. Audubon (1840) says:

A bird stuffer whom I knew at Camden had many fine specimens, all of which he had procured by baiting fishhooks with the common mussel, on a trot-line sunk a few feet beneath the surface, but on which he never found one alive, on account of the manner in which these ducks dive and flounder when securely hooked. It procures its food by diving amidst the rolling surf over sand or mud bars; although at times it comes along the shore and searches in the manner of the spoon-bill duck. Its usual fare consists of small shellfish, fry, and various kinds of seaweeds, along with which it swallows much sand and gravel.

Other writers say that it fed on shellfish which it obtained by diving on the sand shoals, whence it derived the common name of "sand shoal duck."

Mr. S. F. Cheney, of Grand Manan, wrote to William Dutcher (1891) in 1890:

The female Labrador duck I gave to Mr. Herrick was with some oldsquaws or long-tailed ducks when I shot it, and I think there were no others of the kind with it. This one had small shells in its crop. It dove to the bottom with the squaws.

Behavior.—Audubon (1840) wrote:

Its flight is swift, and its wings emit a whistling sound. It is usually seen in flocks of from 7 to 10, probably the members of one family.

Col. Nicholas Pike sent to William Dutcher (1891) the following interesting account of his experiences with the Labrador duck:

I have in my life shot a number of these beautiful birds, though I have never met more than two or three at a time, and mostly single birds. The whole number I ever shot would not exceed a dozen, for they were never plentiful. I rarely met with them. The males in full plumage were exceedingly rare; I think I never met with more than three or four of these; the rest were young males and females. They were shy and hard to approach, taking flight from the water at the least alarm, flying very rapidly. Their familiar haunts were the sandbars where the water was shoal enough for them to pursue their favorite food, small shellfish. I have only once met with this duck south of Massachusetts Bay. In 1858, one solitary male came to my battery in Great South Bay, Long Island, near Quogue, and settled among my stools. I had a fair chance to hit him, but in my excitement to procure it I missed it. This bird seems to have disappeared, for an old comrade, who has hunted in the same bay over 60 years, tells me he has not met with one for a long time. I am under the impression the males do not get their full plumage in the second year. I would here remark, this duck has never been esteemed for the table, from its strong, unsavory flesh.

Probably the Labrador duck was never abundant or even very common throughout its known winter range; certainly we have very little positive evidence to that effect. The statement, so often quoted, of Thomas Morton in his New English Canaan (1637) may not refer to this species at all. In writing of the birds noted by him in New England between 1622 and 1630, he says "Ducks there are of three kinds, pide ducks, gray ducks and black ducks in great abundance." It seems to be taken for granted that by the name "pide ducks" he referred to the Labrador duck. It seems to me much more likely that he referred to the goldeneye, which is still called the "pied duck" all along our northern coasts, or to one of several other species called by that name or, perhaps, to a number of species in general having more or less black and white plumages. Audubon (1840) considered it rather rare, although he says: "Along the coast of New Jersey and Long Island it occurs in greater or less number every year."

Dekay (1844), writing of this species in New York, says:

This duck, well known on this coast under the name of skunk head, and sand-shoal duck on the coast of New Jersey, is not, however, very abundant.

Walter J. Hoxie wrote to me a few years ago, as follows:

During my youthful experience among the ponds and creeks about the mouth of the Merrimac we sometimes got a duck which we called a "black-belly" and many of the gunners considered it a cross with the "sea coots." In the brackish ponds it was commonly found in company with the gadwall, or as we called it, the "gray duck." We rather disdained it, and I remember too it was hard to pick. Lots of down under the feathers that perhaps made us think it was akin to the scoters. One I remember in "Bushy Pond" with a gray duck on a frosty November evening. Did not seem to be as shy as its companion, but kept moving about watching me as I crawled down with a pine sapling for shelter. The old flintlock hung fire a little longer than usual, and though they were both in line when I sighted the gray was too quick. To-day that black-belly would not have been such a disappointment, though I had to wade for it and the water was almost freezing. It must have been in 1862. In 1870 I saw one—perhaps more—in Boston market. But one I know was tied up with an American merganser. I bought the merganser and stuffed it.

George N. Lawrence, in a letter to Mr. Dutcher (1891), wrote:

I recollect that about 40 or more years ago it was not unusual to see them in Fulton Market, and without doubt killed on Long Island; at one time I remember seeing six fine males, which hung in the market until spoiled for the want of a purchaser; they were not considered desirable for the table, and collectors had a sufficient number, at that time a pair being considered enough to represent a species in a collection. No one anticipated that they might be-

come extinct, and if they have, the cause thereof is a problem most desirable to solve, as it was surely not through man's agency, as in the case of the great auk.

Dr. D. G. Elliot (1898) saw a considerable number of Labrador ducks, mostly females and young males, in the New York markets between 1860 and 1870, but full plumaged males were exceedingly rare.

George A. Boardman endeavored to get some specimens for Doctor Elliot from his collectors about Grand Manan, but found that these ducks had all gone; the last one taken in that vicinity was shot by S. F. Cheney in April, 1871; this specimen was sent, by Mr. Boardman, to John Wallace, of New York, to be mounted for the Smithsonian Institution, but, not knowing its value, Wallace parted with the skin, and all trace of it was lost. The last specimen taken and preserved was shot on Long Island in the fall of 1875, purchased, from J. G. Bell, by George N. Lawrence and presented by him to the Smithsonian Institution; it was a young male and possibly its parents or others of the same brood may have survived for a few years; but probably the Labrador duck became an extinct species at about that time.

Since then only one specimen has been recorded as taken; Dr. W. H. Gregg (1879) reported the capture of a Labrador duck, near Elmira, New York, on December 12, 1878; the duck had been eaten before he heard about it and he was able to procure and save only the head and neck; these remnants were preserved for some years, but finally lost; it is unfortuate that this record can not now be verified.

William Dutcher (1891, 1893, and 1894) has made a careful study of the records relating to the Labrador duck and a thorough investigation as to the number of specimens in existence, so far as known, in American and European collections. He published a number of papers on the subject and finally succeeded in locating, up to 1894, only 42 specimens, 31 of which were in American collections at that time. How many specimens have come to this country since, I have made no effort to determine. Many specimens were shipped abroad between 1840 and 1850, which have not been located, and some may turn up later in private collections. J. H. Gurney (1897) recorded a specimen in the museum at Amiens in France, which was apparently unknown to Mr. Dutcher; this, with one since discovered by Winthrop S. Brooks (1912) in the Boston Society of Natural History, brings the published record of specimens up to 44.

There has been considerable speculation among ornithological writers as to the causes which led to the disappearance of this species, which was apparently as well fitted to survive as several other species of ducks. It was a swift flyer, rather shy and difficult to approach in its offshore resorts; it was essentially a maritime species and seldom resorted to inland bays or rivers, though Audubon said that it was known to ascend the Delaware River as far as Philadelphia; it was not particularly popular as a table bird and often proved a drug in the market, when other more desirable ducks were obtainable; for the above reasons it is fair to assume that it was not exterminated by gunners and never was shot in very large numbers. What evidence we have goes to show that it never was a numerous species and that it probably had a very limited breeding range. If this breeding range was, as it appears, restricted to the southeast coast of Labrador, its disappearance may easily be charged to the wholesale destruction of bird life which took place on that coast during the last century. Continued persecution on its breeding grounds, where its nests and eggs were apparently conspicuous and where both young and old birds were easily killed in summer, when unable to fly, is enough to account for it. That certain other species, which are known to have wider breeding ranges, survived the same persecution is no proof that the Labrador duck did not succumb to it.

DISTRIBUTION

Breeding range.—Unknown. Supposed to have bred in Labrador, probably in some very restricted range on the south coast of the Labrador Peninsula.

Winter range.—On the Atlantic coast from Nova Scotia to New Jersey and probably to Chesapeake Bay. Most of the specimens with known data were taken near Long Island, New York.

Casual record.—A specimen, since lost, is said to have been taken at Elmira, New York, December 12, 1878.

POLYSTICTA STELLERI (Pallas)

STELLER EIDER

HABITS

This beautiful and oddly marked duck was first described by the Russian naturalist, Pallas, who named it after its discoverer. Steller obtained the first specimens on the coast of Kamchatka, which is near the center of its abundance and not far from its principal breeding grounds in northeastern Siberia. Illustrating the abundance of this species on the Siberian coast of Bering Sea, Dr. E. W. Nelson (1883) writes:

> The first night of our arrival was calm and misty, the water having that peculiar glassy smoothness seen at such times, and the landscape rendered indistinct at a short distance by a slight mistiness. Soon after we came to anchor before the native village this body of birds arose from the estuary a

mile or two beyond the natives' huts and came streaming out in a flock which appeared endless. It was fully 3 to 4 miles in length, and considering the species which made up this gathering of birds it was enough to make an enthusiastic ornithologist wild with a desire to possess some of the beautiful specimens which were seen filing by within gunshot of the vessel.

Mr. F. S. Hersey's notes of July 26, 1914, state:

As we steamed into St. Lawrence Bay there appeared in the distance a long low, sandy island known as Lutke Island. As we drew nearer we could see a cloud of birds hovering above it which our glasses showed us were Arctic terns. The island itself was very low, hardly above the sea level, and as we looked at it seemed to be strewn with small black rocks. With our glasses, however, we could see some movement among these black objects. At last we made them out to be birds, then suddenly they arose and swept out toward us, their black and white plumage flashing in the sunlight, and we saw that they were eiders. There were many kings and Pacifics among them, and these separated from the main flock and went out to sea, but the remainder, which were Steller eiders, returned to the farther side of the island. A boat was soon lowered and a party of us put off from the ship. When we landed and started to walk across the island the eiders again took flight but soon settled on the water a short distance offshore. They were not at all shy. While we stayed on the island small parties of from 2 or 3 to 8 or 10 were constantly flying back and forth, often close to us, although we were in plain sight at all times, for the island offered no concealment. We had no difficulty in obtaining all the specimens we wanted.

Spring.—From their winter home in the Aleutian Islands the main flight of the spring migration seems to pass westward through the Commander Islands, where they are very abundant in April, to the Siberian coast and northward. There is also a northward migration through Bering Straits to the Arctic coast of Alaska. Mr. John Murdoch (1885) found these eiders common at Point Barrow; he says:

Early in June they are to be found at the "leads" of open water at some distance from the shore, and perhaps the majority of them pass on in this way to their breeding grounds. From the middle to the end of June they appear on land in small parties scattered over the tundra. At this time they are in full breeding plumage, and the males are generally in excess in the flocks. They are generally to be found in small "pond-holes," frequently sitting on the bank asleep, and are very tame, easily approached within gunshot, and generally swimming together when alarmed, before taking wing, so that several can be secured at one discharge. I have stopped a whole flock of five with a single shot.

Mr. Alfred M. Bailey writes to me:

At Cape Prince of Wales, during the spring of 1922, the first Steller eiders were seen May 12. At this time the straits are still choked with pack ice and salt water freezes on the leads. On May 18 a few birds were seen and again on May 29, but the big migration past this westernmost point was on June 3. We had been walrus hunting in the straits for two days and were returning heavily loaded with meat when the wind suddenly died down and a slick calm prevailed—a very unusual occurrence. Immediately great strings of birds appeared on their northward journey, gulls, loons, ducks, and geese, and among them were many of this species. The natives said, "Plenty birds come from south, bime bye—mebbe one, two hours—plenty south wind." It was true; the birds seemed to be going just ahead of the storm from the south. I learned to foretell a change in the wind by the migration of the birds, for invariably a large migration occurred just before a south wind. We feared a south wind, for if caught offshore, we could not sail back to Wales, and would be forced to drift into the Arctic, so the migration of birds was watched with interest.

Nesting.—Nothing seems to be known about the courtship of the Steller eider and very little has been published about its nesting habits, which is not strange considering the remote and inaccessible regions in which it makes its summer home.

The following brief references to the nesting habits of this species are given by John G. Millais (1913):

Middendorff found nests on flat tundra in the moss, and describes them as deep, round, and lined with down. The male keeps in the vicinity of the female, who sits closely and leaves the nest unwillingly, and when disturbed flies off "with a harsh cry reminiscent of our teal, but still more harsh." Steller found a nest in Kamchatka amongst precipitous rocks near the coast.

Personally I have had no experience with this species, and Mr. Hersey never found its nest. I had five sets of eggs sent to me by my correspondent, T. L. Richardson, who collected them near Point Barrow, Alaska, during the summer of 1916. Unfortunately no data came with them, but one of the sets was accompanied by the nest, or rather the nest lining. This nest, which contains 10 eggs, consists of a bulky mass of curly, coarse grasses and various mosses and lichens, such as grow on the tundra, thoroughly mixed with considerable very dark brown down and a few feathers from the breast of the duck. Evidently the female plucks the down from her breast, together with such feathers as casually come with it, and mixes it with the coarser nesting material, as incubation advances. The nest is quite different from any other duck's nest that I have seen, and is easily identified by the peculiar breast feathers of the female Steller eider. The down varies in color from "benzo brown" to "fuscous."

Eggs.—The five sets referred to above consist of two sets of 6 eggs and one set each of 7, 8, and 10 eggs. They are typical eider's eggs in appearance. The prevailing shape is elliptical ovate, some are elongate ovate, and a few are nearly elliptical oval or approaching oval. The shell is smooth, with little or no gloss. The color varies from "light yellowish olive" to "water green" or from "deep olive buff" to "olive buff." Many of the eggs are clouded or mottled with darker shades of the above colors and many are quite badly nest stained.

The measurements of 75 eggs, in various collections, average 61.4 by 42 millimeters; the eggs showing the four extremes measure **70.5** by 45.5; 66.9 by **47.1, 55.5** by 40.5, and 59.2 by **37** millimeters.

Plumages.—The downy young Steller eider is easily recognizable, as it is quite different from the young of other species. The bill, even in the smallest specimens, shows the characteristic shape peculiar to the species, tapering evenly from forehead to tip, slightly compressed in the middle, with an overhanging upper mandible near the base and near the tip. The color is decidedly dark; the upper parts, including the crown, hind neck, back, and rump are very dark, glossy "bone brown" or "clove brown"; slightly lighter shades of the same colors extend downwards on the sides of the head to the chin and throat, on the sides of the body and across the chest; there is a "buffy brown" spot above the eye, a whitish spot below it and a stripe of "buffy brown" behind it; the throat and chin are "light vinaceous cinnamon" or "pinkish buff" in the youngest birds, grayer in older birds, the colors merging gradually into the darker colors above; the breast and belly are dull, silvery, grayish brown, invaded on the sides with darker browns. The bill and feet are black in dry skins.

In the juvenal plumage, during the first fall young males and females are very much alike and somewhat resemble the adult female except that they are lighter colored, redder, and more mottled below; in the young male the breast and flanks are heavily barred with rich reddish brown or "chestnut"; while in the young female the under parts are barred with paler browns; in the young male the wing is much like that of the adult female, with the curved tertials; but in the young female there is less blue in the speculum and the tertials are straighter, less curved. In both sexes the feathers of the back and scapulars have brownish buff edgings; and the under parts are wholly mottled or barred, instead of being uniform dark brown as in the adult female.

During the first winter and spring the sexes begin to differentiate more. The young male becomes lighter colored; the dusky throat patch and the black neck ring begin to show; the breast begins to assume a tawny shade; and in some forward birds some of the white-edged scapulars and long curved tertials appear before summer. But, on the whole, there is not much change until the summer molt occurs in July and August. This produces what might be called a first eclipse plumage, relatively similar to that of the common eider. The plumage is completely changed during this molt, after which old and young birds are practically indistinguishable.

Adults have one complete molt each year, which produces in the male a fairly complete double molt and eclipse plumage of the head, neck, and upper parts. It apparently occurs in July and August, as I have in my collection males in full nuptial plumage up to the end of June and a series of nine adult males, taken July 26, showing various stages of the eclipse plumage. In full eclipse the striking colors of the head and neck—white, green, and black—are wholly replaced by "bister" or "mummy brown," darker above and lighter below, with only a trace left on the hind neck of the purplish black collar. The back becomes dull black and the showy scapulars are replaced by plain "clove brown" feathers which overhang the showy wings. The wings are still further concealed by "clove brown" feathers on the flanks and by a suffusion of dusky and brown barred feathers on the shoulders and chest, some of which invade the breast. The remainder of the under parts and the wings remain as they were and are apparently molted only once. Specimens showing the change into the full plumage are not available.

Food.—Referring to their feeding habits, Mr. Murdoch (1885) says:

When open water forms along shore, that is, in the latter part of July and early part of August, they are to be found in large flocks along the beach, collecting in beds at a safe distance from the shore, feeding on marine invertebrates, especially gephryean worms.

Mr. Bernard J. Bretherton (1896) says that at Kodiak Island in winter they feed largely on decapods and mollusks, which they obtain in deep water, seldom feeding near the shore. Mr. Millais (1913) writes:

They feed on fish spawn, young fish, crabs, and possibly on vegetable growths, but principally on conchylia and mussels. These they obtain by diving, and their favorite resorts are mussel banks lying at the same depths as those frequented by eiders and long-tailed ducks.

Behavior.—Referring to their behavior on the Siberian coast, Doctor Nelson (1883) writes:

Flocks of thousands were found about Cape Wankarem during our stay there the first of August, 1881, and, in company with an equal number of king eiders and a few of the Pacific eider, were seen passing out and in each evening to and from the large estuary back of the native village. This village was built upon the spit cutting this estuary from the sea at this place, and lay directly in the track of flight followed by these eiders as they passed to or from the sea. As these flocks passed back and forth the birds were being continually brought down by the slings thrown into the midst of the passing birds by the natives; yet, notwithstanding this, the birds continued from day to day the entire season to pass and repass this place. Their heedlessness in this respect may be accounted for from the fact that these people were without guns of any kind, and were thus unable to frighten them by the noise of the discharge. The birds were easily called from their course of flight, as we repeatedly observed. If a flock should be passing a hundred yards or more to one side, the natives would utter a long, peculiar cry, and the flock would turn instantly to one side and sweep by in a circuit, thus affording the coveted opportunity for bringing down some of their number. These flocks generally contained a mixture of about one-twentieth of the number of Pacific eiders, and the remainder about equally divided of stellers and the king eiders. At times the entire community of these birds, which made this vicinity their haunt, would pass out in a solid body, and the flock thus formed exceeded in size anything of the kind I ever witnessed.

Fall.—At Point Barrow, according to Mr. Murdoch (1885), the fall migration, or rather the movement away from their breeding grounds begins early.

Birds that have bred, judging from the looks of the ovaries, begin to come back from the first to the middle of July, appearing especially at Pergniak and flying in small parties up and down the coast. They generally keep to themselves, but are sometimes found associating with small parties of king ducks. They disappear from the first to the middle of August, and when gathered in large flocks are exceedingly wild and hard to approach.

The main migration route in the fall is southward along the Siberian coast of Bering Sea to their winter homes in the Kurile, Commander, and Aleutian Islands. But Doctor Nelson (1887) says that—

In autumn, as they pass south, stray individuals and parties are found in Norton Sound. Those taken there are usually young of the year. When found at St. Michael they usually frequented outlying rocky islets and exposed reefs, and fed in the small tide rips. The shallow turbid water of Norton Sound seems to be offensive to the majority of these birds, as their chosen haunts are along coasts where the water is clear and deep close to the shore.

Winter.—Steller eiders are almost as abundant in their winter resorts about the Aleutian Islands as they are in summer on the Siberian coast. Here they gather in large flocks, associated with king eiders, about the harbors which are free from ice. They resort to the vicinity of sunken ledges and rocky islets where they can obtain their food by diving to moderate depths, although they can dive in deep water if necessary. They are rather shy at this season when in large flocks. The winter range extends eastward to Kodiak Island, where this species is said to be abundant. Chase Littlejohn, in some notes sent to Major Bendire in 1892, writes:

These ducks are by far the most numerous of any duck during the winter, and a few were nesting at Morzhovia Bay in June. They are known locally as soldier ducks, from their habit of swimming single file and then as if by a given signal they all disappear beneath the surface in search of food, where they remain for some time, but when they arise they usually form a solid square or, in other words, a compact bunch, and then single file and repeat. Such chances are taken advantage of by men in search of game; if near shore they run to the nearest point where the ducks disappeared, and when they come to the surface shoot into the flock, sometimes killing a large number. The same tactics can be employed using a boat. They are not bad eating if the skin is removed.

Westward the winter range extends at least to the Commander and Kurile Islands. Probably all the birds which breed in eastern Siberia and Alaska winter in some of the resorts named above. But there is evidently a westward migration also, along the Arctic coast of Europe to a well-known winter resort in the unfrozen waters off the coast of Norway; this flight is probably made up of birds which breed in western Siberia or northern Europe.

DISTRIBUTION

Breeding range.—Coasts of northwestern America and northern Asia. East on the Arctic coast of America to Point Barrow, Alaska, and perhaps farther. South regularly, in the Bering Sea region, to St. Lawrence Island, Anadyr Bay, and Kamchatka; recorded as breeding on the Aleutian and Shumagin Islands and on the Alaska Peninsula, (Morzhovia Bay), but probably only sparingly and irregularly. The main breeding range is on the Arctic coast of Siberia from Bering Straits westward, at least as far as the Taimyr Peninsula, and perhaps on Nova Zembla.

Winter range.—The vicinity of the Aleutian Islands, eastward on the south side of the Alaska Peninsula to the Shumagin Islands and the Kenai Peninsula. Westward to the Commander and Kurile Islands. North in Bering Sea as far as open water extends. A few winter in northern Europe, as far west as Scandinavia, Heligoland, Denmark, and the Baltic Sea.

Spring migration.—Arrivals have been noted at Point Barrow, Alaska, as early as June 5 and at Nijni Kolymsk, northern Siberia, June 9. First seen at Cape Prince of Wales, May 12, and a heavy flight on June 3. The last birds leave the Commander Islands from May 25 to 31, and leave Nushagak, Alaska, about May 20.

Fall migration.—Early dates of arrival in the Bering Sea region: St. Michael, September 21; Nushagak, October 8; Commander Islands, November 1. Late dates of departure: Point Barrow, September 17; St. Michael, October 15; Ugashik, November 28. No dates are available for the migrations to and from the European winter range, which is probably occupied by birds breeding in western Siberia.

Casual records.—Accidental in Greenland (Disco Bay, August, 1878), Quebec (Godbout, February 17, 1898), England (Norfolk.

February 10, 1830 and Yorkshire, August 15, 1845), France, Germany and Japan (Yezzo, March 9, 1894 and May 3, 1894).

Egg dates.—Alaska: Eleven records, June 17 to July 10; six records, June 22 to July 6.

ARCTONETTA FISCHERI (Brandt)

SPECTACLED EIDER

HABITS

If the preceding life history was unsatisfactory, this will be more so, for still less is known about the habits of the oddly marked spectacled or Fischer eider, which occupies such a restricted breeding range in northwestern Alaska and northeastern Siberia. Few naturalists have ever seen it in life. Dr. E. W. Nelson (1887), to whom we are indebted for most of our knowledge of the habits of this species, says on this point:

Its restricted range has, up to the present time, rendered this bird among the least known of our waterfowl. Even in the districts where it occurs it is so extremely local that a few miles may lead one to places they never visit.

In Mr. Dall's paper upon the birds of Alaska he limits the breeding ground of the spectacled eider to the marshes between the island of St. Michael and the mainland. This, with the statement made to him by natives that they are never found north of St. Michael, is not borne out by my observations, for these eiders breed from the head of Norton Bay south to the mouth of the Kuskoquim, at least. St. Michael may be noted as the center of abundance. The spectacled eider is so restricted in its range and so local in its distribution, even where it occurs, that, like the Labrador duck and the great auk, it may readily be so reduced in numbers as to become a comparatively rare bird. A species limited in the breeding season to the salt marshes between the head of Norton Bay and the mouth of the Kuskoquim River occupies but a very small territory, and a glance at the map will show this coast line not to exceed 400 miles, even following its indentations. The width of the breeding ground will not exceed 1 or 2 miles, and there are long stretches where it does not breed at all.

In addition to the natural struggle for existence the species has to contend against thousands of shotguns in the hands of the natives. The diminution in all the species of waterfowl breeding along the coast is more and more marked each season; and while this may mean a desertion of one region for another in the case of the great majority of geese and ducks, yet for such narrowly limited species as the spectacled eider, and to a less extent the Emperor goose, this diminution is but the beginning of extermination. Moreover, the present scarcity of large game along the coast is having great effect in causing the natives to wage a continually increasing warfare upon the feathered game.

Apparently Doctor Nelson's fears have been realized, as the spectacled eider has nearly disappeared from the vicinity of St. Michael and from the Yukon delta. My assistant, Mr. Hersey, spent the season of 1914 at the mouth of the Yukon and the summer of 1915 in the vicinity of St. Michael with this species as one of the things especially wanted; and during the two seasons he succeeded in securing only one pair of the birds and did not find a single nest. I doubt very much if they breed there at all at the present time, for he saw only a few in the canal early in June, at which time they seemed to be already mated; they soon disappeared and were not seen again during the season. Wherever the center of abundance may have been in Doctor Nelson's time, it is now to be found somewhere in northeastern Siberia, where it is one of the commonest eiders.

Spring.—Doctor Nelson (1887) says:

> Although living so far north, yet it is one of the last among the waterfowl to reach its breeding ground at the Yukon delta and the coast of Norton Sound. My observations show this species to be strictly limited to the salt marshes bordering the east coast of Bering Sea, and thus favoring the shallow, muddy, coast waters, which appear to be so distasteful to Steller's eider. Very soon after reaching their destination the flocks disband and the birds quietly pair, but the first eggs are rarely laid earlier than the first days of June.
>
> When first paired the birds choose a pond on the marsh, and are thenceforth found in its vicinity until the young are hatched. Their love-making is very quiet. I have never heard any note uttered except by the female while conducting the brood out of danger. As the grass commences to show green and the snow and ice are nearly gone, although the other denizens of the marsh are already well along in their housekeeping, these ducks choose some dry, grassy spot close to the pond, and making a slight hollow with a warm lining of grass, they commence the duties of the season.

Nesting.—The same writer gives us the following account of the nesting habits of the spectacled eider:

> One nest found on June 15 was on a bed of dry grass within a foot of the water on the border of the pond, and when the female flew off the single egg could be seen 20 yards away. Tussocks of dry grass, small islands in ponds, and knolls close to the water's edge are all chosen as nesting places, and as a rule the nest is well concealed by the dry grass standing about. If the nest contains but one or two eggs the female usually flies off and remains until the intruder is gone; but if the set is nearly completed or incubation is begun she will soon return, frequently accompanied by the male, and both circle about, showing the greatest uneasiness. The female will sometimes alight in the pond, within easy range, and both parents may be obtained by watching near the nest.

A set of 9 eggs of this species, sent me by Rev. A. R. Hoare, was collected at Point Hope, Alaska, on June 15, 1917, on a small islet, about 3 feet square, in a tundra pond, in which the water was from 3 to 4 feet deep; the nest was concealed in the long grass at the edge of the islet and was composed of grass and very little down; the eggs were fresh and more down would probably have been added later.

Mr. T. L. Richardson sent we several sets of spectacled eider's eggs from Point Barrow, Alaska. The nest shown in the accompanying photograph was evidently in plain sight, in a depression in the tundra moss and grass, about 10 feet from the shore of a small pond; it was lined with a little moss and down; the 5 eggs that it contained were collected on June 26, 1917. He says that down is added to the nest as incubation advances, so that there is a heavy lining of it before the eggs hatch. The down is soft and closely matted; it varies in color from "bister" to "sepia," with inconspicuous, slightly lighter centers; small mottled breast feathers and dusky tipped belly feathers are usually found in it.

Eggs.—The spectacled eider lays from 5 to 9 eggs, the smaller sets being apparently commoner than the larger ones. In shape they vary from ovate to elongate ovate, but the prevailing shape is elliptical ovate. The shell is smooth with a slight gloss. The color varies from "deep olive buff" or "olive buff" to "water green" or "yellowish glaucous." The measurements of 101 eggs in various collections average 65.4 by 44.6 millimeters; the eggs showing the four extremes measure **73** by 45.7, 70.5 by **47.8**, and **59.5** by **40.5** millimeters.

Young.—Doctor Nelson (1887) says of the young:

> The male is rarely seen after the young are hatched, but the female shows the greatest courage in guarding her brood, as the following incident will show: A brood was swimming away from me, and the female tried to protect them by keeping between the young and myself. I fired two charges of No. 12 shot, killing all the young, yet, in spite of the fact that the parent received a large share of the charge each time, she refused to fly, and kept trying to urge her dead offspring to move on, until a charge of larger shot mercifully stretched her among her offspring. Upon removing the skin her back was found to be filled with fine shot, and her desperate courage in defense of her brood shows the strength of parental feeling. Other similar instances attest the courage and devotion of this species.

Mr. Koren, while collecting for me, near the Kolyma Delta in northeastern Siberia, on July 21, 1916, saw a female spectacled eider swimming in a tundra pool followed by two downy young white-fronted geese, which she had evidently adopted and was carefully guarding; she allowed him to come near enough to photograph them, after which he shot all three of them and sent them to me.

Doctor Nelson (1887) says:

> The middle of August young birds are frequently seen from a few days old to those nearly ready to take wing. During this month the adult birds pass through the summer molt, and with the half-grown young desert the marshes and tide creeks for the seacoast and outlying rocky islands.
>
> By September 1 scarcely a single individual can be found on the marshes, and by the 20th they are scarce along the coast.

Plumages.—The downy young is easily recognized by the shape of the bill and the feathering at its base, which are just as they are in the adult; the bill slopes gradually to a point, with straight edges; the nail at the tip is light colored, but the bill is black in dried skins; the feathering extends to the nostrils and beyond them to a point above. The "spectacles" also are conspicuous. The colors on the upper parts shade from "warm sepia" on the crown and rump to "snuff brown" on the mantle, hind neck, and flanks; the dark color of the crown extends down over the lores and auriculars; a circular space around the eye is "wood brown," surrounded by a broken circle of "cinnamon buff," forming the "spectacles"; the lower cheeks, chin, and throat are pale "cinnamon buff," shading off to dull grayish buff on the breast and belly, into which the darker colors of the upper parts blend.

In the juvenal plumage the sexes are much alike, but they are quite different from the adult female. In the young male the head and neck are much like those of the adult female, with the "spectacles" only indicated; but the upper parts are darker, the feathers of the back and scapulars being "warm sepia" or "bister," edged with "clay color" or "cinnamon buff"; the under parts are uniformly, but rather faintly, barred with dusky, not strongly barred, as in the adult female; the wings are brownish black, with brownish buffy edgings on the greater and lesser coverts, secondaries and tertials. In the young female the juvenal plumage is much the same, except that the under parts are spotted rather than barred; the wings are like those of the adult female, but more brownish, with more buffy-brown edging in the coverts.

Specimens are lacking to show the progress toward maturity during the first winter, but probably it is similar to what takes place in the young common eider. The young male assumes during the following summer a first eclipse plumage, quite different from that of the adult male. In this plumage the "spectacles," lower cheeks, and throat are pale buff fading off to grayish buff on the neck, faintly mottled with dusky; the rest of the head and neck are "hair brown" or "mouse gray," becoming "fuscous" on the crown and occiput and mottled with buffy shades on the sides of the head; the back, scapulars, and flanks are "hair brown" or "deep mouse gray"; the wings are like the juvenal wings until they are molted in August or September; and the under parts are as in the juvenal plumage.

I have not been able to trace the immature plumages beyond this stage, but probably the second winter plumage, as in other eiders, is not fully adult, but very much like it. The perfection of the adult plumage is probably acquired after the second eclipse, when the young bird is over 2 years old.

The adult male apparently has but one complete molt each summer, at which most, if not all, of the contour plumage is molted twice, involving a nearly complete eclipse plumage. The adult eclipse, and probably the second eclipse, can easily be distinguished from the first eclipse by the wings, which are molted but once, and by the under parts, both of which remain as in the fully adult plumage. During the spring the plumage of the male becomes very much worn and in June it begins to molt into the eclipse. The brilliant plumage of the head and neck entirely disappears; the "spectacles" become "mouse gray," mottled with buff, and the rest of the head and neck become mottled and variegated with various shades of gray, buff, and dusky; the white mantle is entirely, or nearly all, replaced by plain "wood brown" or "deep mouse gray" feathers; many feathers barred with dark brown and buffy shades appear on the chest and shoulders; the white rump spots disappear; the conspicuous white wing coverts and white curving tertials are concealed by the dark scapulars and flank feathers while the bird is not in flight.

Food.—All that I can find published as to the food of this species is the short statement by Doctor Nelson (1887) that: "Their food in summer consists of small crustacea, grass, seeds, and such other food as the brackish pools afford."

Behavior.—The same writer says:

They fly in small compact flocks, rarely exceeding 50 birds in a flock, and skim close along the surface of the ice or marsh with a flight very similar to that of other heavy-bodied sea ducks.

Winter.—The winter home of the spectacled eider does not seem to be well known, but, as it has been recorded in winter in both the western and the eastern Aleutian Islands, its main winter range is probably in the vicinity of these islands, where so many other northern sea fowl are known to spend the winter in the comparatively mild open water, tempered by the Japan current.

DISTRIBUTION

Breeding range.—Arctic coasts of Alaska and Siberia. East to Point Barrow at least. South to the Bering Sea coast of Alaska to the mouth of the Kuskokwim River. Westward along the north coast of Siberia to the mouth of the Lena River and to the New Siberia Islands.

Winter range.—Mainly in the vicinity of the Aleutian and Pribilof Islands, and more sparingly eastward along the south side of the Alaska Peninsula to Sanakh Island.

Spring migration.—Early dates of arrival in Alaska: St. Michael, May 6; Point Hope, May 4; Cape Prince of Wales, May 16; Wainwright, May 28; Point Barrow, May 26.

Fall migration.—Latest date of departure from Point Barrow is September 17.

Egg dates.—Alaska: Twelve records, June 8 to July 4; six records, June 15 to 26.

SOMATERIA MOLLISSIMA BOREALIS (Brehm)

NORTHERN EIDER

HABITS

Spring.—Winter lingers on the outer coast of Labrador well into the summer months; all through the month of June and part of July the northeast winds and the Arctic current drive the drifting pack ice onto these exposed and barren rocky coasts. Long before the icy barriers yield to the soft west winds and as soon as the lanes of open water begin to break up the fields of ice, flocks of these heavy-bodied sea ducks may be seen wending their way northward in the opening leads, flying with slow and labored wings beats close to the cold, dark waves or resting in flocks on the larger pans of ice until the way opens for further progress. Many of them have been wintering just beyond the ice floes and are seeking the first opportunity to find open water near their northern breeding grounds.

Regarding their arrival in Cumberland Sound, Kumlien (1879) says:

As soon as there is any open water they are found in spring; still they are not common at Annanactook till the latter days of May. Eskimos from the south reported them on the floe edge near Niantilic early in May, and I saw a few on an iceberg near the Middliejuacktwack Island on the 30th of April. They can stand almost any temperature if they can find open water.

W. Elmer Ekblaw writes to me of their arrival in northern Greenland, as follows:

The all-winter residents are probably the first eiders to appear along the mainland shore in early spring, wherever open water may be found off the outermost capes and islands, usually about April 20. The number of eiders frequenting these open places gradually increases, but slowly, until the last week in May, when the immigration begins in earnest and continues until mid-June, when apparently the last comers have arrived. The females come later than the males, but the last females come with the last males. They are usually rather shy and wary and will not permit near approach.

By mid-June the mating season is usually at its height, but in years of heavy snow when the islets are covered until late, the season is retarded. The summer of 1914 was a summer of late melting of snow and the nesting season of the eiders had hardly begun by the 20th of June.

Courtship.—John G. Millais (1913) describes the courtship of the European subspecies, as follows:

The courtship of the eider is a very simple one, and somewhat undemonstrative. It is essentially in accordance with the gentle disposition of the bird. The female seems to be at least as amorous as the male, and pays considerable court to the object of her affections. Having selected a mate, she follows him round and round in all his movements, stretching her neck out and sinking low in the water, calling and pushing herself against his side until he responds. The male, on his part, makes a very slight "lift" in front, the bill being lowered and the neck drawn up. At the same time he inhales, and on releasing the air as he slightly sinks forward, he utters a gentle "*Pu-whoo*" or "*Aa-u*," almost a dove-like cry. At the moment the call is emitted, the mouth is slightly opened. The call of the male is repeatedly uttered and is often made without "lifting" in front. At such times the head is held forward, then erected to the normal position as the cry is given. At the moment of calling, the whole throat is somewhat distended. When a general display is in progress amongst a flock of eiders the males and females are in a constant state of movement and activity. The males often make half turns and bows toward their inamorata, and utter a high soft note like the syllable "whoop."

Lucien M. Turner found northern eiders very abundant in Hudson Straits; his notes say:

They were by far the most abundant duck, probably exceeding all others together. The islands of Ungava Bay are crowded with them. During the mating season the males are irascible and when the mate is chosen he carefully resents an intrusion from another male. Severe and, often fatal, encounters take place between rival males, resulting in complete defeat to the one or the other. They fight by seizing with the beak and slapping with the wings; more of a kind of wrestle in which they endeavor to get the head of the adversary under the water. When enjoying quiet the male is fond of uttering a cooing sound *Oo oo*, spreading one wing out while he rolls on his side, then recovering and kicking rapidly through the water that makes it fly on both sides. This note with a *curring* sound made in their contests are the only ones I have heard the males produce.

The immature males, during the breeding season, do not associate with the adults. They keep aloof and are usually solitary. Not until the fully adult plumage of the male is assumed does he enter into contest for the female.

Nesting.—The same observer describes two interesting nesting localities as follows:

A few miles below Mackay's Island, about 18 miles up the Koksoak River, is a deep cove on the left bank and nearly opposite "Pancake" point. I gave the name of Eider Cove to that locality from the number of eider's nests I discovered in it during my first visit there—June 17, 1883. The cove is about 400 yards deep and 75 yards wide, preserving a nearly uniform width to the head, where a lively stream dashes down over the jagged rocks. The south side is inaccessible, formed by a steep wall of granite sloping very slightly to the summit, which is about 400 feet high. The northern side or wall is composed of ledges and projections covered with rank grasses, weeds, and ferns. On these ledges and rocks 14 nests of the common eider were found. The first nest was at the base of the rock on a flat scarcely above high-tide mark. This nest contained 5 eggs. Near by were 2 other nests, one of 3, and the other of 1 egg. Farther within were 11 nests each containing from 3 to 11 eggs. Only in the nests containing the greater number of eggs were they unfitted for food. I secured 49 eggs perfectly fresh and about a dozen that were too far advanced to be eaten.

On that same trip I visited the islands off the mouth of Whale River. Here James Irvine and myself collected, in less than an hour and a half, over 500 eggs from a single island and could, doubtless, have obtained many more, but a storm was near by and and we had to make for a larger island where we could secure the heavy whaleboat we had with us. As we approached that island the number of male eiders in the surrounding water and occasional females flying from the water and settling on the land gave promise of a great nesting place. We hauled the boat on a shelving ledge and quickly scrambled to the top of the bank. Here an immense ice cake and drifted snow had collected on the edge of the bank and extended for several hundred feet in length and over 30 yards wide. The height of the seaward edge was then, June 29, over 4 feet. The dripping water and slippery rock made it difficult to surmount in our anxiety to get at the eiders, which had taken alarm and were scurring in hundreds by wing and walk from the land to the sea. In a moment a nest was found and then another and so on until hundreds were discovered. Some with 1 or 2 eggs, others with 6 or 7, these being the more numerous; others with as many as 12. Every grass patch in the depressions of the rocks was examined and the eggs put into piles to be taken to the boat. Several small ponds surrounded by high grasses which were given a luxuriant growth by the droppings of these birds where they had come to bathe or drink for many successive seasons. Among these patches were also the nests of a few Phalaropes, *Phalaropus lobatus*, which twittered and flitted before us. A single nest of a gull was also found. The nests of the eiders were so differently constructed even on this one island that it would be impossible to describe them all. The materials of which they were composed were grasses, weeds, stalks, and down. The amount of the vegetable matter depended on the particular situation of the nest, for if in the midst of plenty of such material, the nest was often several inches high, resting on a mound formed from the decayed mass of material used as a nest many years ago. At times merely a slight depression was cleared of vegetation and on the bare earth the egg was deposited and covered with down.

On my trip down the coast of Labrador in 1912 I found eiders common all along the coast from the Straits of Belle Isle northward, but generally they were so shy that it was impossible to shoot any. The largest breeding colony I saw was on one of the outer islands off the coast near Hopedale, which we visited on July 22. It was a small, low, rocky island with a very little grass and a few mosses growing in the hollows and crevices between the rocks. No male eiders were seen on or about the island, but the females began flying off as we landed and we flushed many from their nests as we walked over the flat rocks. We found between 20 and 30 nests with eggs, varying in number from 1 to 5. Some natives had visited the island a fortnight or more previously and had collected about 150 eggs; there must have been between 30 and 40 nests on the island at that time. The nests were on the ground, in the grass or moss, or in hollows between the rocks; some of them were well made, with a generous supply of pure down, but in most of them the down was mixed with grass and rubbish, and in some of the nests the supply of down was very scanty. Apparently these nests were second or third attempts at raising broods, and evidently the supply of down was becoming exhausted. A drizzling rain was falling all the time that we were on the island, so my attempts at photographing the nests were not as successful as they might have been. We shot five of the ducks as they flew from the nests, all of which proved to be the northern eider. A pair of great black-backed gulls were breeding on the island; we saw the old nest and a young gull running about, as well as the old birds flying overhead. There is generally a pair of these gulls on every island where the eiders are breeding. The natives, who rob the duck's nests regularly, never disturb the gull's nest, for they believe that if the gulls are driven off the ducks will not return to the island to lay again. They say that the black-backed gulls are good watch dogs, to warn the eiders of approaching danger and to keep away the ravens and other gulls which might rob the nests. The great black-backed gulls are notorious nest robbers and destroyers of young eiders elsewhere, but perhaps they do not indulge in highway robbery and murder so near home. Perhaps, however, the gulls do levy their toll in eggs and young eiders, which the latter are too stupid to avoid.

On the coast of Greenland the northern eider frequently nests on cliffs, according to J. D. Figgins, as the following quotation from his notes, published by Dr. Frank M. Chapman (1899) will illustrate:

It prefers the small islands lying some distance offshore, but also breeds on the mainland. Its nest is usually well up the cliffs, and in some cases quite a distance from shore. One nest containing 4 eggs was at an altitude of about 450 feet, and more than three-quarter of a mile from shore.

Dalrymple Rock is the favorite breeding place of this species, it is much broken, and the many ledges offer fine nesting sites. There is a heavy growth of grass on these ledges, and the nest, when it has been used for many years, is a depression in the sod, lined with the down from the breast of the female. As soon as incubation begins the male birds form into flocks of from 4 or 5 to 20, and seem to be always on the wing. There is a constant line of the male birds flying around Dalrymple Rock, all going in the same direction. As soon as incubation is completed, the young are transferred to the water, where they seem perfectly at ease, even when there is a heavy sea running.

Mr. Ekblaw thus describes, in his notes, a visit to one of the great breeding resorts of these birds in northern Greenland:

On June 23 and 25, 1914, we went in a whaleboat to the Eider Islands, between Wolstenholme Island and Saunders Island. These are favorite nesting places of the eiders. In normal years the islands are covered with thousands of nests, but we found a relatively small number of the birds at this time.

As we approached the largest islet of the group flock after flock of eiders flew about us, skimming fast and low over the water. The males and females seemed about equal in number. The bright-colored plumage of the male contrasts vividly in the sunlight with the dark, uniformly barred coat of the female, both awing and alight.

Numerous pairs of the eiders were swimming about in the sea or idly preening their feathers or wooing on the ice pans. Their wooing "song" very closely resembles the cooing of our domestic pigeons. The noise from a flock together in the mating season might readily be mistaken for the "music" from a dovecot. The sound is audible for a long distance; from the time we left Saunders Island until we returned we were not beyond reach of these "love songs."

As we set foot upon the islet hundreds of eiders flew about us. The snow still left on the ice foot and on the slopes of the low, rocky hills was beaten down by their footprints. We scrambled hastily over the islet in search of eggs, but found only a dozen all told, where last year on June 15 several thousand were collected. The fact that no nest held more than one egg indicated that the lateness of the season and the heavy snow on the islet had retarded the nesting season. On the other hand, it may have been that the heavy inroad upon the nests last year discouraged the return of the birds to the islet this year. They may have gone to some other, more remote islet to nest.

The eiders nest on all the small skerries and islets and many favorable places along the mainland. The Eider Islands, Dalrymple Rock, Lyttleton Island, McGarys Rock, the small islets in the bay south of Cape Hatherton, Sutherland Island, Hakluyt, and the Cary Islands are all frequented by large numbers of nesting eiders. They may nest in large numbers close together where conditions are favorable, but many nest alone far from any other companions, either on the mainland, sometimes far inland, though more frequently near the shore, or on the larger islands. Safety from depredations by foxes, jaegers, and ravens is a factor in the choice of nesting sites.

The nest of the eider is profusely lined, around, under, and sometimes over the eggs with a thick bed of soft, fluffy down, densely matted, a famous product of considerable commercial value; in color it is "drab," "light drab," or "drab gray," with poorly defined lighter centers and light tips; mixed with it are occasional bits of pure white down, dusky belly feathers, and barred breast feathers.

Eggs.—The northern eider raises but one brood in a season and lays ordinarily from 4 to 6 eggs; larger numbers have been found, even as many as 19, in one nest, but probably the larger numbers are the product of more than one female. In this connection the following notes from Mr. Turner are of interest:

The number of eggs in a nest is not always a safe index to the number of birds having made the deposit. It frequently happens that a single female will be attended by as many as five males; although it is scarcely probable that they all enjoy equal rights. One of the males is always the leader and the others appear to be entirely under his guidance. I have again, and on repeated occasions, known a single male to have as many as three females under his charge. It is, of course, difficult with these wild birds to determine whether under such circumstances two or more of these ducks have a nest in common or whether they make separate nests.

The eggs of all three of the large eiders, *mollissima*, *dresseri*, and *V-nigra*, are indistinguishable in size, shape, or color, so one description will serve for all three. In shape they vary from ovate to elliptical ovate. The shell is smooth, with only a slight gloss, which increases with incubation. The color varies from "olive" to "deep olive buff," or from "yellowish glaucous" to "vetiver green." The eggs are often mottled or clouded with darker shades of green, olive, or buffy olive, through which the ground color sometimes shows in washed-out spots. The measurements of 76 eggs, in various collections, average 75.4 by 50.4 millimeters; the eggs showing the four extremes measure **83** by **53**, **67.8** by 47, and 73.2 by **46** millimeters.

Young.—The period of incubation of the European eider has been ascertained by several observers to be about 28 days. It is performed wholly by the female. The males are said to desert the females at this time and form small flocks by themselves; but the following observation by Kumlien (1879) is interesting:

I have often lain behind a rock on their breeding islands and watched them for a long time. On one occasion we disturbed a large colony, and the ducks all left the nests. I sent my Eskimos away to another island, while I remained behind to see how the ducks would act when they returned. As soon as the boat was gone they began to return to their nests, both males and females. It was very amusing to see a male alight beside a nest, and with a satisfied air settle himself down on the eggs, when suddenly a female would come to the same nest and inform him that he had made a mistake—it was not his nest. He started up, looked blankly around, discovered his mistake, and with an awkward and very ludicrous bow, accompanied with some suitable explanation, I suppose, he waddled off in search of his own home, where he found his faithful mate installed.

Mr. Turner's notes state that:

The young leave the nest when about 36 hours old and immediately accompany the parent to the edge of the water. The distance to be traveled varies from a few feet to half a mile. I have not on the Atlantic coast found the nests so far from the water as were found at St. Michael, Alaska. In some instances the nest is placed on ledges that have no path by which the delicate young can reach the water excepting by plunging several feet to the next ledge or else be assisted by the parent. The latter I have not seen the old ones do or have I seen it so recorded. The young remain with the mother during the summer and probably do not leave her at all, but join with other broods to compose the flocks seen in the fall of the year. As soon as the young are hatched the males separate from the females and do not join them again until fall.

Mr. Ekblaw says in his notes:

The first nesting birds hatch their eggs soon after July 1. The most are hatched about July 15 to 20. On Sutherland Island as late as August 16, 1912, I found a nest of four eggs still hatching in a single isolated nest. In this nest one little fellow had hatched and dried, and when I flushed the mother he followed her to the water. Another little fellow just hatched was not yet dry, but even so, he sensed the alarm and tried to hide in the down with which the nest was lined. A third duckling was just cutting the shell; later in the evening when I came back to the nest he had come safely into the world. The fourth egg was not even cracked, but a vigorous peeping within the shell indicated that on the morrow the last of the brood would follow the mother into the water.

The downy little ducklings get into the water just as soon as they can. They paddle after their mothers like animated little black balls of fur, keeping always close to her. Even when tiny, they dive like a flash, and come up like

a bubble. Like the oldsquaws, the eider mothers join their little flocks, and one mother leads the file while the other brings up the rear. From the time they are hatched until they leave for the South, they keep to the open sea, coming in to the shore sometimes to rest, but more frequently resting on ice floes.

While the young birds are developing in strength and ability to swim, the mothers constantly lead them southward. From mid-August to mid-September a constant procession of eiders swims outward from the bays and fjords and goes on southward. The southward flight migration is late, beginning about the 1st of September and continuing until the ice freezes, often in October. Even so, many of the young birds are still unable to fly when the fjords freeze over, and frequently flocks of them are caught by sudden freezes and imprisoned in little pools where they finally freeze to death or fall prey to the foxes, bears, or Eskimo.

Plumages.—The downy young of the three eiders, *mollissima*, *dresseri*, and *V-nigra*, are all practically indistinguishable. The upper parts are "olive brown," deepening to "clove brown" on the crown and rump, and paling to "light drab" on the sides and to "pale drab-gray" on the throat and belly; there is a broad superciliary stripe of "light drab" or dull "wood brown" above the eyes and the cheeks, and lores are abruptly darker than their surroundings. Before the young bird is half grown the first dusky, brown-tipped feathers of the juvenal plumage appear on the flanks and scapulars; the tail feathers start next, then the breast plumage and the head; the young bird is nearly fully grown before the last of the down disappears from the upper back and rump; the wings appear last of all, so that both old and young birds are flightless together in August.

In the juvenal plumage the sexes are much alike; this plumage is worn by the young male through the fall and resembles that of the young female, but not that of the adult female. Sometimes as early as October, but more often not until December, the young male begins to differentiate by acquiring a few black feathers in the flanks and scapulars and a few white feathers on the lower neck or chest. These two colors increase in purity and extent by a practically continuous molt throughout the winter and spring, with considerable individual variation. The tail is molted during the winter, the time varying with different birds, and some white feathers appear in the scapulars and rump, in early spring or before that. By April many young males have the throat, neck, and chest almost wholly white, the back almost wholly black and white, and the head nearly all black above, but the wings and the under parts are still wholly immature; at this stage, the sea-green patch over and behind the ear coverts is often completely developed or intermixed with black and brown feathers of the first stage.

The next change takes place at the summer molt, in July and August. This is a complete molt involving a double molt of much of the contour plumage and producing a first-eclipse plumage, which Mr. Millais (1913) describes as follows:

> The feathers of the whole head and neck are shed and replaced in a few days by a plumage resembling, but somewhat darker than, that of the juvenile; eye-stripe dull white with blackish markings; crown, upper parts of cheeks, and back of head and neck black; rest of cheeks and throat grey-brown; mantle and scapulars blackish-brown. In a bird killed on July 6 at Fitfulhead, Shetland, which has effected the above change, the wings, tail, and nearly all the lower parts are still in juvenile plumage, much worn and faded; the white-and-buff shield on the upper chest and its sides is replaced by a new set of feathers—white with brown-black bars, and edged with reddish-brown; the long faded scapulars are still unshed and sandy-yellow as well as the primaries.

During the fall this first eclipse plumage is replaced by the second winter plumage, which is not completed until November. This resembles the adult plumage in a general way, but it can be easily recognized by its imperfections. The center of the crown is mottled with grayish brown; the green areas on the head are paler and more restricted; the white of the back is broken by scattering dusky feathers; the lesser wing coverts are brownish and the greater wing coverts are edged with dusky, both of which are pure white in adults; the curving tertials are less developed and edged with dusky, instead of being pure white; and the under parts, which are clear, deep black in adults, are now dull, brownish black, with the anterior border broken and mottled. This plumage is worn without much change until the second eclipse plumage appears the next summer. This is less complete than the adult eclipse plumage and can be distinguished from the first eclipse by the wings. The fall molt out of this plumage produces the adult winter plumage, characterized by the pure white back, wing coverts, and curved tertials. The young bird thus becomes adult at an age of 28 or 30 months. A few birds, otherwise adult, retain signs of immaturity during their third winter, chiefly in the form of dusky-edged feathers in the cream-colored breast.

The adult male has one complete molt each year, reaching its climax in August; the plumage of the head and neck is all molted twice to produce and to replace the eclipse plumage, that of the breast and back partially twice and the rest of the plumage only once. The eclipse plumage is very striking and very interesting, as it is beautifully adapted to conceal the brilliant colors of the male during the time when he is quite incapable of flight and obliged to seek refuge by swimming and diving on the open sea. The bright colors of the head and neck are completely replaced by blacks and browns in mottled effect, a complete molt of these parts beginning in July; the white back is screened by a new growth of grayish white feathers, broadly tipped with dusky; and the breast is completely concealed by new feathers, subterminally barred with the black and tipped with brown. In the fall the winter plumage is reproduced by a complete new growth of feathers on the head and neck; the plumage of the back and the breast is restored partly by molt and partly by wearing away of the dark tips; the whole process is a beautiful illustration of the maximum of concealment with the minimum of molt.

Mr. Millais (1913) says the young female can be distinguished from the male, even in the downy stage, being "generally darker on the under parts and the eye stripe narrower and shorter." In the juvenal plumage she has a "smaller eye stripe, paler upper parts and darker upper breast." The young female remains largely in the juvenal plumage through much or all of the first winter; specimens collected in February show the molt into the first spring plumage in various stages; but by March most of the birds have acquired a semiadult plumage. In this the dull-brown feathers, with narrow sandy-brown edges, of the juvenal plumage have been replaced by the dusky or dusky-barred feathers, with broad edges of deeper and richer browns, of the adult plumage. But birds in this plumage can always be distinguished from adults by their juvenal wings, which still retain the old, worn, dusky secondaries, tertials, and long scapulars; the long, curved, brown-edged tertials and the white-tipped secondaries and secondary coverts of adults are lacking; the belly plumage also remains largely immature. At the next summer molt, which is complete, a second winter plumage is assumed, which is nearly adult; but the white tips of the secondaries and secondary coverts are smaller and narrower; and the birds are usually more heavily barred above and more uniformly dark brown below. At the next molt, when a little over 2 years old, the fully adult plumage is acquired. Some females probably breed during their second spring, but probably most of them do not do so until they are nearly 3 years old.

The foregoing account of the molts and plumages of this species will suffice equally well for the American eider and the Pacific eider, as the sequence is the same in all three species, or subspecies, and the immature plumages of all three are practically indistinguishable.

Food.—Eiders obtain their food almost wholly by diving to moderate depths; almost any kind of marine animal life is acceptable and easily digested in their powerful gizzards; most of it is found on or about the sunken ledges or submerged reefs off rocky shores, which support a rank growth of various seaweeds and a profusion of marine invertebrates. They prefer to feed at low tide when the food supply is only a few fathoms below the surface; they often dive to depths of 6 or 8 fathoms and sometimes 10 fathoms, but when forced by the rising tide to too great exertion in diving, they move off to some other feeding ground or rest and play until the tide favors them again. They are usually very regular in their feeding habits, resorting to certain ledges every day at certain stages of the tides, as long as the food supply lasts. They seem to prefer to feed by daylight and to roost on some inaccessible rock to sleep at night. Many other ducks are forced to feed at night, as they are constantly disturbed on their feeding grounds during the day; but the eider's feeding grounds are so rough and inaccessible that they are seldom disturbed. Even in rough weather these tough and hardy birds may be seen feeding about the ledges white with breakers; they are so strong and so expert in riding the waves and in dodging the breakers that they do not seem to care how rough it is. I have seen them feeding, off our eastern coasts in winter, in water so rough that no boat could approach them.

Their favorite food seems to be the common black mussel (*Mytelis edulis*), which grows in such extensive beds as to furnish abundant food for myriads of sea fowl; the eiders devour these in such large quantities that their crops are most uncomfortably distended. Periwinkles, limpets, and a great variety of other univalve and bivalve mollusks are eaten; their stomachs are crammed full of such hard-shelled food, mixed with pebbles, all of which is ground up by the strong muscular action of the stomach, assisted by the chemical action of the gastric juices; the soft parts are digested and the pulverized shells pass out through the intestines. They are said to eat small fish occasionally, as well as fish roe and that of crustaceans. Starfish, sea urchins, and crabs are eaten, even the great spider crab and other large crabs measuring 2 inches across the carapace. Mr. Millais (1913) says:

I remember once, in Orkney, running down to a flock of feeding eiders that for the moment had vanished beneath the waves. One rose near the boat with something like a thick stick projecting 5 or 6 inches from its mouth, which it was unable to close. I shot the bird, an old female, and found that the obstruction, when drawn out, was a razor shell (*Ensis siliqua*), 10 inches long and 3 inches in circumference. How any bird, even with the digestion of a sea duck, could assimilate so tough a morsel with a hard and thick shell seemed a marvel, but it is doubtless the case that they are able to break them up and eject the shells as pellets.

Kumlien (1879) writes:

Their food in autumn consists almost entirely of mollusks. I have taken shells from the oesophagus more than 2 inches in length; from a single bird I have taken out 43 shells, varying from one-sixteenth to 2 inches in length.

The adult birds in spring did not seem to be quite so particular; in them I found almost all the common forms of marine invertebrates, and sometimes even a few fish (*Liparis*, and the young of *Cottus scorpius*).

Mr. Andrew Halkett (1905) found the following material in the gizzards of some 20 eiders:

Numerous shells of *Acmaea testudinalis*, numerous fragments of valves of *Tonicella marmorata*, a few shells of *Margarita cinerea*, a number of shells of other small gastropods, a few opercula of a gastropod, egg capsules of a gastropod, numerous valves of *Crenella*, fragments of valves of various small and medium sized lamellibranchs, various parts of the shells of *Hyas* and other crustaceans, a few pieces of the arms of an ophiurian, a few bones of a very small teleost, fragments of alga, numerous small stones.

Behavior.—The flight of the northern eider is apparently slow, heavy, and labored, but in reality it is much stronger and swifter than it appears and exceedingly straight and direct. Its heavy head is held low, with the bill pointing slightly downward, a characteristic and diagnostic attitude. Eiders usually fly in small flocks, in Indian file, close to the water, often following the indentations of the shore line, but very seldom flying over the land. In rough weather a flock of eiders is apt to follow the trough of the sea and is often lost to sight between the waves. I have seen one, when shot at and perhaps wounded, dive out of the air into the water and not show itself again. It is an expert at diving and hiding below the surface; if there is only the slightest ripple on the water it can conceal itself and swim away with only a portion of its bill protruding and almost invisible.

As stated above, eiders are capable of diving to depths of 8 or 10 fathoms if necessary. In diving the wings are partially opened and used to a limited extent in swimming under water, but the wings are not wholly spread; progress seems to be made mainly by the use of the feet, and there is nothing like the full subaqueous flight practiced by some of the Alcidae. Mr. Millais (1913) relates the following interesting incident:

Personally I have the gravest doubts of the truth of the statement made by many writers that the eider and other sea ducks "hold on" to the seaweed at the bottom of the ocean rather than allow themselves to come to the surface and be shot. One morning in February, 1866, I pursued an old male eider, which I had winged from a flock, into some shallows off the island of Reisa Little, in the Orkneys. The white back of the bird could be plainly seen under water entering some dark weeds amongst small rocks near the shore. Presently it disappeared in the tangle, and as the bird did not again come to the surface I leaned over the side of the boat and made search for it. I had seen it enter a comparatively small area of dark ground round which there were sand spaces, so I concluded it must be hidden amongst the fronds, and after a short search I saw the white back gleaming beside a small rock, the head and neck being concealed under the seaweed. It occurred to me that it would be interesting to see whether the bird would voluntarily leave this position or not; so after waiting for a quarter of an hour, during which it did not move, I gave it a lift with my long seal gaff, when it at once floated to the surface quite dead. The mouth was half open: some thin weeds encircled the neck. Doubtless this bird allowed itself to be drowned, as its half-open bill showed; but that it was actually holding on to the weeds I could see no sign. I could narrate several instances of a similar character which would only tend to show that whilst the birds both voluntarily get into positions under water from which they will not move until death overtakes them and also into crannies and encircling weeds from which they can not escape owing to lack of strength, yet there is not actual proof that they hold on to the weeds at the bottom of the sea, as Naumann suggests.

Outside of the vocal performances indulged in during the mating season, eiders, particularly males, are decidedly silent birds. Mr. Turner says:

The females utter a grating croak while flying to or from their nests and a hiss while on the nest. This hissing sound gives rise to the Eskimo name of this species, Mitik.

Mr. Millais (1913) says:

In winter eiders are very silent birds, like all the sea ducks except the long-tail, and their voice is not often heard except when single individuals are searching for their friends. The male when swimming occasionally utters a hoarse, grating call like the words "*Kor-er-korkorr-kor,*" and the female a slightly higher note, "*Kar-er-karkar-kaa.*" The female also utters this call when she is flying.

Referring to the enemies of eiders, Mr. Millais (1913) writes:

In the winter eiders have few enemies except man, though sea eagles often attack them along the coast line in Norway, whilst the great black-backed gull has a wonderful eye for a "picked" bird and will hunt it until it falls a prey owing to exhaustion. In the summer eiders have many enemies in their Arctic home, and a few in our islands. Even in the Orkneys and Shetlands a few of the young fall a prey to both lesser and greater black-backed gulls; whilst Richardson's skua is not wholly above suspicion. In Unst the great skua has been seen to attack and swallow young eiders. In Iceland numbers of young eiders are killed by Richardson's skua, sea eagles, and a few by the Iceland falcons. Arctic foxes are not numerous here, as they are in the Russian islands, Greenland, and Labrador, where these animals levy heavy toll on the old birds on the nests as well as the young. Polar bears also kill quantities of young eiders, and will break and eat their eggs. In West Greenland the harp seal is said to catch eiders on the water, coming up and seizing them from below, and it is possible that the small whale, *Orca gladiator*, kills a few. In northeast Greenland the chief marauder of all sea birds is the glaucous gull, which creates much havoc amongst young ducks.

Dr. I. I. Hayes (1867) gives the following graphic account of the predatory habits of the glaucous gulls:

A rugged little ledge, which I named Eider Island, was so thickly colonized that we could hardly walk without treading on a nest. We killed with guns and stones over 200 birds in a few hours; it was near the close of the breeding season. The nests were still occupied by the mother birds; but many of the young had burst the shell, and were nestling under the wing, or taking their first lessons in the water pools. Some, more advanced, were already in the ice-sheltered channels, greedily waiting for the shellfish and sea urchins which the old bird busied herself in procuring for them. Near by was a low isolated rock ledge, which we called Hans Island. The glaucous gulls, those cormorants of the Arctic seas, had made it their peculiar homestead; their progeny, already fully fledged and voracious, crowded the guano-whitened rocks; and the mothers, with long necks and gaping yellow bills, swooped above the peaceful shallows of the eiders, carrying off the young birds, seemingly just as their wants required. The gull would gobble up and swallow a young eider in less time than it takes me to describe the act. For a moment you would see the paddling feet of the poor little wretch protruding from the mouth; then came a distension of the neck as it descended into the stomach; a few moments more and the young gulls were feeding on the ejected morsel.

J. D. Figgins (1902) says that, in northwest Greenland, "eider ducks are much prized by the natives and are killed by spearing from the kayak. The spear is simply a sharpened rod of iron set into the end of a light shaft. At 15 or 20 yards the hunter seldom misses his mark."

Referring to the food value of eider's eggs to the Eskimos, Dr. Donald B. MacMillan (1918) writes:

How impatiently we awaited the discovery of those first golden nuggets in the nests. Can we ever forget those annual pilgrimages to the shrine at historic Littleton and Eider Duck Islands and McGarys Rock. Here, among a laughing, jolly company of men, women, and children, we pitched our tents among the nests; we boiled eggs, and we fried eggs, and we scrambled eggs, and we shirred eggs, and we did everything to eggs. In a few hours 4,000 delicious fresh eggs were gathered from one small island alone. Cached beneath the rocks, away from the direct rays of the sun, they remain perfectly fresh; they become chilled in August; and freeze hard as so many rocks in September—a much-appreciated delicacy during the long winter months. The shells are often broken and the contents poured or squirted from the mouth of the Eskimo into the intestinal sheath of the bearded seal or the walrus, a most nutritious sausage to be eaten on the long sledge trips.

The Moravian missionaries of northern Laborador showed me some beautiful eider-down blankets which were made by the Eskimos of Greenland for sale in the Danish markets; they were made of the breasts of eiders from which the feathers had all been plucked, leaving the down on the skins, which had been cured so that they were very soft and pliable; the edges of the blankets were trimmed with the cured skins of the heads of many northern and king eiders, making very attractive borders. They were the softest, lightest, warmest, and most beautiful blankets I had ever seen, and I was told that they brought such fancy prices that they were beyond the reach of ordinary mortals. I believe the natives also use these plucked skins for winter underwear, wearing them with the down side next to the skin; eider-down underwear and Arctic-hare stockings must be very soft and warm.

The eider-down industry has never been so highly developed on the American side of the Atlantic as it has on the other side. It would undoubtedly prove a profitable industry and would also serve to protect the birds if properly conducted. The following account of how it is done in Iceland, written by C. W. Shepard, is published by Baird, Brewer, and Ridgway (1884):

The islands of Vigr and Oedey are their headquarters in the northwest of Iceland. In these they live in undisturbed tranquility. They have become almost domesticated, and are found in vast multitudes, as the young remain and breed in the place of their birth. As the island (Vigr) was approached we could see flocks upon flocks of the sacred birds, and could hear their cooing at a great distance. We landed on a rocky, wave-worn shore. It was the most wonderful ornithological sight conceivable. The ducks and their nests were everywhere. Great brown ducks sat upon their nests in masses, and at every step started from under our feet. It was with difficulty that we avoided treading on some of the nests. On the coast of the opposite shore was a wall built of large stones, just above the high-water level, about 3 feet in height, and of considerable thickness. At the bottom, on both sides of it, alternate stones had been left out so as to form a series of square compartments for the ducks to nest in. Almost every compartment was occupied, and as we walked along the shore a long line of ducks flew out, one after the other. The surface of the water also was perfectly white with drakes, who welcomed their brown wives with loud and clamorous cooing. The house itself was a marvel. The earthen walls that surrounded it and the window embrasures were occupied by ducks. On the ground the house was fringed with ducks. On the turf slopes of its roof we could see ducks, and a duck sat on the door scraper. The grassy banks had been cut into square patches, about 18 inches having been removed, and each hollow had been filled with ducks. A windmill was infested, and so were all the outhouses, mounds, rocks, and crevices. The ducks were everywhere. Many were so tame that we could stroke them on their nests; and the good lady told us that there was scarcely a duck on the island that would not allow her to take its eggs without flight or fear. Our hostess told us that when she first became possessor of the island the produce of down from the ducks was not more than 15 pounds in a year; but that under her careful nurture of 20 years it has risen to nearly 100 pounds annually. Most of the eggs are taken and pickled for winter consumption, one or two only being left in each nest to hatch.

Fall.—By the middle of the summer, or as soon as the egg-laying season is over, the adult eiders desert their mates and begin to move away from their breeding grounds. This might be called the beginning of the fall migration. The immature males of the previous year keep by themselves all summer in large flocks and do not even now mingle with the old males; they spend the summer well out at sea near the drift ice. The young eiders often are late in hatching and are slow in developing, so that it is often quite late in the fall before they are able to fly away with their mothers and join the mixed flocks in their winter resorts. All have to undergo the annual summer double molt, which lasts well into the fall and delays migration, for they are absolutely flightless for a few weeks while the wing quills are molting. So the fall migration is very irregular and much prolonged; many birds spend the winter not far away from their breeding grounds, moving only far enough away to find open water and good feeding grounds.

Winter.—On the coast of Greenland many northern eiders spend the winter in the open waters of the fjords. Near Ivigtut, Hagerup (1891) observed that:

In October, 1886, the females began to come into the fjord singly, and in November they came in small flocks. As the weather grew colder the number increased, and it became still larger after Christmas, the period of greatest abundance being March and April. The males did not come in as great numbers into the fjord that winter. I saw, indeed, none at Ivigtut until March, while they were quite numerous at Christmas of the following year.

In the evening these birds generally go as far inland as there is open water, and during the night they are almost constantly on the move. Then their cries may be plainly heard, as also their splashing near the shore; but if a match be lit, they fly aloft with a great uproar.

Eiders are at all seasons essentially sea ducks, but especially so in winter. Low temperatures have no terrors for them; their winter resorts extend from Greenland to the coast of Maine, wherever they can find open water and plenty of food. Even in the roughest winter storms they brave the rigors of the open sea, riding at ease among the white caps, diving for food among the surf-swept ledges, safe from the molestation of the hardiest gunners, and retiring at night to rest on some lonely rock or drifting iceberg.

Mr. Ekblaw writes:

A few eiders stay all through the four months winter night in the open waters of Smith Sound. Like the guillemots, the eiders find sufficient food in the upwelling of the tidal currents about the Gary Islands to maintain themselves throughout the coldest winters. These strong swift tidal currents running back and forth through Smith Sound between Baffin Bay and Kane Basin prevent the formation of widespread ice, and are the controlling factor which permits a luxuriant far-Arctic plant and animal life, including a pleasant homeland for the polar Eskimo, in this habitat a thousand miles within the Arctic Circle, far beyond the usual northern limit of life.

The far northern coasts of Greenland afford a sanctuary for the eider, where this splendid species of the duck family may save itself from total extermination. To the lonely, inaccessible rocks and islets of these far northern shores, the egg hunters, and down gatherers are not likely to come in numbers enough or often enough to destroy the species. The natives are too few and the value of the eider in their economic needs is too small to constitute any serious menace to the species. Farther south, in Greenland, and elsewhere, the eider is threatened with extinction, though in Danish Greenland the Danes are vigilantly safeguarding the birds as far as it lies in their power to do so. It is fortunate that the eiders have their far northern habitats, relatively safe from man's devastation.

DISTRIBUTION

Breeding range.—Coastal islands of Greenland and northeastern America. South on the Atlantic coast of Labrador to the vicinity of Hamilton Inlet and, in the regions north of Hudson Bay, to Baffin

Land, Ungava Bay, Southampton Island, and Cape Fullerton, intergrading at these points with *dresseri*. West about 100° west longitude, where it may intergrade, somewhere in the Arctic Archipelago, with *V-nigra*. North to northern Ellesmere Land (81° 40′) and northern Greenland (82° on the west coast and 75° on the east coast). Represented in Iceland and Europe by closely allied forms.

Winter range.—South along the coast to Maine and rarely to Massachusetts. North to southern Greenland, as far as open water extends.

Spring migration.—Early dates of arrival: Labrador, Battle Harbor, May 1; Cumberland Sound, April 30; northern Greenland, Etah, April 20; Wellington Channel, latitude 76°, May 17; Cape Sabine, latitude 79°, May 28; Thank God Harbor, latitude 81°, June 4. Late dates of departure: Massachusetts, April 3; Maine, April 6.

Fall migration.—Early dates of arrival: Maine, October 19; Massachusetts, late October. Late dates of departure: Thank God Harbor, November 4; Etah, November 1; Cumberland Sound, November 17.

Egg dates.—Labrador: Eight records, June 15 to August 2. Greenland: Two records, June 23 and July 2.

SOMATERIA MOLLISSIMA DRESSERI (Sharpe)

AMERICAN EIDER

HABITS

Contributed by Charles Wendell Townsend

The eider is a duck of which the Americans should well be proud. Large and splendid in plumage, interesting in courtship display, pleasing in its love notes, susceptible to kind treatment by man, and capable of furnishing him with a product of great value; notwithstanding all this, the bird is so incessantly persecuted, especially on its breeding grounds, that it is rapidly diminishing in numbers. If the senseless slaughter is not stayed the eider will continue to diminish until it is extinct. Happily, even now there are signs of a better era. On the Maine coast—the bird's most southern breeding station—there were less than a dozen pair breeding in 1905. As a result of protection, however, through the efforts of the Audubon Society, their numbers are now increasing, and Bowdish (1909) in 1908 reported as many as 60 eiders breeding on Old Man Island alone. Farther north the persecution still goes on; on the Nova Scotia coast not more than two or three remain to breed, while on the coast of Newfoundland and of the Labrador Peninsula south of Hamilton Inlet, where they formerly bred in immense numbers, but a remnant is left. All the ornithologists from the time of Audubon to the present day who have visited the Labrador coast have bewailed the fact that the eider was singled out for destruction. In "A Plea for the Conservation of the Eider" (Townsend, 1914) occurs the following:

There is no reason why the eider, which furnishes the valuable eider down of commerce, should not be made a source of considerable income without any reduction of its natural abundance. The principle of conservation can as well be applied to the eider as to a forest. The conservation of the common eider of Europe (*Somateria mollissima*), a species that differs but very slightly from the American bird, has been practiced for many years in Iceland and Norway. The birds are rigidly protected during the nesting season and offered every encouragement. They are not allowed to be shot, and even the discharge of a gun in their vicinity is forbidden by law. Suitable nesting sites are furnished close to the houses and the birds become semi-domesticated, losing all fear of man. The people are allowed to take the eggs and down during the first of the season, but the birds are permitted to hatch out and rear a few young in order to keep up the stock. The last down is taken after the birds have left.

Many quotations are given from various authors showing what is being accomplished in Iceland and Norway. For example (Annandale, 1905):

The one offense against the Icelandic bird laws which a native can not commit with impunity is the slaughter of the eider duck. What is more important than many laws, namely public opinion, protects the species, and there seems to be a sentimental interest in it. Probably it is due to the great tameness of the bird, which appears actually to seek the vicinity of a human dwelling for its nesting place and to frequent those parts of the coast which are more frequented by man. The Icelandic eider farms are frequently situated on little islands off the coast. Small circular or oblong erections of rough stones are made among the hummocks, to protect the brooding ducks from wind and driving rain. All the sea fowl in these farms become exceedingly tame, as no gun is allowed to be fired and everything liable to disturb the ducks is carefully banished. Those who know how to handle them can even stroke the backs of the ducks as they sit on their eggs. On such farms there is a separate building or large room entirely devoted to cleaning the down. It was formerly the custom to take away all the down supplied by the female; but this practice was said to lead to great mortality among the ducks through exhaustion, and nowadays each nest is generally rifled only once before the eggs are hatched, and then again after the young have left it.

Townsend (1914) then goes on to say:

Eiderdown is not only extremely light and elastic but is also one of the poorest conductors of heat. It is therefore an ideal substance for preserving warmth and is the best material for coverlets, puffs, cushions, etc. Its money value is considerable, and there is always a demand for it in the markets of the world. The down obtained from dead eiders, however, soon loses its elasticity and is of little value. The retail price in Boston at the present time of well-cleaned Iceland or Norwegian eiderdown is $14 a pound. It is probable that each nest furnishes—as a very conservative estimate—from an ounce to an ounce and a third of down; therefore 12 to 16 nests or breeding females are needed for each pound. Burton states that the annual supply of down in Iceland rose from 2,000 pounds in 1806 to 7,000 pounds in 1870. One can easily understand the great value of this product even if the producer receives only one-half of the retail price. He could count on at least 50 cents a season for each breeding female in his eider fold.

Imagine the pleasure as well as the profit that could be obtained along the coast of Labrador, Newfoundland, Nova Scotia, and Maine if these birds were treated in the manner above described and flocked and nested about the habitations of man. Then, each dweller in suitable localities by the sea could have his own flock of these beautiful birds, for the female is as beautiful in her modest dress of shaded and pencilled brown as is the male in his striking raiment of jet black and cream and snow white, delicate sea green and dark navy blue. The cooing notes, so long few or absent in many places, would again resound over the waters, and best of all, to the practical minded, the birds would pay well for their protection by gifts of eggs and of valuable eider down.

How can the present senseless habit of destruction be stopped and this desirable state of affairs brought about? As a preliminary step in Labrador and Newfoundland, I would suggest that a few islands scattered along the coast should be made bird reservations, and carefully guarded by one or two families who live on or near the islands. These people should be allowed to take the first set of eggs and down, as well as the down left behind after the duck has hatched out the second set and has left for the season, but should not be allowed the use of firearms, and their Eskimo dogs must be confined during the nesting season. In other words, these people must not frighten the birds and must treat them kindly. The object of the experiment should be spread broadcast along the coast with the request for fair play, so as to restrain others from poaching and frightening the ducks on the reservation.

The rapidity with which the birds will respond to this treatment and the intelligence they will display in the recognition of the safety spots will surprise the people. This is the case wherever bird reservations are established. At Ipswich, Massachusetts, the shores of a small, protected pond are thronged with shore birds of many species which display almost no fear of man, while on the neighboring beaches, where they are shot, they are very wary. In the city of Boston the Charles River Basin and Jamaica Pond are the resort of numerous ducks that pay but little attention to the people, while in the sea and ponds near by, where shooting is allowed, the ducks show their usual wildness.

It is useless to pass laws if they are not observed or if the sentiment of the community is against them. This reform, which will be of such great value to our northern seacoast, can only be accomplished by education, and these bird reservations with their eider farms will be one of the best means to that end.

Spring.—The spring migration occurs on the New England coast in the latter part of March or early in April. The birds have been wintering south to Nantucket and in rare cases to Delaware and Virginia. The adult males go north two or three weeks or a month ahead of the females and immature. In the latter part of May and early in June we found them abundant on the southern coast of the Labrador Peninsula. Some of them were nesting, but it was probable that many of them were on their way farther north along the coast.

Courtship.—The courtship of the eider has a certain resemblance to that of the goldeneye, but it is not so spectacular. It can be observed during the latter part of May and in June along the southern coast of the Labrador Peninsula. It is thus described by Doctor Townsend (1910):

> The actual courtship of the eider may be recognized from afar by the love note of the male, which can be expressed by the syllables *aah-ou* or *ah-ee-ou*, frequently repeated, and, while low and pleasing in tone, its volume is so great that it can be heard at a considerable distance over the water. On a calm day when there were many eiders about, the sound was almost constant. While the syllables *aah-ou* express very well the usual notes, there is much variation in tone, from a low and gentle pleading to a loud and confident assertion. In fact, the tones vary much as do those of the human voice, and there is a very human quality in them, so much so that when alone on some solitary isle I was not infrequently startled with the idea that there were men near at hand.
>
> But the showy drake eider does not depend on his voice alone; he displays his charms of dress to best advantage and indulges in well-worn antics. It always seemed to me a pity that the magnificent black belly should disappear when the drake is swimming on the water, and the bird evidently shares my sentiment, for during courtship he frequently displays this black shield by rising up in front, so that at times in his eagerness he almost stands upon his tail. To further relieve his feelings he throws back his head and occasionally flaps his wings. The movements of the head and neck are an important part of the courtship, and although there is considerable variation in the order and extent of the performance, a complete antic is somewhat as follows: The head is drawn rigidly down, the bill resting against the breast; the head is then raised up until the bill points vertically upward, and at this time the bill may or may not be opened to emit the love notes. Directly after this the head is occasionally jerked backward a short distance, still rigidly, and then returned to its normal position. All this the drake does swimming near the duck, often facing her in his eagerness, while she floats about indifferently, or at times shows her interest and appreciation by facing him and throwing up her head a little in a gentle imitation of his forcefulness.

In walking around the shore of Eskimo Island, one of the Mingan group of southern Labrador, on June 3, 1909, we found the water everywhere dotted with these splendid birds, all intent on courtship. In one cove there were 104 birds, in another 80. Only one eider was in partially immature plumage. The birds were for the most part in pairs, but there were frequent groups of 3 or 4 males ardently courting 1 female. There were also occasional coteries of 15 or 20 where the sexes were about evenly divided.

Nesting.—It is evident that the eider prefers to nest in communities, but where diminished in numbers from persecution they nest singly. On some small islands off the southern coast of Labrador we found, in 1909, 20 or 30 nests in the space of an acre; this, too, when the birds were much harassed and evidently less numerous than in former years. The nests are placed on the ground, generally close to salt water and almost always on islands; we have found them, however, a hundred yards or more from the water. The nesting site may be open to the sky in a depression among the rocks of a barren island, but it is often partially or wholly concealed among and under spruce, alder, and laurel bushes or in the grass and rushes.

Mr. Harrison F. Lewis (1922) refers to a nest which he found in an unusual situation, as follows:

> On June 24 I found an eider's nest with six eggs on a bush-covered rock in the midst of the second falls of the Kegashka River, more than a mile from the sea. In an expansion of the river near by I saw seven female eiders swimming about. Residents of the vicinity informed me that considerable numbers of young eiders descended this river each autumn. It would be interesting to know just how and at what age the young eiders from the nest which I found left their birthplace, situated as it was in the midst of a foaming cataract.

The nest itself is made of seaweeds, mosses, sticks, and grasses matted together, but is chiefly distinguished for the famous eider down which is plucked by the mother from her breast. The down is of a dull gray color, very soft, light, and warm, and is supplied in such liberal amounts that the eggs can be entirely covered when the sitting bird is absent. If the departure of the mother is sudden and forced, the eggs are left exposed. The female can supply plenty of down for two sets if the first is stolen, but the story that the male is called upon to supply down for the third set is not true, for he does not go near the nest. The down is rarely clean, as it generally contains bits of moss, twigs, and grasses. If the nests are repeatedly robbed of their down the poor bird is obliged to use other material in its place, and some of the nests under these circumstances at the end of the season are practically destitute of down.

Eggs.—Under normal circumstances only one set of eggs is laid. Five eggs constitute a setting which, however, varies from 4 to 6 and in one instance, that came under our observation, to 7 eggs.

The eggs are nearly oval in shape, of a rough and lusterless exterior, as if the lime had been put on with a coarse brush. Their color is a pale olive green with patches and splashes of dull white. Sometimes their color is pale brownish or olive. The measurements of 59 eggs in various collections average 76 by 50.7 millimeters; the eggs showing the four extremes measure **83.5** by **54.8**, **65** by 44.5, and 66.4 by **41.5** millimeters.

[AUTHOR'S NOTE.—As the eggs, young, and plumage changes of the common eider are exactly like those of the northern eider, no further attempt will be made to describe them here, and the reader is referred to what has been written on this subject under the foregoing subspecies.]

Young.—The period of incubation is 28 days, and the incubation is performed only by the female. The young as soon as hatched are led to the water by the female, who is also said to help them over difficult places by carrying them in her bill. From the first they are expert divers in shallow water, and are assiduously tended by the mother who draws them from danger, or acts the part of the wounded duck to distract attention while they hide among the rocks or in the grass. The downy young are of grayish brown color, lighter on the belly. There is a pale line along the side of the head over and under the eye.

Plumages.—In his full nuptial plumage the male eider is a splendid sight, a very conspicuous object. It has been said by Thayer that the drake's plumage is in reality concealing, as the white matches the snow, the green and dark blue the ice and the water, and the black the rocky cliffs. But to one who is familiar with the bird, either among snow and ice, on the open sea, or under the beetling crags, on rocks or among mosses and dwarf spruces of the northern bogs, such an idea does not appeal, for the bird is always conspicuous. The display he makes of his plumage in courtship, and the fact that he retires after this season is over into the eclipse plumage which is similar to that of the female, is good proof—if proof were needed—that the courtship dress is for show only and not for protection. The eclipse plumage worn by the males and the nearly similar brown plumage of the females and young is indeed an inconspicuous one, and the birds wearing this are to a large extent protectively colored. I have almost stepped on the nesting female and did not see her until she ran from the nest; and at a distance on the ocean, one may see a band of eiders of which only the males are visible until a nearer approach, when one is surprised to find an equal number of females.

After the nesting season is over the males retire to the outer islands and rocks, where they are for a time unable to fly owing to the extensive molt into the eclipse plumage. According to Audubon the sterile females molt at the same time, but the females with broods do not molt until fully two weeks later.

With rare exceptions all the eiders in the region of the Mingan Islands in southern Labrador we found to be in full adult plumage in the last half of May and the first half of June. In the first half of July, on the southern part of the eastern coast, many birds were seen that were molting from the nuptial drake plumage into the eclipse, while in the last part of July and early in August, in the Mingan region, we found nearly all the eiders to be in the brown plumage; only a very few showed traces of the brilliant nuptial dress of the drakes. Native hunters say that the drakes leave the coast about the last of July, but it is probable that this is apparent only, and that the brown birds are females and immature birds of both sexes, as well as adult drakes in the eclipse plumage. The change from the eclipse to the nuptial plumage occcurs in November and December and from the immature to the first nuptial dress in the spring, after the 1st of March. It probably takes three years before the full drake plumage is acquired, although there is considerable variation in this.

Eiders with V-shaped marks under the chin suggestive of *Somateria v-nigra* of the Pacific coast have been reported on the Atlantic coast by W. A. Stearns (1883). Arthur H. Norton (1897) says that "the black lancet is a character of frequent occurrence in the young drakes of *S. dresseri*, and there are strong reasons for the belief that it occurs in *S. mollisima borealis*." He describes four specimens where the mark occurs in *S. dresseri*. He states that they are all immature birds and "show nothing than can be considered as of a hybridic nature."

Food.—The favorite food of the eider is the edible mussel (*Mytelis edulis*), although various other mollusks, crustaceans, echinoderms, and worms are taken with avidity. Mr. Mackay (1890) reports the finding of sculpin spawn in the stomach of the eider. Such food is particularly abundant around rocky ledges, and the birds gather there from all sides during the day, but toward evening they fly out to sea to spend the night.

Behavior.—The eider is an expert diver and uses the wings under water. This is evident from the fact that it flaps open the wings for the first stroke. If alarmed when diving it often comes out of the water flying, merely changing from subaqueous to aerial flight.

Floating or swimming on the water, the head and neck are generally drawn down as if resting at ease. When on the alert the neck is stretched up and is much elongated. The tail is often cocked up at an angle.

The flight of the eider is generally close to the water, swift and powerful, but in the absence of a head wind the bird often flaps along the surface of the water for several yards before it is able to rise. The neck is stretched out and the bill is pointed obliquely downward at an angle of 45°, a field mark of some value in the recognition of the bird.

The speed of flight was estimated by Cartwright (1792), who recorded it in his Labrador Journal, as follows:

> In my way hither I measured the flight of the eider ducks by the following method, viz, on arriving off Duck Island, 6 miles distant from Henry Tickle, I caused the people to lie on their oars, and when I saw the flash of the guns,

which were fired at a flock of ducks as they passed through, I observed by my watch how long they were in flying abreast of us. The result of above a dozen observations ascertained the rate to be 90 miles an hour.

The male, aside from his courtship notes or love song already described, appears to be a silent bird. The female utters at times a rolling quack or a succession of sharp *kuk kuk kuks;* the latter is heard when she is suddenly disturbed at the nest.

The great black-backed gull is probably the greatest enemy of the eider, aside from the arch enemy, man. Nesting in the same region and having a voracious progeny to support this gull takes frequent opportunities of pillaging the nest and capturing the downy young. The raven also eats both eggs and young. Another enemy, which is now entirely extirpated from the Labrador region south of Hamilton Inlet, is the polar bear. Cartwright (1792) gives several instances of their depredations. He says in his Journal, under date of June 18, 1777:

On examining the paunches of the bears (an old bitch polar bear and her cub), found them well filled with eggs. I had often heretofore observed that all the nests upon an island had been robbed and the down pulled out, but I did not know till now how those things happened.

Winter.—In the fall migration eiders arrive on the New England coast late in November or early in December. The eider winters throughout its range wherever there is open water and as far south regularly as Nantucket, rarely as far as Delaware and Virginia. Along the Maine coast it is still abundant in winter, although its numbers are much reduced over those of former days. On the Massachusetts coast the eider may be found off Cape Ann and Cape Cod, and especially in the tempestuous and shallow seas about Nantucket.

Mackay (1890) records the shooting of 87 eiders in a December day in 1859 by one man near the Salvages, small rocky islands off Rockport, on the end of Cape Ann. I was told that in 1875 a hundred eiders were shot there by a gunner in January. Of late years one is lucky to find any, but on March 14, 1909, I saw a flock of 17 near the Salvages. Mackay (1890) says that as the birds come in to feed in the morning they alight some distance outside these rocks and swim in in a compact body. They dive for the mussels outside the breakers. On March 18, 1875, he saw a flock between Muskeget and Nantucket Islands that he estimated contained 12,000 birds, and a flock near the harbor of Nantucket in 1890 of about 1,500. In March, 1894, he estimated about 200 eiders near Muskeget and 2,000 near Cape Poge, Marthas Vineyard. Also between 4,000 and 5,000 near Woods Hole, attracted by the great beds of the edible mussel.

Game.—Eiders are shot in winter off the Maine coast from blinds among the rocks off the islands and occasionally from boats. Wooden decoys are used and the sport requires much energy and endurance in the winter seas. The birds are very shy, easily taking alarm, and with their strong flight and well-made armor of thick feathers and down are very difficult to kill. Cripples are rarely recovered, as they escape by rapid diving and long swimming under water. Their value for down should, however, preserve them from persecution at all seasons.

DISTRIBUTION

Breeding range.—Islands along the coasts of Labrador (south of Hamilton Inlet), Newfoundland, eastern Quebec (north shore of the Gulf of St. Lawrence), Nova Scotia, and Maine (west to Penobscot Bay). Also in the southern half of Hudson Bay and James Bay at least as far north as Richmond Gulf, Southampton Island, and Cape Fullerton. It intergrades with *borealis* at the northern limits of its breeding range.

Winter range.—Northeastern coasts, from Newfoundland and the Gulf of St. Lawrence south regularly to Massachusetts (Vineyard Sound and Nantucket), rarely to New Jersey, and casually farther south.

Spring migration.—Northward movement starts late in March or early in April. Usual date of departure from Massachusetts is about April 20; unusually late dates are, Connecticut, Milford, May 29; Massachusetts, May 18.

Fall migration.—Early dates of arrival in Massachusetts: Essex County, September 20; Cohasset, September 18; usual date of arrival is early in November.

Casual records.—Inland wanderings have occurred as far west as Wisconsin (Lake Koshkonong, November, 1891), Iowa (Sioux City, November 1, 1901), and Colorado (Loveland).

Egg dates.—Labrador Peninsula: Twenty-five records, May 26 to July 9; thirteen records, June 5 to 20. Maine, New Brunswick and Nova Scotia: Ten records, May 25 to July 5; five records, June 12 to July 1.

SOMATERIA V-NIGRA Gray

PACIFIC EIDER

HABITS

The common eider of the Pacific coast is closely related to the eiders of the Atlantic coast, perhaps more closely than its present status in nomenclature would seem to indicate. The ranges of the Pacific eider and the northern eider on the Arctic coasts of North

America come very close together and perhaps overlap, with a good chance for hybridizing or intergrading as subspecies.

Some interesting specimens have been taken in eastern waters which are worth considering. Hagerup (1891) mentions specimens of *Somateria mollissima*, taken by Holboell in Greenland, which showed the black lancet-shaped figure on the throat so characteristic of *S. v-nigra.* Holboell supposed that these were hybrids with *S. spectabilis* which also has this black V. Mr. Arthur H. Norton (1897) obtained several specimens of immature males of *S. dresseri* on the coast of Maine, from which he inferred "that the black lancet is a character of frequent occurrence in the young drakes of *S. dresseri;* and there are strong reasons for the belief that it occurs in *S. mollissima borealis.*" Mr. W. A. Stearns (1883) obtained similar specimens on the coast of Labrador and even recorded the Pacific eider as of regular occurrence there. The occasional appearance of this mark in immature males might perhaps indicate an occasional cross between two species, but it seems more reasonable to regard it as a reversion to an ancestral type, which would mean that at some date, probably not very remote, these three eiders belonged to a single species and perhaps even now they are merely intergrading subspecies. In this connection I would refer the reader to Dr. Charles W. Townsend's (1916*a*) interesting paper and plate showing intergradation between *S. mollissma borealis* and *S. dresseri.*

Whatever the systematic status of the Pacific eider may prove to be, its habits and its lift history are practically the same as those of the Atlantic species, and nearly everything that has been written about the latter would apply equally well to the former. Therefore I shall not attempt to write a full life history of this species, which would be largely repetition.

Nesting.—By the time that we reached the Aleutian Islands, early in June, the vast horde of eiders that winter in the open waters of this region had departed, to return to their extensive breeding grounds farther north. But we found the Pacific eider well distributed, as a breeding bird, on all of the islands west of Unalaska. They were particularly abundant about Kiska Harbor in small flocks and mated pairs. They frequented the rocky beaches at the bases of the cliffs, where they sat on the loose rocks, fed in the kelp beds about them, and built their nests among the large boulders above high-water mark. Here on June 19, 1911, I examined two nests of this species; one, containing 5 fresh eggs, was concealed in a hollow under or between two tufts of tall, rank grass which grew back of a large boulder on the beach at the foot of a high grassy cliff; the other, containing 4 fresh eggs, was hidden in the long grass at the top of a steep grassy slope; both nests were well supplied with down.

A pair of Peale falcons were flying about some cliffs, and probably had a nest, not very far from where we found these eiders' nests. This reminds me of what Mr. Lucien M. Turner (1886) says, on this subject:

> Another peculiarity that was brought to my notice by a native, was that these birds (the eiders) usually seek some slope where the duck hawk has its nest on the high point forming one end of the slope. This was true in three instances that came under my observation. The eiders were more numerous in such localities than otherwise. The natives always are glad when the hawk comes screaming overhead, as the canoe is being paddled along the shore, for they know the nest of the hawk is near and that many nests of the eider will be found close by.

About St. Michael the nesting habits of the Pacific eider are somewhat different. Dr. E. W. Nelson (1887) says:

> Their courtship must be conducted before the birds reach the breeding ground, as I have never seen any demonstrations such as are usual among mating birds. The small flocks seen at first glance give place at once to solitary pairs, which resort to the salt marshes. The nesting site is usually a dry spot close to a small pond or a tide creek and not often in close proximity to the seashore. The moss-grown slope of some small knoll, a grassy tussock, or a depression made on an open flat, but hidden by the thin growth of surrounding vegetation, are all chosen as nest sites.
>
> The first evening after my arrival at St. Michael I walked back on the flat about 200 yards from the fort and put up a female from 5 fresh eggs. The nest was thickly lined with down and concealed by dwarf willows and other low Arctic vegetation. This was the only instance noted by me where the nest was so near human habitations. The nest is usually lined with dead grasses and sometimes fragments of moss when the first egg is laid, and the down is added as the eggs multiply. The male is a constant attendant of the female until her eggs are nearly all deposited, when he begins to lose interest in family affairs, and dozens of them may be found at all hours sunning themselves upon the long reefs about shore, and if we are behind the scenes on the marshes they may be seen flying silently back to their partners as the dusky twilight of night approaches from 8 to 10 in the evening.

Mr. F. Seymour Hersey contributes the following notes:

> Unlike the spectacled, Steller, and king eiders, which spend considerable time and frequently nest among the tundra ponds some distance back from salt water, I found the Pacific eider to be almost exclusively a bird of the seacoast during the breeding season. About St. Michael Bay, portions of the shore of which is covered by volcanic rocks, these birds were quite abundant and were often seen about the rocky island at the entrance to the "canal," swimming in the surf or resting on the rocks and preening their feathers. They were also frequently met with in pairs flying up or down the canal near the entrance, but did not seem to follow it farther than the point where it begins to narrow, a mile or so from the bay. The land at this point is 6 or 7 feet above the reach of normal high tides, level, and quite dry, and with very few ponds. The ground is thickly and softly carpeted with a growth of mosses, creeping vines, and such Arctic vegetation as is common to this region. Here the birds nest, making a deep cup-like hollow in the thick mosses, the edge flush with the surface and abundantly lined with a thick wall of soft gray down.

Apparently the most extensive breeding grounds of this species are in the vicinity of Franklin and Liverpool Bays, where MacFarlane (1891) collected and sent to Washington over a thousand eggs. He says:

The nest is usually a shallow cavity in the ground, more or less plentifully lined with down. We found some nests on a sloping bank at a distance of 300 or more feet from the sea. Others were also on the mainland, but the bulk of those secured by us were obtained from sandy islets in the bays.

Eggs.—The Pacific eider has been credited with laying anywhere from 5 to 10 eggs, but probably the larger sets are exceptional. The eggs are indistinguishable in every way from those of *mollissima borealis* and *dresseri*, so I will not attempt to describe them here. The measurements of 85 eggs in the United States National Museum and the writer's collections average 75.9 by 50.4 millimeters; the eggs showing the four extremes measure **86.5** by 52, 74.5 by **55.5**, **70** by 48.5 and 71.5 by **47** millimeters.

Young.—Although incubation is performed wholly by the female, and although the male usually deserts the female after the eggs are all laid, he sometimes remains near her during the process of incubation and may help to guard the nest or young.

Doctor Nelson (1887) says:

From the 15th to the 20th of June nearly all the males desert their partners and are thenceforth found at sea or about outlying reefs and islands in large flocks, as already described. Toward the end of June the first young appear, but the majority are not hatched until the first of July. As the young are hatched they are led to the nearest large pond or tide creek and thence to the sheltered bays and mouths of streams on the seacoast. About this time the females lose their quill feathers and, like the young, are very expert in diving at the flash of a gun. At this time the Eskimo amuse themselves by throwing spears at the young, but the latter are such excellent divers that they are rarely hit. As a rule the young do not fly before the 10th of September, and broods with the female are often seen unable to fly even later.

Mr. Hersey's notes on the young of this species say:

When the young are hatched the female leads her brood to some small pond or lagoon just back from the coast, and apparently they do not take to the open sea or even the outer bays until well grown. I never saw any young birds on St. Michael Bay until they had become strong on the wing. A female with a brood of young was found in a small lagoon just back from the beach on Stuart Island on July 8, 1915. In her endeavors to lead her brood to safety the mother bird was absolutely fearless. I stood on the bank within 30 feet of her and watched for about 20 minutes. She splashed about in the water, making quite a commotion, all the while calling in a low guttural tone. Some of the young swam around her while others dove. Those that went under seemed unable to stay down more than the smallest fraction of a minute and reappeared almost instantaneously, bobbing up as buoyantly as corks. After considerable difficulty she got her brood around her and started to swim away, but the young did not follow and, after swimming some 12 or 15 feet and calling, she returned and the performance was repeated.

Whether the actions of the young were due to panic or inability to realize the presence of danger I could not tell, but they appeared to me to be decidedly stupid. I know of no other duck, not excepting the oldsquaw, which at times seems to be devoid of any sense of danger, that would not have led her young to safety in a fraction of the time it took this eider.

Plumages.—The downy young are absolutely indistinguishable, so far as I can see, from those of either *mollissima* or *dresseri*, and the molts and plumages of all three are practically identical.

Mr. Hersey secured for me a fine specimen of an adult male in full eclipse plumage, of which he says, in his notes:

> The molt of the males into eclipse plumage takes place during July, and by August 1 the birds are flightless. This flightless period is probably spent on the open sea where they are practically safe. On August 11, 1915, a bird in full eclipse plumage was secured on St. Michael Bay. The new primaries were about 3 inches long, but the bird was still unable to fly. Its diving and swimming powers fully compensated for its loss of flight, and it was captured only after more than four hours' pursuit, when a mistake in judgment brought it to the surface, after a long dive, within range of our boat.

Food.—The Pacific eider apparently does not differ from its Atlantic relative in its food and feeding habits.

Behavior.—Pacific eiders have similar enemies and as meekly submit to their depredations as do their eastern relatives. MacFarlane (1891) saw a snowy owl eating the eggs in an eider's nest. Mr. Turner (1886) writes:

> The bird is very shy except when on land during boisterous weather. At that time the natives of the western islands of the Aleutian chain used small handnets to throw over the birds as they sat stupidly on the shore. A bright night with a hard gale of wind was the best time to secure them. The birds then sit in a huddle and many are caught at one throw of the net. The natives assert that the common hair seals catch these birds when on the water and drag them under to play with them; hence, these birds are constantly on the alert for seals and take flight as soon as a seal is discovered near.

Winter.—The Pacific eider has a more decided migration than the Atlantic species, for, though it breeds abundantly on the Arctic coast of Alaska and eastward to the Coppermine River, it is not known to winter north of Bering Straits, as there is no open water to be found in this portion of the Arctic Ocean. The main winter resort of the Pacific eider is in the vicinity of the Aleutian Islands, though it has been detected in winter as far north as the Diomedes. This means a migration route of about 2,000 miles from the remotest breeding grounds. On the other hand, the birds which breed in the Aleutian Islands and south of the Alaska Peninsula probably do not migrate far from their breeding grounds.

DISTRIBUTION

Breeding range.—Coasts of northwestern America and northeastern Asia. East to Coronation Gulf and west to Cape Irkaipij, Siberia. South on both coasts of Bering Sea to the Commander and Aleutian Islands, and eastward along the south side of the Alaskan Peninsula to Kodiak Island and Cook Inlet (Chugachick Bay). North to Banks Land (Cape Kellett) and Victoria Land (Walker Bay). It may intergrade with *borealis* at the eastern extremity of its breeding range, in the vicinity of 100° west longitude.

Winter range.—Mainly in the vicinity of the Aleutian Islands and the Alaska Peninsula, extending but little south of its breeding range and north as far as open water extends, sometimes as far as the Diomede Islands.

Spring migration.—Early dates of arrival: Alaska, Point Hope, April 15; Point Barrow, May 16; and Demarcation Point, May 26; northeastern Siberia, May 28. Usual date of arrival at St. Michael, Alaska, is from May 10 to 20.

Fall migration.—Birds which breed in Coronation Gulf leave in September and migrate about 2,000 miles. Last seen at Point Barrow November 4.

Casual records.—Has wandered as far south on the Pacific coast as Washington (Tacoma, January 6, 1906), and has been recorded a number of times in the interior of Canada, as follows: Severn House, 1858; Fort Resolution, 1861; Fort Good Hope, June 14 and 30, 1904; and Manitoba (Giroux, November 11, 1911, and Lake Manitoba, October 23, 1911).

Egg dates.—Arctic Canada: Thirty records, June 6 to July 15; fifteen records, July 2 to 8. Alaska: Sixteen records, June 10 to July 12; eight records, June 20 to 29.

SOMATERIA SPECTABILIS (Linnaeus)

KING EIDER

HABITS

Although generally regarded as one of the rarer ducks and although comparatively few naturalists have ever seen it in life, this beautiful and showy species is astonishingly abundant in certain portions of its northern habitat. Among the vast hordes of wild fowl which migrate in the spring from Bering Sea, through the straits, to their Arctic summer homes the king eider takes a prominent and a conspicuous place, as the following observations by Mr. John Murdoch (1885), at Point Barrow, will illustrate:

This is by all means the most abundant bird at Point Barrow. Thousands hardly describes the multitudes which passed up during the great migrations, within sight of the station, and yet equally great numbers passed up along the "lead" of open water several miles off shore. They appear in the spring before there is any open water except the shifting "leads" at a distance from the shore, and travel steadily and swiftly past Cape Smythe to the northeast, following the coast. Some flocks cross to the eastward below Point Barrow, but the majority follow the barrier of grounded ice past the point. It is probable, however, that they turn to the east after passing Point Barrow, because all the returning flocks in the autumn come from the east, hugging the shore of the mainland.

The first ducks in the spring of 1882 were seen on April 27, a comparatively warm day, with a light southerly wind blowing. They were flying parallel to the coast over the barrier of grounded ice. The natives said they were all "kingaling," "nosy birds," or males (referring to the protuberance at the base of the bill), and the first flocks of the migration appear to be composed exclusively of males. There were six great flights in 1882, the first on May 12 and the last on June 11, and five in 1883, the first on May 17 and the last on June 4.

As a rule, these flights took place on comparatively warm days, with light westerly or southwesterly winds. On one day each year, however, there was a large flight with a light breeze from the east. A warm southwest wind is pretty sure to bring a large flight of eiders. The flight seldom lasts more than two or three hours, beginning about 8 or 9 in the morning, or between 3 and 4 in the afternoon. More rarely a flight begins about 10 in the morning and lasts till afternoon. During the flights, the great flocks in quick succession appear to strike the coast a few miles from the station, probably coming straight across from the Seahorse Islands, and then follow up the belt of level ice parallel to the coast toward Point Barrow, going pretty steadily on their course, but swerving a little and rising rather high when alarmed. Their order of flight was generally in long diagonal lines, occasionally huddling together so that several could be killed at one discharge. A few flocks in a great flight usually followed up the line of broken ice a mile or two from the shore, and a flock occasionally turned in at the mouth of the lagoon and proceeded up over the land.

He further says:

The majority of them are paired by the middle of May, and the flocks are made up of pairs flying alternately, ducks and drakes. If a duck is shot down, the drake almost immediately follows her to the ice, apparently supposing that she has alighted.

Mr. W. Elmer Ekblaw contributes the following notes on his observations in northern Greenland:

The most strikingly beautiful of all the Arctic birds is undoubtedly the male king eider. His regal plumage warrants fully his royal name. He is almost as rare as he is royal. Only a relatively small number of his gorgeous family appears to be left, if ever it was plentiful. His mate is soberly dressed, quite undistinguishable from her cousin, the female of the common eider.

The king eiders reach the Smith Sound shores somewhat later in the season than do the common eiders. Because the males are so conspicuous they attract immediate attention; and since my first record for them is June 22, I feel confident that they do not arrive much earlier than that date. Probably June 15 would be a reasonable date for their first appearance in the land of the polar Eskimo. I can not say with certainty that the females arrive as early as the males, but I am inclined to think that they do, because the first males that we recorded were apparently paired with females. I believe that the king eiders are paired when they arrive, and that the sexes arrive together.

When the king eiders arrive, the inland ponds are already open, and to these fresh-water pools the king eiders go at once. They do frequent the open sea somewhat, but their favorite haunts through the mating and nesting season are the inland lakelets.

Courtship.—Mr. W. Sprague Brooks (1915) describes the courtship of this species as follows:

Once I found this species courting. On June 14, when approaching a small lagoon, but still unable to see it owing to a slight elevation of the tundra before me, I heard a strange sound on the other side of the elevation. This peculiar noise came in series of three "*Urrr-urrr-URRR*," the last being the loudest, a sort of drumming call as when one expels air forcibly through the mouth with the tongue lightly pressed against the palate. I had heard this noise once before during the winter made by an Eskimo and used with indifferent results for encouraging his dog team. I thought this call was an invention of his own at the time, but when in sight of the lagoon I found that the disturbance came from a small flock of king eiders, three females and five males. They were on the beach, and three males were squatted in a triangle about a female, each about a yard from her. They did much neck stretching, as many male ducks do in the spring, and frequently bowed the head forward. The males constantly uttered the above drumming note. During this time the female was very indifferent to the attentions of her suitors, doing nothing more than occasionally extending her head toward one of them. After a brief period of these tactics, one or more of the males would enter the water and bathe vigorously, with much bowing of heads and stretching of necks, to return to the beach in a few moments and repeat the foregoing performance. Finally they all took wing, uttering the croaking sound similar to the Pacific eider.

Nesting.—Although the eggs of the king eider are not rare in collections, I am surprised to find that remarkably little has been published regarding the details of its nesting habits. I have three sets in my collection from Point Barrow, Alaska, but unfortunately no particulars came with them. MacFarlane secured over 200 eggs from Franklin Bay and Liverpool Bay, and says in a letter that when on Island Point, as he was walking along the sea beach, a female of this species got up and flew violently away to a short distance, where she alighted on the ground. He at once discovered her nest, which was a mere hole or depression in the ground, about 50 yards from the beach, wholly composed of eider down, and containing 6 eggs. Other nests were found on the coast during several seasons, and also among the islands of the Arctic Sea. All appear to have been similar to the one described, and 6 is the largest number of eggs mentioned as having been found in any one nest.

Messrs. Thayer and Bangs (1914) report that:

On June 26 a nest was found (by John Koren) in a tuft of grass 10 feet from the edge of a small lake on one of the islands in the delta of the Kolyma (northeastern Siberia). It contained the broken shells of two fresh eggs, evidently destroyed by a pair of glaucous gulls that were nesting nearby. The pair of eiders were swimming about in the lake.

A. L. V. Manniche (1910), on his explorations in northeastern Greenland, found two nests "on the slopes of the low rocks by the coast not far from the mouth of the river. The fresh down and eggshells proved that they had lately been inhabited." Again he states that the king eider—

always stayed in the fresh waters on the mainland, on which it undoubtedly exclusively nests. I did not succeed in finding nests with eggs, but all the old nests I found at Stormkap proved that this bird nests singly. The nests were placed on the lower slopes, with luxuriant vegetation, or on small hills in the lowland, with large stones surrounded by grass. None of the observed nests were far from the bay (as a maximum 1 kilometer).

According to Rev. C. W. G. Eifrig (1905), this species breeds commonly on the islands north of Hudson Bay and "places its soft, down-lined nest on tussocks of grass along the shores and on islands of inland ponds."

In northern Greenland Mr. Ekblaw noted that:

About July 1 the breeding season is at its height. In one day's tramp over the lake-dotted valleys about North Star Bay I saw some 20 pairs swimming about the ponds or resting on the grassy banks, feeding on the abundant life of the pools, preening their feathers, or sleeping either on land or water with their bills tucked under their wings. The birds were all paired. The females were very shy, indeed, and both females and males were shier than the oldsquaws and common eiders.

The nests are placed at some distance from the pools; the nearest I found was about 200 yards from any water and 2 miles from the sea; the farthest at least a mile from any pool and 4 miles from the sea. Some of the nests were concealed in the sedges and grasses of the wet swales, usually on a hummock; some were placed in small hollows in the gravel of old glacial moraines. The nest is well-lined with down, much lighter in color than the down from the oldsquaws, and somewhat lighter than the down from the common eiders.

Rev. F. C. R. Jourdain writes to me as follows:

It is very remarkable that, although the king eider is found in fair numbers in several districts in Spitsbergen, hardly anything was known as to its breeding habits there until quite recently. The reason seems to be that it was surmised that its breeding habits were similar to those of the common eider, whereas they are really very different. While the common eider generally nests in thickly populated colonies on low flat islands in the fjords or off the coast, the king eider prefers to nest among the moss and lichen-covered expenses of flat tundra on the mainland. Instead of 100, or several hundred, nests being found crowded together on an acre or two of ground, the female king eiders are scattered over the tundra, perhaps half a mile of monotonous moorland intervening between the nests, and, as they squat closely in the moss, in most cases they are invisible till approached within a few yards. On the eider holms, too, the drakes, with their boldly contrasted plumage of black and white, stand close to the hens and always accompany them when they leave the nest to wash and feed, while, on the other hand, the male king eiders are never seen in the vicinity of the nest, but after the ducks have begun to sit they congregate in flocks and haunt the open fjords. When the young are hatched out the female king eiders lead them to some fresh-water lake in the vicinity of the nest, unlike the common eiders, which at once take to the sea. On a moderate-sized lagoon or fresh-water lake on the tundra scores of broods of young eiders may be found congregated together late in the summer, accompanied by the ducks, and Doctor Van Oort observed that the ducks, by adroit splashing and feigned attacks, were able to preserve their young from the powerful sledge dogs which haunted the camp.

Summing up what little data we have on the subject, it seems that, whereas the other eiders of the north Atlantic nest on the seacoast exclusively, preferably on islands and often in densely populated colonies, the king eider seems to prefer to nest near the shores or on the islands of fresh-water ponds and streams, and its nests are usually widely scattered. Some writers have said that the nest resembles that of the common eider, and others have called attention to the darker down in the nests of the king eider. This character is very well marked in the only nest I have of this species, in which the down is dark " bone brown " or dark " clove brown "; mixed with the down are numerous bits of moss, lichens, and grass from the tundra, a few dry leaves of willows, and a few breast feathers of the duck.

Eggs.—The king eider lays from 4 to 7 eggs, usually 5. These resemble the eggs of the common species, but they are decidedly smaller, somewhat more elongated, and rather more pointed. The shape is elliptical ovate or elongate ovate. The shell is smooth, but without much gloss. The color varies from " dark olive buff " to " deep olive buff " or even " olive buff." The eggs are often clouded or mottled with darker shades of olive or brown and are frequently much nest stained, giving them a darker appearance.

The measurements of 152 eggs, in various collections, average 67.6 by 44.7 millimeters; the eggs showing the four extremes measure **79.5** by 47, 78.5 by **52**, **61.3** by 45, and 62.5 by **41.5** millimeters.

Incubation is performed by the female alone. The males desert the females at this season and fly out to sea, where they form in large flocks, often far from land and about the edges of the sea. Mr. Manniche (1910) writes:

The females would, in the breeding season, sometimes leave the nest for a short while and fly to the nearest pond for the purpose of bathing and seeking food. Like many other birds, the king eider is irritable and quarrelsome at this period. One evening I observed a female which had just left her nest. She flew quickly straight toward me and so low that she seemed to touch the earth with the tips of her wings. I was standing on the beach of a pond with shallow water. Uttering an angry grunting she circled around and quite near to me and then flew to the pond. Having quenched her thirst and by a pair of quick bounds under the surface put her feathers in order, she swam straight toward me, all the while uttering a peculiar growling and hissing; the feathers on her head were erected, and she seemed to be very much displeased at my presence; now and then she cackled in the shallow water like a domestic duck, again to show her displeasure.

Young.—Referring to the care of the young, the same observer says:

On the arrival of the expedition at the ship's harbor on August 17 several females, accompanied by their still downy young ones, were lying in the small openings in the ice. Three days later I met with 5 broods of ducklings at the mouth of Stormelven; one of these broods was scarcely 1 week old. These broods all contained 5 ducklings—in one case 6 were seen. The old birds behaved very anxiously when I approached and swam, grunting, around quite near me and the coast, while the young ones, with a surprising rapidity, moved outwards, swimming and diving till they at last disappeared far out in the bay.

Immature birds do not breed during their first spring and perhaps not during their second, as the full plumage is not acquired until the third winter, or when the young bird is about 2½ years old. These young birds flock by themselves until they become of breeding age, and frequent different resorts. They do not go as far north in the summer, are never seen on or near the breeding grounds, and they usually winter farther south. Most of the straggling inland records are made by birds of this class. Kumlien (1879) saw in July large numbers of these immature birds in Cumberland Sound, on the west coast of Davis Straits, and around Disko Island. They were in various stages of immature plumage, and the sexual organs of those he examined were not developed. Apparently these are the summer resorts of the immature birds not yet ready to breed.

Plumages.—The downy young king eider bears a superficial resemblance to the young of the common eider, but it can be easily recognized by the shape of the bill, head, and feathered tracts at the base of the bill. The bill is longer and more slender; the forehead is more rounded and prominent, less sloping; the feathered points on the sides of the bill do not extend so far forward; and there is a long, slender feathered point extending out onto the culmen to the outer end of the nostrils. The under parts are more extensively lighter, and the upper parts are tinged with more yellowish buff. The color pattern of the head is also different; the crown is "bister," paling to "Saccardo's umber" on the hind neck; there is a broad superciliary stripe of "pinkish buff," and a paler tint of the same color extends down the sides of the head and neck, paling almost to white on the throat; there is a dark postocular stripe of "sepia," and the cheeks and lores are washed with a paler tint of the same, leaving the lower half of the pointed feather tract nearly white, as on the throat. The colors on the upper parts shade from "Saccardo's umber" anteriorly to "bister" on the rump. The under parts are grayish white.

The progress toward maturity in the young male king eider advances by similar stages and at the same rate as in the common eider.

The plumage changes are parallel, and yet the two species can be easily recognied at any age in both sexes. In the brown stage, during the first fall and early winter, the young male king eider is much darker above and the shape of the bill and its feathered borders are distinctive. During the late winter and early spring the back, scapulars, and flanks become nearly black; the crown and neck become darker brown; a variable amount of white appears in the chest, each feather tipped with dusky; and some white, dusky-bordered feathers appear in the rump patches. Some forward birds show considerable white on the neck and throat, with a suggestion of the black V during the first spring. The under parts remain dull mottled brown and the immature wings, with dusky, light-edged coverts, are retained until the complete summer molt.

This molt involves the first eclipse plumage, which does not entirely disappear until November. The young bird is then in its second winter plumage, which is similar to the adult winter plumage, but duller and less complete. This plumage can be easily recognized, however, by the wing; in the adult male the lesser wing-coverts, except for a dusky border around the bend of the wing, and the median wing-coverts are pure white; but in the second-winter male these white feathers are more or less margined or shaded with dusky. At the next summer molt these wings are shed and the young male becomes adult as soon as the second eclipse plumage disappears in the fall, when about 28 months old.

The eclipse plumage of the adult male king eider is very similar in appearance, extent, and duration to that of the common eider; it begins to appear early in July or even in June, with the growth of dull-brown feathers in the head and neck; dark-brown or blackish feathers in the white area of the back, and buffy brown feathers barred with dusky on the breast. The brilliant colors of the head, including the black V, disappear entirely; the white rump spots are nearly or quite obliterated; there is very little white left in the back; and the buff breast shield is nearly concealed by the barred feathers. In this plumage, which is complete in August, the wings and the rest of the plumage are molted. The adult winter plumage is completely renewed again by the end of November.

The plumage changes of the female king eider are similar to those of the common eider. This species can always be easily recognized by its smaller size, smaller head and bill, and by the pattern of the feathering at the base of the bill, the central feathered point extending down to the nostrils. The colors of young females are always duller, particularly on the under parts. Adults are richer brown in winter and paler in summer or spring, owing to the wear and fading of the brown edgings, particularly on the scapulars, wings, and flanks.

Food.—Referring to the food of this species on its breeding grounds, Mr. Manniche (1910) says:

In the season in which the king eider lives in fresh water its food consists principally of plants. In the stomachs which I examined I found, however, many remnants of insects, especially larvae of gnats. In the stomachs of downy young ones I found indeterminable remnants of crustaceans, plants, and small stones.

At other seasons king eiders are essentially salt-water birds and spend much of their time out on the open sea. They obtain their food by diving, generally at moderate depths, to the rocky shoals and ledges where they find the bulk of their food, which consists mainly of mollusks, conchylia, bivalve, and univalve shellfish of a great variety of species. They are also said to eat a number of crustaceans, shrimps, starfish, small fishes, and fish spawn. On the coast of Maine in winter, Arthur H. Norton (1909) found that one "had its gullet filled with large specimens of *Gammarus locusta*, the common sea flea of our shores. Another was similarly filled with young crabs (*Cancer irroratus*), in both instances to the exclusion of other food." Ora W. Knight (1908) says: "They are said to feed in rather deeper water than the other eiders, and Mr. Norton has recorded the fact that certain individuals had been eating sea cucumbers (*Pentacta frondosa*) to the practical exclusion of other material. While a few I have examined also evidenced some fondness for such a diet, they also had been eating great quantities of mussels." Apparently almost any kind of animal food to be found in the sea is a welcome addition to the food of this bird.

Behavior.—The flight of the king eider is similar to that of the other eiders, but the male can be easily recognized, even at a long distance, by the larger amount of black in the back and wings. The adults are usually very shy, but the immature birds are often very unsuspicious. As this species has to fly inland to its breeding places, it has less fear of the land than the common eider and does not object to flying over points of land which lie in its line of flight.

F. Seymour Hersey says, in his notes:

The flight of the king eider, like all the eiders with which I am familiar, is swift. The wings are moved very rapidly and the large heavy body is propelled with a speed and directness that is bulletlike. A bird shot on the wing will frequently strike the water and bound along the surface for a considerable distance before coming to a stop. A bird which I shot at Cape Dyer just as it was about to round a turn in a small inlet it was flying over at the time, continued straight ahead from the momentum it had acquired and struck the tundra several yards from the water, where I found it nearly hidden from sight in the tundra mosses into which the impact of its fall had embedded it.

It always seemed to me that with the exception of the yellow-billed loon the king eider is the strongest flyer of any of the northern waterfowl I met.

As a diver it is an expert and can penetrate to great depths; birds are said to have been taken in gill nets at a depth of 150 feet. In diving it partially opens its wings and probably uses them, as well as its feet, in swimming under water.

Referring to the behavior of this species in Greenland, Mr. Manniche (1910) writes: "Every day they used to fly from the lakes and ponds inland down to the bay, and especially to the mouth of Stormelven, in which they would lie and dive for food. They used to lie for hours on the grass-clad beaches of the lake in order to rest or to sleep, with their heads hidden under their wings. During their excursions in the field, they always flew very low and sometimes uttered a slight growling or grunting sound." Hagerup (1891) says that "its only note is a single cooing sound, heard especially at night."

Fall.—Mr. Murdoch (1885) says of the fall migration at Point Barrow:

By the second week in July, before the ice is gone from the sea or from Elson Bay, the males begin to come back in flocks from the east, and from that time to the middle of September there is a flight of eiders whenever the wind blows from the east. The flocks are all males at first, but mixed flocks gradually appear, and the young of the year were first observed in these flocks on August 30, 1882.

Most of the flight birds make no stay, but continue on to the southwest, generally a couple of miles out at sea, though they occasionally stop to rest, especially when there is much drifting ice. Between the regular flights they continue to straggle along, coming off the land, and occasionally sitting apparently asleep on the beach. Small flocks and single birds are to be seen till the sea closes, about the end of October, and in 1882 many were seen as late as December 2, when there were many holes of open water.

When the birds are flying at Pergniak, it is quite a lively scene, as there is a large summer camp of Eskimos close to the point where the ducks cross when the conditions are favorable. When the wind is east or northeast, and not blowing too hard, the birds come from the east and strike the land at a point which runs out on the shore of the bay about half a mile from Pergniak, close to where the lagoons begin.

They would be apt to turn and fly down these lagoons were it not for a row of stakes, set up by the natives, running round the semicircle of the bay to the camp. As soon as the flock reaches this critical point, all the natives, and there may be 50 of them on the watch, with guns and slings, just at the narrowest part of the beach above the tents, immediately set up a shrill yell. Nine times out of ten the flock will waver, turn, follow round the row of stakes, and naturally whirl out to sea at the first open place, where, of course, the gunners are stationed. With a strong wind, however, the ducks do not follow the land, but come straight on from the east and cross wherever they happen to strike the beach, so that the shooting can not be depended on.

The flocks during the fall flight are not so large and do not follow one another in such rapid succession as in the spring; and though they arrive from the east in the same stringing order, they huddle into a compact body as they whirl along the line of stakes and out over the beach.

Mr. Hersey contributes the following notes on the fall migration:

No one who has not observed the migration of the king eider along the Arctic coast of Alaska can realize the enormous abundance of this species in the north. The southward migration was well under way when we left Kotzebue Sound early in August, 1914, and started northward for Point Barrow. When we reached Icy Cape we encountered ice, and by the time we were off Wainwright Inlet the ice conditions were so bad and the wind so unfavorable that the captain decided it was not prudent to go further until the wind changed. We spent the time from August 10 to 20 lying off Wainwright Inlet waiting for a change of wind, and during these days migrating flocks of king eiders were constantly passing. The birds travel mostly in large flocks of 75 to 350 birds, following the shore line but keeping at least a mile from land. They spread out in a long line, the birds flying nearly abreast, but in the larger flocks quite a few will be bunched toward the center, or sometimes two or three small parties of 10 to 20 birds will follow directly in back of the main line. There is an undulating motion to these flocks when seen at a distance similar to that of a flock of Canada geese. They fly some 30 or 40 feet above the water or ice and follow in what appears to be the exact course of the flock that had passed a few moments before. During the 10 days that this migration was under observation there appeared to be no diminution in the numbers of birds coming out of the north. A flock would appear on the horizon to the northeast, fly steadily toward us to a certain point where they always swerved away from the ship, pass at a distance of a quarter of a mile, and a moment later disappear in the southwest. Turning our faces to the northeast again another flock would be seen coming into view at the same point on the horizon where the last birds had appeared and pressing steadily on along the same flight line. Throughout the entire 10 days there was hardly a quarter hour in which a flock of birds was not passing, and often more than one flock was in sight. The migration moved on without interruption from daybreak, which at this time of year takes place between 3 and 4 in the morning, until the sun sets, about 9 p. m. The flight of one of these migrating flocks seems slow, probably on account of the wavy motion of the line of birds, but when they finally sweep past it is seen that they are really flying swiftly, and there is a roar of wings audible for a long distance.

Winter.—To visit the winter haunts of the king eider on the New England coast, one must be prepared to brave the rigors of the cold, rough sea in the most exposed places; for these hardy birds do not come until wintry conditions have made offshore boating far from comfortable, and they prefer to frequent the outer ledges which at that season are almost always unapproachable. I can well remember a December morning on the coast of Maine, the first chance after a week of waiting for a day smooth enough to reach the outer islands, when we started long before daylight for a little eider duck shooting. Fifteen miles or more we had to go in our little launch to reach the ledges where we were to shoot. With the first signs of daylight and for an hour before sunrise we could see small flocks of scoters, darkly painted on the lightening sky, flying from their bedding grounds at sea up into Jericho Bay to feed on the mussel beds in shallow water With the coming of the dawn the gulls became active, and and their shadowy forms could be made out against the rosy clouds. The black figure of an occasional cormorant was seen flying high in the air, and scurrying flocks of oldsquaws flew past us at safe distances. We soon realized, as we began to reach the outer islands, that it was none too smooth; a heavy ocean swell was rolling in and breaking on the ledges; and the west wind, coming up with the sun, was stirring up a troublesome cross chop. As we approached Spirit Ledge, where we intended to do our shooting, all hope of landing was dispelled, for the waves were breaking over it with clouds of spray and all around it the submerged ledges were white with combing breakers. It was no place for us, this wild scene of ocean fury, but for the birds it held no terrors. There, just beyond our reach were hundreds of American eiders, surf and white-winged scoters, flocks of oldsquaws, and a few of the black-backed king eiders; flocks were going and coming, settling in the water among the breakers or circling about the rocks. It was a wild and attractive scene, but we could only view it from a distance, and we were finally obliged to retire to a more sheltered ledge where we succeeded in landing and setting out our decoys in the lee. Here only occasional flocks, pairs, or single birds came in to us, as we lay concealed among the rocks while our boatman was anchored at a distance. Off around the outer ledges we could still see the flocks of eiders feeding in the surf, riding at ease among the angry waves, paddling backwards or forwards to avoid the breaking crests, or diving under a combing breaker. There were both old and young birds in the flocks, but the latter decidedly in the majority; the old birds were too shy to come to us, but we secured young males in various stages of plumage. Before long it became too rough to stay even here, and our boatman insisted on our leaving before it was too late; as it was we lost one oar and nearly lost our skiff; we were glad to leave the sea ducks alone in their glory.

On the southern coasts of New England and Long Island the king eider is an irregular winter visitor, and during some seasons it is quite common. I have a small series of immature males and females taken in midwinter about Hen and Chickens Reef, partially submerged ledges a few miles off the coast of Westport, Massachusetts; here these birds are known as "cousins," owing to their resemblance to the common eiders, which are known as "wamps." On Long Island they are known as "Isle of Shoal ducks"; William Dutcher (1888) received, in January, 1887, a female king eider from

Capt. J. G. Scott, keeper of the lighthouse at Montauk Point, who reported them:

> Living off the Point since early in November (11), when I saw a flock of 4; the next day I saw 10 at one time. They appear less shy than the other wild fowl and will permit a nearer approach in a boat. In this locality it is seen occasionally in the winter months on the ocean from one-quarter to one-half mile from shore. It is not a common duck, and I believe it is only a few years since they have been seen off Montauk Point, but this winter they have been more than usually common. There is a shoal, with a depth of water from 15 to 20 feet, about one-quarter of a mile off the Point, where I go to shoot ducks, but can only do so when the surf will permit. Every time I have visited this spot this winter I have seen from 4 to 20 king eiders.

The king eider spends the winter as far north as it can find open water; in southern Greenland it associates with the northern eider in the open water in the fjords, but probably the greater number spend the winter at sea on the edges of the ice packs or in the open leads. On the western side of the continent the principal winter resort of this species is in the vicinity of the Aleutian Islands, where so many other sea ducks find congenial surroundings and abundant food. Some few birds winter as far north as the Diomede Islands, and many resort to the islands south of the Alaska Peninsula.

DISTRIBUTION

Breeding range.—Arctic coasts of both hemispheres. On both coasts of Greenland, north to 82° 30′, less abundantly in southern Greenland. In North America south to northern Labrador (Nachvak), Hudson Straits (Wales Sound), northern Hudson Bay (Southampton Island and Cape Fullerton), and the entire Arctic coasts of Canada and Alaska. On St. Lawrence and St. Matthew Islands, in Bering Sea. All along the Arctic coast of Siberia, on Nova Zembla, and on Spitsbergen. North on Arctic islands to Melville Island and probably others up to at least 76° N.

Winter range.—South on the Atlantic coast more or less regularly to Massachusetts (Vineyard Sound) and New York (Long Island), more rarely to New Jersey, and casually farther south. South in the interior frequently to the Great Lakes (Ontario, Erie, and Michigan) and casually beyond. On the Pacific coast south to the Aleutian, Kodiak, and Shumagin Islands. North as far as open water extends in Bering Sea and around southern Greenland. In the Eastern Hemisphere it visits Iceland, the Faroes, Norway, Denmark, Great Britain, Holland, and the Baltic Sea, and has occurred in France (twice) and Italy (four times).

Spring migration.—Early dates of arrival: Greenland, Igloolik, latitude 69°, April 16, and Etah, May 1; Wellington Channel, latitude 76°, June 9; Fort Conger, latitude 82°, June 11. Late dates of departure: Georgia, Brunswick, May 5; New York, Long Island, June 8, and Waterford, April 30; Maine, May 29; Ontario, Ottawa, May 8. Dates of arrival in Alaska: Point Hope, March 17; Point Barrow, April 27; Humphrey Point, May 15. Arrival at Banks Land, Mercy Bay, June 1.

Fall migration.—Early dates of arrival: Massachusetts, October 21; New York, Cayuga, November 3; Pennsylvania, Erie, November 13; Ontario, Toronto, November 25; Ohio, Columbus, November 4: North Carolina, Dare County, December 3; Virginia, Cobb Island, December 19. Late dates of departure: Alaska, Point Barrow, December 2, and Wainwright, November 9; Mackenzie, Fort Simpson, October 25.

Casual records.—Has wandered south on the Atlantic coast to Georgia (Ossabaw Island, December 1, 1904, St. Catherine Island, December 3, 1904, and Brunswick, April 25 and May 5, 1890), on the Pacific coast to central California (San Francisco, winter of 1879–80), and in the interior to Alberta (Calgary, November 4, 1894), and to Iowa (Keokuk, November 18, 1894).

Egg dates.—Arctic Canada: Sixteen records, June 27 to July 8. Alaska: Twelve records, June 10 to July 5; six records, June 20 to July 3. Greenland: Four records, July 11 to 25.

OIDEMIA AMERICANA Swainson

AMERICAN SCOTER

HABITS

This is the least known and the rarest of the three species of scoters or "coots" which migrate up and down our coasts on both sides of the continent; it is also the hardest of the three to identify in life, as it has no distinctive marks visible at a distance; hence it is commonly referred to by nearly all observers in connection with the other two, and it is very difficult to separate much that is applicable to this species alone. This group as a whole seems to be rather unpopular among naturalists, as it is among sportsmen; consequently comparatively little effort has been devoted to the study of its habits.

Spring.—The spring migration of the American scoter, on the Atlantic coast, is eastward and northward along the seacoast; I can find no evidence of an overland flight to the interior, which is so conspicuous in the white-winged scoter. During April and the early part of May large numbers may be seen migrating eastward through Vineyard Sound and around Cape Cod. On the south coast of Labrador we saw them flying eastward during the latter part of May and the first half of June. While migrating the three species usually keep in flocks by themselves, but mixed flocks are occasionally seen; such mixed flocks are more apt to contain American and surf scoters than white-winged scoters.

Nesting.—Dr. E. W. Nelson (1887) writes:

At St. Michael these ducks are never seen in spring until the ice begins to break offshore and the marshes are dotted with pools of open water. May 16 is the earliest date of arrival I recorded. Toward the end of this month they leave the leads in the ice and are found in abundance among the salt and fresh water ponds on the great marshes from the Yukon mouth north and south. The mating is quickly accomplished and a nesting site chosen on the border of some pond. The spot is artfully hidden in the standing grass, and the eggs, if left by the parent, are carefully covered with grass and moss. At the Yukon mouth Dall found a nest of this species on June 17. The nest contained two white and rather large eggs, and was in a bunch of willows on a small island, and was well lined with dry grass, leaves, moss, and feathers.

Edward Adams (1878) found this species breeding in the same general region, and says of its nesting habits:

These birds were rather late in their arrival; I met with none until the 19th of May. Toward the end of the month several pairs had taken possession of the larger lakes near Michalaski; here they remained to breed, seldom going out to sea, but keeping together in small flocks in the middle of the lake. Their nests were well secreted in the clefts and hollows about the steep banks of the lakes, close to the water; they were built of coarse grass, and well lined with feathers and down. They had not laid when I last examined the nests.

Audubon's (1840) historic account of the nesting of the American scoter on the south coast of Labrador is interesting as showing that it once bred farther south than it now does and as illustrating its method of nesting in Labrador, where it probably still breeds abundantly in the more remote sections; he writes:

On the 11th of July, 1833, a nest of this bird was found by my young companions in Labrador. It was placed at the distance of about 2 yards from the margin of a large fresh-water pond, about a mile from the shore of the Gulf of St. Lawrence, under a low fir, in the manner often adopted by the eider duck, the nest of which it somewhat resembled, although it was much smaller. It was composed externally of small sticks, moss, and grasses, lined with down, in smaller quantity than that found in the nest of the bird just mentioned, and mixed with feathers. The eggs, which were ready to be hatched, were 8 in number, 2 inches in length, an inch and five-eighths in breadth, of an oval form, smooth, and of a uniform pale yellowish color.

We did not observe this species on the south coast of Labrador except as a migrant, but on the east coast we found it fairly common all summer all along the coast, at least as far north as Hopedale. Flocks made up entirely of males were seen in many of the inner bays and in the mouths of rivers in July and August; probably their deserted mates were incubating on their eggs or tending broods of young about the inland ponds a few miles back from the coast.

It is surprising that none of the numerous ornithologists who have visited Labrador have ever found and identified a nest of the American scoter since Audubon's time, but anyone who has ever attempted to explore into the interior of this discouraging country will appreciate why. The region is so vast, so hopelessly impassable, and so exceedingly poor in bird life that one soon gives it up in despair. This and all other species are so widely scattered that the chances of finding their nests are very small. I doubt if the American scoter migrates very far north to breed; we did not see it north of Hopedale, though we did see the surf scoter; it has never been found breeding abundantly about Hudson Bay in the northwest territories or on the Arctic coast, and Turner reported it as very scarce about Ungava Bay. Its main breeding grounds have apparently never been found. One would naturally infer then that the large numbers of this species which winter on our eastern coasts must breed in the interior of the Labrador peninsula, probably in the southern half of it, and perhaps near the marshy coasts of James Bay and the southern half of Hudson Bay; all of which regions are sadly in need of further exploration.

The American scoter undoubtedly breeds regularly, but not abundantly in Newfoundland. My friend, J. R. Whitaker, told me that he had seen this duck on Grand Lake with a brood of young, though his attempts to find a nest have proved unsuccessful. This is another vast region, difficult to travel in and largely unexplored.

I have a set of 9 eggs in my collection, said to be of this species, taken by Rev. C. E. Whitaker on Gary Island, Mackenzie Bay, on June 10, 1910; the nest is described as made of down in a tussock of grass. The down is rather dark in color, varying from "bone brown" to "dusky drab" and is flecked with bits of whitish down, uniformly mixed with the dark down; the down is mixed with bits of dry leaves, pieces of grass, and small sticks.

Eggs.—The American scoter is said to lay from 6 to 10 eggs. They vary in shape from ovate to elliptical ovate. The shell is clean and smooth, but without gloss. The color varies from "light buff" or "pale pinkish buff" to "cartridge buff." The measurements of 58 eggs, in various collections, average 61.9 by 41.7 millimeters; the eggs showing the four extremes measure **72.5** by 46, 63 by **46.2**, and **53** by **33.6** millimeters.

Young.—Nothing seems to be known about the period of incubation. This duty is performed solely by the female, who is entirely deserted by the male at this season. Doctor Nelson (1887) writes:

As the set of eggs is completed, the male gradually loses interest in the female and soon deserts her to join great flocks of his kind along the seashore, usually keeping in the vicinity of a bay, inlet, or the mouth of some large stream. These flocks are formed early in June and continue to grow larger until the fall migration occurs. Males may be found in the marshes with females all through the season, but these are pairs which breed late. A set of fresh eggs was taken on August 3, and a brood of downy young was obtained on September 9. The habits of these flocks of males are very similar to those of the male eiders at this season. They are good weather indicators, and frequently, 10 or 20 hours in advance of a storm, they come into the sheltered bays, sometimes to the number of a thousand or more. At such times they show great uneasiness, and frequently pass hours in circling about the bay, sometimes a hundred yards high and again close over the water, the shrill whistling of their wings making a noise which is distinctly audible nearly or quite half a mile. Until the young are about half grown the female usually keeps them in some large pond near the nesting place, but as August passes they gradually work their way to the coast and are found, like the eiders of the same age, along the reefs and about the shores of the inner bays until able to fly.

Regarding the care of the young, Audubon (1840) says:

> I afterwards found a female with seven young ones, of which she took such effectual care that none of them fell into our hands. On several occasions, when they were fatigued by diving, she received them all on her back, and swimming deeply, though very fast, took them to the shore, where the little things lay close among the tall grass and low tangled bushes. In this species, as in others, the male forsakes the female as soon as incubation commences.

Plumages.—The downy young, when first hatched, is dark colored above, varying from "Prout's brown" or "verona brown" to "bister," darkest on the crown and rump; the throat and cheeks, below the lores and the eyes, are white; the under parts are grayish white centrally, shading off on the flanks into the color of the upper parts; the bill is broadly tipped with dull yellow. The plumage appears first, when about half grown, on the breast and scapulars; the tail appears next and the wings are the last to grow.

The following remarks are based largely on two papers by Gurdon Trumbull (1892 and 1893) and one by Dr. Jonathan Dwight (1914) on the molts and plumages of the scoters, all of which are well worth reading, and to which I would refer the reader for details. In the juvenal plumage the sexes are practically alike, the female averaging slightly smaller. The upper parts, including the crown, back, wings, and tail, are deep, rich brown, varying from "Prout's brown" to "mummy brown," darkest on the scapulars and tertials and palest on the neck, chest, and flanks, where it fades into the light color of the under parts; the lower half of the head and the belly are grayish white, mottled with lighter browns. This is the plumage in which the birds are known to the gunners as "gray coots"; it is worn during the fall and often well into the winter without change. Sometimes as early as November, but more often not until January or later, the sexes begin to differentiate; a growth of black feathers begins in the head and neck of the young male and a similar growth of brown feathers appears in the young female. The growth of black feathers in the male increases during the winter and spring until some of the most advanced birds become nearly all black except on the belly and wings. Doctor Dwight (1914) says:

> Shortly after new feathers appear, the bill of the young male begins to take on the colors of the adult and still more gradually assumes its shape. The colors may closely approximate, by the end of the winter, those of the adult, but the shape is not perfected for at least a year, the swelling of the hump not being marked in the first winter birds, although the yellow color may be brilliant. The bill of the female and the legs and feet of the male remain dusky, adults differing very little from young birds.
>
> No matter how black the plumage may be nor how bright the colors of bill or feet, young males may infallibly be told from adults by the shape of the first primary, which is not replaced until the first postnuptial molt. The iris in *americana* is always brown in both sexes at all ages.

At the first complete postnuptial molt, the following summer, the young bird becomes practically adult; the plumage is wholly black in the male and wholly dark brown in the female. At this molt the adult wing is acquired, in which the outer primary is deeply emarginated; the broad tipped outer primary is worn by the young bird for one year only.

Adults have two molts each year, a partial prenuptial molt in March and April, involving the body feathers and the tail, and a complete postnuptial molt in August and September. There is no evidence of anything like an eclipse plumage in this or in the other scoters. The plumages described as such by European writers are probably produced by wear and fading or by left-over traces of a former plumage.

Food.—Writing of the feeding habits of the three species of scoters on the Massachusetts coast, George H. Mackay (1891) says:

> These scoters are the most numerous of all the sea fowl which frequent the New England coast, collecting in greater or less numbers wherever their favorite food can be procured—the black mussel (*Modiola modiolus*), small sea clams (*Spisula solidissima*), scallops (*Pecten concentricus*), and short razor shells (*Siliqua costata*), about an inch to an inch and a half long, which they obtain by diving. Mussels measuring 2½ inches by 1 inch have been taken from them; but usually they select sea clams and scallops varying in size from a 5-cent nickel piece to a quarter of a dollar. They can feed in about 40 feet of water, but prefer less than half of that depth. As these mussels are frequently difficult to detach, and the sea clam lives embedded endwise in sand at the bottom with only about half an inch above the sand, the birds are not always successful in obtaining them, it requiring considerable effort on their part to pull the mussels off or to drag out the clams. Eight or ten of these constitute a meal, but the number varies according to the size. I have heard of a mussel closing on a scoter's tongue, which was nearly severed at the time the bird was shot (Muskeget Island, about 1854). The fishermen frequently discover beds of shellfish (scallops) by noticing where these birds congregate to feed. In the shoal waters adjacent to Cape Cod, Nantucket,

and Marthas Vineyard, these mollusks are particularly abundant, and consequently we find more of the scoters in those localities than on any other part of the coast or perhaps than on all the rest of the coast combined.

E. H. Forbush (1912) writes:

Its food consists largely of mussels, and when feeding on fresh water it prefers the *Unios* or fresh-water clams to most other foods. Thirteen Massachusetts specimens were found to have eaten nearly 95 per cent of mussels; the remaining 5 per cent of the stomach contents was composed of starfish and periwinkles. It is a common belief that all scoters feed entirely upon animal food, but this is not a fact. Along the Atlantic coast they appear to subsist mostly on marine animals, but, in the interior, vegetable food also is taken. Mr. W. L. McAtee found the scoters in a Wisconsin lake living almost exclusively for a time on the wild celery, but he does not state definitely what species of scoter was represented there.

Dr. F. Henry Yorke (1899) says that, while on the lakes and ponds of the interior, this species eats minnows and small fish, slugs, and snails, larvae of insects, fish spawn, crawfish, small frogs, and polliwogs; also a variety of vegetable food such as duckweed, pondweed, flags, water milfoil, bladderwort, and several other water plants. Probably the young are fed largely on insect food.

Behavior.—The American scoter is not easily recognized in flight; its size, shape, gait, and general appearance are all much like those of the surf scoter, from which only the adult males can be distinguished by the head and bill markings at short range; the females and young of these two species can not be distinguished in life at any considerable distance, and many gunners do not recognize them in the hand. Its flight is not quite so heavy as that of the white-winged scoter. All the scoters fly more swiftly than they appear to be going, but at nothing like the speed at which they have been reported to fly; I doubt if they ever fly at over 60 miles an hour or even attain that speed except under the most favorable circumstances. Migrating flocks in the fall usually fly high in fair weather, but in stormy or very windy weather the flocks sweep along close to the water and usually well in shore, following the indentations of the coast line and seldom flying over the land, except on their occasional visits to inland ponds. The flocks vary greatly in size and form, some are great irregular masses or bunches, others are strung out in long straight or curving lines, and sometimes they form in more or less regular V or U shapes. The wings make a whistling sound in flight, which the gunners imitate to attract the attention of passing flocks.

All the scoters are strong, fast, and tireless swimmers, either on the surface or below it; they dive quickly and neatly and can remain under water for a long time. Mr. Mackay (1891) writes:

In these shallow waters the tide runs rapidly over the shoal ground and sweeps the scoters away from where they wish to feed, thus necessitating their flying back again to it; consequently there is at such times a continual movement among them as they are feeding. When wounded and closely pursued, they will frequently dive to the bottom (always using their wings as well as feet at such times in swimming under water) and retain hold of the rockweed with the bill until drowned, prefering thus to die than to come to the surface to be captured. As an instance of this, I may mention that on one occasion I shot a scoter when the water was so still that there was not even a ripple on its surface; after pursuing the bird for some time I drove it near the shore, when it dove and did not reappear. I knew it must have gone to the bottom, as I had seen the same thing repeatedly before. As the occasion was a favorable one for investigation, the water being clear and not more than 12 or 15 feet in depth, I rowed along carefully, looking continually into the water near the spot where the bird was last seen. My search was at last successful, for on getting directly over where the bird was I could look down and distinctly see it holding on to the rockweed at the bottom with its bill. After observing it for a time I took one of my oars, and aiming it at the bird sent it down. I soon dislodged it, still alive, and captured it. I have often seen these birds, when wounded and hard pressed, dive where the water was 40 to 50 feet deep, and not come to the surface again. I therefore feel much confidence in stating that it is no uncommon occurrence for them under such circumstances to prefer death by drowning to capture. This they accomplish by seizing hold of the rockweed at the bottom, holding on even after life has become extinct. I have also seen all three species when wounded dive from the air, entering the water without any splash. All are expert divers, it requiring considerable experience to retrieve them when wounded.

Scoters are usually silent birds; I can not remember having heard any notes from any of them, but Mr. Mackay (1891) says: "The American scoter makes a musical whistle of one prolonged note, and it can frequently be called to the decoys by imitating the note." Rev. J. H. Langille (1884) says: "The note of the scoter in spring is like *whe-oo-hoo*, long drawn out."

Maj. Allan Brooks (1920) writes:

In British Columbia this scoter is an exclusively maritime duck; at least I have not come across a single reliable inland record. Not only is it a maritime bird, but it is seldom found in the small bays and inlets where the other species swarm, but frequents the exposed shores and outer reefs together with the harlequin. It has many points in common with that duck, rising easily from the water and doing much flying about in small lots of four or five—mostly males—seemingly for the pleasure of flying, usually returning to the point they started from. In flight the silvery undersurface of the primaries, in both sexes, is very conspicuous. In fine, calm weather they call a great deal and their plaintive *cour-loo* is the most musical of duck cries, very different from the croaking notes of most diving ducks.

Fall.—The fall migration of the American scoter is somewhat earlier than that of the other two species. On the Massachusetts coast the flight begins in September, and during the latter half of that month there is often quite a heavy flight which consists almost entirely of adult birds. The young birds, which are known as "gray coots," come along with the other scoters in October. Each species usually flocks by itself, and flocks of adults are often separated from flocks of young birds; but mixed flocks are often seen, particularly of young American and surf scoters. The American scoter is more often seen in fresh-water ponds a few miles back from the coast than are the other two, though all three are often seen on our large inland lakes.

Game.—From the sportsman's standpoint the American scoter is a more desirable game bird than the other two scoters. The young birds particularly, when they first arrive from their northern feeding grounds in the interior, are fatter, more tender, and less strongly flavored than are the others. The value of "coots" as food has been much maligned and in my opinion unjustly; if young and tender birds are selected and if they are properly cooked their flesh is much more palatable than is generally supposed; the popular prejudice against them is largely due to an erroneous impression that they must be parboiled, which is a pernicious practice and will render any oily seaduck unfit for food by saturating the flesh with the oily flavor. There are only two proper ways to cook a sea duck; one is to skin it and broil it; and the other is to scrape as much oil out of the skin as possible and then roast it quickly in a hot oven, letting the oil run off.

The Massachusetts method of "coot shooting," in which I have often indulged, is described under another species, so I shall quote from Walter H. Rich (1907) as to the methods employed on the Maine coast; he writes:

> Probably the least wary of the duck family, they may be approached quite readily as compared with other members of the tribe. Gunners use many methods for capturing the coots, but the greater number are killed over decoys. A string of "tolers" is set in a promising place just off some rocky point or ledge in the deep water, the gunner is well hidden, and if the birds are flying there is every prospect of good shooting, for the coot is one of the best of birds to decoy. Often in the early part of the season, before the birds have become shy from constant peppering, the gunner may set his decoys on a line from his boat, only keeping below the gunwale when the flocks are coming in. And they *will* come in. I have often seen them fly close enough to be struck with an oar—I may say that they make it an invariable rule to do this when the gunner has taken the shells out of his gun or laid it aside to pick up his decoys after a morning's cootless waiting in the cold. One oddity in the gentle art of duck shooting is the practice of "hollerin' coots"—that is, of making a great noise when a flock is passing by out of shot—when they will often turn and come to the decoys. The report of a gun sometimes has the same effect, but we New Englanders are too thrifty to waste powder and lead where our vocal organs will serve as well.
>
> Next to decoying, the use of the "gunning float" is the most effective method of killing coots. The "gunning float" is a long, low craft, drawing but little water and showing only a foot or so above the surface when properly trimmed down with ballast. In the fall, for use in the open water, they are "trimmed" with "rockweed"; in the marshes with "thatch." In the spring and winter months the proper thing is snow and ice to represent a drifting ice cake. It takes sharp eyes to detect the dangerous one among the many

harmless pieces of ice when the gunner, clad in his white suit, is working his cautious way along toward the feeding flocks. The deception is so complete that I have known that crafty old pirate, the crow, to almost alight on the nose of a float when it was being pushed after a flock of sea fowl. This float gunning is the method most used for all duck and goose shooting on the eastern New England coast line.

Winter.—There is not a month in the year during which scoters may not be seen on the Massachusetts coast; straggling birds, crippled, sick, or nonbreeding birds are present more or less all summer; the heavy migration flights last all through the fall and during much of the spring; and in winter this is one of the main resorts of all three species. The waters lying south of Cape Cod and in the vicinity of Nantucket and Marthas Vineyard are particularly congenial to these birds in winter, where the numerous islands, bays, and reefs offer some shelter from the winter storms, where they are not much disturbed by gunners at that season, and where they can find extensive beds of mussels, scallops, clams, and quahogs within easy reach on the numerous shoals and ledges. Here they congregate in enormous numbers, the three species associated together and often with eiders and oldsquaws. They have their favorite feeding grounds, to which they resort regularly every day at daybreak, feeding, playing, and resting during the day and flying out again at night to sleep on the bosom of the ocean, or on some more sheltered portion of the sound, far enough from land to feel secure.

DISTRIBUTION

Breeding range.—Northern North America and northeastern Asia. East to the coast of Labrador (north of the Straits of Belle-Isle) and Newfoundland (Grand Lake). South nearly or quite to the Gulf of St. Lawrence, to James Bay, to an unknown distance in the interior of Canada and Alaska and to the base of the Alaska Peninsula (Lake Clark). On the Aleutian and Kurile Islands. West to the Bering Sea coast of Alaska and to northeastern Siberia (Gichiga). North to northern Alaska (Kowak River), northern Canada (Mackenzie Bay) and probably north in the Labrador Peninsula to Ungava Bay and perhaps Hudson Straits.

Winter range.—Mainly on the seacoast. On the Atlantic coast regularly south to Long Island Sound and New Jersey, rarely to South Carolina, and occasionally to Florida. North regularly to Maine and more rarely to the Gulf of St. Lawrence and Newfoundland. On the Pacific coast from the Pribilof and Aleutian Islands south to southern California (Santa Barbara Islands) and from the Commander Islands south to Japan and China. In the interior it winters on the Great Lakes more or less regularly and has occurred irregularly or casually as far west and south as Wyoming (Cheyenne), Colorado (Fort Collins), and Louisiana (Lake Catherine).

Spring migration.—Early dates of arrival: Gulf of St. Lawrence, March 25; Ontario, Ottawa, May 4; Alaska, St. Michael, May 16, and Bering Straits, May 8. Late dates of departure: South Carolina, Bulls Bay, May 7; Virginia, Cobb Island, May 19; New York, Shelter Island, June 5; Massachusetts, Woods Hole, June 10; Alaska, Admiralty Island, June 10.

Fall migration.—Early dates of arrival: Ontario, Ottawa, September 1; Massachusetts, September 8; Minnesota, Heron Lake, October 5; Colorado, Denver, October 2. Main flight passes Massachusetts in October. Late date of departure: Alaska, St. Michael, October 15.

Egg dates.—Arctic Canada: Five records, June 10 to 21. Alaska: Four records, June 2 to August 3. Labrador: Two records, June 10 and 17.

MELANITTA FUSCA (Linnaeus)

VELVET SCOTER

HABITS

This is strictly an Old World species which owes its somewhat questionable place on the American list to the fact that it has been recorded as a straggler in Greenland. I have never been able to understand why the birds of Greenland should be included in our North American fauna, while those of Cuba and the Bahamas, which are both geographically and faunally much closer to us, are excluded. Greenland both geographically and faunally is but a little nearer North America than Europe; it is intermediate. If we exclude cosmopolitan species, common to both hemispheres, about three-fifths of the breeding birds of Greenland are also North American and about two-fifths are also European.

As the velvet scoter is practically unknown, as an American bird, I can not do better than quote its life history from one of the best of the European writers, an eminent authority on ducks, Mr. John G. Millais (1913) as follows:

Nesting.—Velvet scoters arrive on the lakes of Norway and Sweden about the end of April, in fact, as soon as the ice breaks up, and even earlier on the lake swamps of Lithuania, which seems to be about the southern limit of nesting birds. The male and female are much devoted to one another and keep close together during the early part of the nesting season. It has often been noticed that if one of the pair is shot the other will fall to the water and dive or stay close to its fallen mate.

They seem to prefer inland lakes and small ponds on which to breed. Collett has found them breeding in large numbers in the hill lakes of Gudbrandsda, Valders, Osterdal, and north to Finmark, and I have myself seen females and young birds on the lakes of Valders and Trondhjem in September, the males having departed.

The nest is often found in a depression of the dry ground in the open; at other times sheltered by brushwood such as salix or juniper. C. E. Pearson found one nest in a clump of marram grass amongst sand hills. Others were placed deep down in cracks of the peat, overgrown by *Empetrum nigrum*, so that the sitting duck was carefully concealed. Seebohm found several nests in the Siberian tundra far from the water, whilst Knobloch says the velvet scoter sometimes breeds in forests. The nest is usually a deep hollow lined with grass and leaves. The earliest clutches are to be met with in the Baltic, and are to be found from May 25 onward, but in Lapland it is more usual to find eggs in June, and generally in the second half of that month. H. F. Witherby took a clutch of eggs on July 22 in Russian Lapland, and Seebohm found eggs on the Petschora in July. Six to ten eggs are usually laid, and incubation is by the female alone. As to the period of incubation no data are available.

Eggs.—Six to ten, as a rule, but clutches of 11 have been recorded. Simonson says that clutches of 10 to 14 may be met with. Oval in shape, creamy white with a warm "apricot" tinge when fresh, which fades after a time. Average size of 90 eggs, 70.8 by 47.9 mm. (2.78 by 1.88 inches). Max., 76.5 by 49.5 and 71.2 by 51.5; min., 64.3 by 46.9, and 68.3 by 44.8 mm. (F. C. R. Jourdain).

Young.—The males appear to desert the females about the time the young are hatching. E. F. von Homeyer says: "I have often seen flocks of 60 to 100, consisting of old males only, in the months of July and August, and these spent the day on the high sea and at dusk came to the shallower water on the coasts in the bays of the island of Rügen."

According to Pleske, they nest in such numbers in the island of Rugoe in Esthland (Russian Baltic Provinces), that the inhabitants make ornaments for their rooms with the blown eggs. All the habits of this duck, the upbringing of the young, and the early departure of the males for the sea, seem to be similar to other true sea ducks. Late in September, when the young are able to fly, the female takes them to the nearest seacoast, where she stays with them until the migration commences in late October.

Plumages.—The sequence of plumages from the downy stage to maturity and the subsequent molts and plumages are so similar to those of our white-winged scoter that it seems hardly necessary to describe them here. For a full account of the molts and plumages of the velvet scoter, I would refer the reader to the excellent work by Mr. Millais (1913).

Food.—On this he says:

The food of this species consists chiefly of conchylia and crustacea, which they gain from a considerable depth. I have found their stomachs filled with large numbers of the common mussel, which seems to be their principal food, mixed with quantities of sand and small pebbles. They are also very partial to the razor shell in Orkney. I have also seen them bring to the surface quite large crabs, which they break up before swallowing. The great black-backed gull often waits on in attendance of feeding velvet scoters, and I have more than once seen these clever robbers swoop down and steal the crab, the duck merely gazing round in surprise when he finds his treasure gone.

Behavior.—I have never found the velvet scoter a very wild bird, except in rough weather, when it is easy for them to take to wing, and this is probably accounted for by the fact that their bodies are very heavy, and they seem to experience considerable difficulty in taking to flight if there is little or no headwind. They are as a rule much tamer than either the surf or common scoter; and if a boat is carefully maneuvered so as not to press them at first, a shot is certain. They rise head to wind with the usual run-up, and can not turn away from a boat until they have traveled some 30 to 50 yards. The flight is at first accompanied with much noise and flapping, and usually performed at a very low elevation. Unlike the other scoters, they usually adopt a "string" formation, and seldom move about in large flocks. It is most common to see single birds or flocks of from 3 to 15, each bird following the leader at a yard or so apart, and only 2 or 3 feet above the water. In the morning and evening these flocks or single birds may often be seen coming up the tideway from the deep sea, where they have been resting, preening, or sleeping during the hours of high tide, and moving toward their regular feeding grounds. On settling they seem to sink into the water with a heavy splash and glide for some distance over the element before coming to rest.

The velvet scoter has no superior in swimming and diving. Its powerful legs and feet enable it to pass rapidly beneath the water, and reach the bottom at depths of 40 feet, and even more. They seem to prefer to search for their food in deep places, probably because mussel beds situated in such spots are far offshore, and consequently safe. I do not think they use the wings under water, at any rate to the same extent as the eider.

Both the male and female velvet scoter make a hoarse guttural cry like the words "*kra-kra-kra.*" The male probably had a distinct call during courtship, but no one, so far as I know, has ever seen the mating display of these birds.

In our islands the velvet scoter is strictly a sea duck, is only very rarely killed on fresh water, and then only on migration. As a rule these birds frequent the neighborhood of mussel banks at some distance offshore, apparently caring little whether these situations are exposed or protected, for they come with the utmost regularity to the same places year after year. Most of the places known to me in Scotland and the islands where these birds spend the winter months are more or less protected by outlying islands or headlands, but in some cases, such as St. Andrews Bay and the Tay estuary, their feeding grounds are usually exposed to north and easterly winds. They seem to be capable, however, of standing as much buffeting by wind and weather as the hardy eiders and long-tailed ducks, and will ride out great storms at sea without coming in for protection.

Winter.—Herr Gätke (1895) gives the following account of the winter home of this and other sea ducks about Heligoland:

During the severe winter all the flocks of birds which, during ordinary winters, are in the habit of staying in the Gulfs of Bothnia and Finland, and under the shelter of the west coast of Holsten, now congregate on the open sea outside of this ice field. Wherever the eye roams it alights upon sea ducks of all possible species, near and far, high and low, in smaller or larger flocks, singly or in pairs. These consist of myriads of common and velvet scoters, flights of from 5 to 50 gay-colored red-breasted mergansers, smaller companies of the beautifully colored goosander, mixed with bands of from 20 to 100 or more scaups (*A. marila*), which flights again may be crossed by from 3 to 5 of the brilliant white, green-headed males of the goldeneye, and the still rarer and elegant eider duck; and travelling high overhead long chains of whooper swans send forth their loud and resonant trumpet calls. The wide surface of the sea presents a scene of aquatic bird life equally rich and varied. Velvet and common scoters assemble in dense crowds near the ice, while large flocks of scaups, all keeping close together, dive and swim about among the rocks off the eastern and western sides of the island.

DISTRIBUTION

Breeding range.—Northern Europe and Asia, from Norway eastward to northeastern Siberia (Marcova and Gichiga) and on Nova Zembla.

Winter range.—Temperate Europe and Asia, south to Spain, Morocco, Egypt, northern Persia, and Turkestan.

Casual records.—Accidental in the Faroe Islands and Greenland.

Egg dates.—Lapland: Five records, May 25 to July 22. Norway and Sweden: Two records, June 18 and July 4.

MELANITTA DEGLANDI (Bonaparte)

WHITE-WINGED SCOTER

HABITS

Spring.—The northward movement of scoters on the New England coast begins early in March, but the main flight comes along during the first half of May and continues in lessening numbers all through that month. It has long been known to gunners that a local westward flight of white-winged scoters takes place on the south coast of New England in May, consisting wholly of fully plumaged adult birds, recognized by the gunners as "May white wings." This undoubtedly indicates an overland migration route to their breeding grounds in the Canadian interior. I have seen several thousand of these birds gathered in large flocks in the waters about Seconnet Point, Rhode Island, early in May, preparing for this flight.

Mr. George H. Mackay (1891) refers to this flight as follows:

This movement is a peculiar one, inasmuch as it takes place about the middle of May, and after the greater portion of the migration of this group has passed by, as also ignoring the coast route accepted by all the rest. My attention was first directed to this unusual movement during the spring of 1870, while shooting at West Island, off Seconnet Point, Rhode Island, and it has occurred regularly every year since that date, as was undoubtedly the case earlier. These birds are apparently all adults and do not seem to heed the regular migration to the eastward of many of their own kind, which has no effect in hastening their own departure for the north. When the time arrives for them to set out on their migration, and the meteorological conditions are favorable—for it must be clear at the westward—they always start late in the afternoon, from 3 to 5 o'clock, and continue the flight during the night, passing by Marthas Vineyard, Woods Hole, Seconnet Point, Point Judith, and Watch Hill, quite a number frequently going over the land near the coast, they being very erratic at such times in their movements. This flight lasts for from three to seven days, according to the state of the weather. I have never heard of their starting before the 7th of May, which is unusually early; the customary time being from the 12th to the 15th, and the latest the 25th. They usually fly at a considerable altitude, say, from 200 to 300 yards, fully two-thirds of them being too high to shoot. They prefer to start during calm warm weather, with light southerly, southeasterly, or easterly winds; though they will occasionally fly when the wind is strong. They never fly in the forenoon; but when once they have determined to migrate they leave in large flocks, some of which number from five to six hundred birds, while as many as 10,000 have been estimated as passing in a single day, I have never heard of, or seen, any similar flight to the eastward after this western flight has taken place. A few of the other two scoters are seen with the white wings during this western movement. No perceptible difference is noted in their numbers from year to year, and I have never heard of a year when such a flight as above described did not take place.

On the south coast of Labrador we saw migrating flocks of scoters flying eastward all through the month of June, but probably some of these were not breeding birds. Flocks of nonbreeding scoters are frequently seen in summer on the coast of New England and from California northward. Probably the bulk of the breeding birds arrive on their nesting grounds early in June, although the nesting season does not begin until the middle or last of the month.

What becomes of the vast hordes of scoters that migrate along our coasts has long been a mystery to the gunners. Although they are widely distributed over an extensive breeding range, they have never been found breeding abundantly anywhere; probably their main breeding grounds have never been discovered. On the south coast of Labrador we saw no evidence of their breeding, and Audubon found them there but sparingly. Judging from what I saw and what I learned from other observers on the northeast coast of Labrador in 1912, I am inclined to think that this species breeds more or less commonly in the interior of that great peninsula. Among the vast flocks of scoters, seen all along that coast in summer, numerous flocks of this species were observed, but they were not nearly as abundant as the other two species. These flocks were composed almost entirely of adult males, which probably meant that the females were incubating or tending their broods of young in inland ponds. I was told that they breed far inland and at long distances from any water.

Nesting.—In the Devils Lake region, in North Dakota, Herbert K. Job found white-winged scoters breeding quite commonly; on June 27, 1898, he found several nests of this species on some small islands in Stump Lake, containing from 1 to 14 fresh eggs and one empty nest. I visited these islands with him in 1901, and on May 31 we did not find a single egg, although we saw a few of the birds flying about in pairs; evidently they had not yet laid. On June 15 we again explored the islands quite thoroughly, finding only one incomplete set of 5 eggs, cold and fresh. This nest was in the center of a small patch of rosebushes, where a hollow had been scraped in the ground and the eggs buried under a lot of dry leaves, sticks, soil, and rubbish, so as to be completely concealed from view. No attempt had been made to line the nest with down which is generally added after the set is complete. The scattered clumps of rosebushes on these islands, where they grew tall and thick among masses of large boulders, formed excellent nesting sites for the scoters and doubtless concealed several nests. One nest we certainly overlooked, which on June 22, was found to contain 12 eggs.

In the Crane Lake region in southwestern Saskatchewan we found a few pairs of white-winged scoters breeding in 1905 and 1906, but only one nest was found on June 28, 1906. While walking through an extensive patch of wild rosebushes near a small slough the female was flushed from the nest almost underfoot and shot by my companion, Dr. Louis B. Bishop. The nest consisted of a hollow in the ground under the rosebushes, profusely lined with dark-gray down; it contained 9 fresh eggs. All of the nests that Mr. Job and I have seen were placed under wild rosebushes or other small deciduous shrubs, but others have found them in somewhat different situations.

Macoun (1909) records a nest found by Walter Raine as follows:

> On June 26, 1893, Mr. G. F. Dippie and myself found a nest containing 9 eggs on an island at the south end of Lake Manitoba. The nest was built between loose bowlders and consisted of a hollow in the sand lined abundantly with dark down. The eggs were very large and of a deep, rich, buff color. The bird sat very close upon the nest and did not fly up until I almost trod upon her. It appears to be a late breeder, nesting late in June on the islands of Lakes Manitoba and Winnipeg. Mr. Newman sent me an egg of this bird which he took from a female he had shot at Swan Lake, northern Alberta, on June 25, 1897.

The nest down of the white-winged scoter is larger than that of the American scoter; in color it varies from "clove brown" to "olive brown," with small and inconspicuous whitish centers.

Eggs.—As this scoter is a late breeder, probably some of the smaller sets referred to were incomplete. I think that the normal number of eggs in a full set varies from 9 to 14. The eggs are elliptical ovate in shape. The shell is smooth but not glossy; in some eggs it is very finely granulated or minutely pitted. When first collected, even after being blown, the color is a beautiful "pale ochraceous salmon" or "sea shell pink," but this color fades to "pale pinkish buff" or "cartridge buff" in cabinet specimens. The pitted eggs are minutely dotted with "pinkish cinnamon," giving them a darker appearance. The measurements of 71 eggs, in various collections, average 65.3 by 45.7 millimeters; the eggs showing the four extremes measure **72.5** by 47, 68.5 by **49, 55.4** by 37.7, and 58.9 by **35.7** millimeters.

Plumages.—The downy young of the white-winged scoter is thickly covered with soft, silky down. The upper parts, including the upper half of the head, down to the base of the lower mandible and a space below the eye, are uniform "clove brown," shading off to "hair brown" on the flanks and into a broad collar of "hair brown" which encircles the lower neck. The chin and throat are pure white, which shades off to grayish white on the lower cheeks and the sides of the neck. The under parts are silvery white, and there is an indistinct, tiny white spot under the eye. The feather outline at the base of the bill is much like that of older birds. Doctor Dwight (1914) mentions a "white patch of down, foreshadowing the white wing patch," but I can find no trace of it in my one specimen.

In the juvenal plumage the sexes are alike, dark brown above, lighter and more mottled brown below; there are conspicuous whitish patches on the lores and on the auriculars, varying in intensity and extent; the white secondaries, forming the speculum, are tipped with dusky, which often invades much of the inner web. This plumage is often worn without much change all through the winter and into the spring. But usually in December, or a little later, the sexes begin to differentiate by the growth of black feathers in the male and brown feathers in the female, starting in the head, obliterating the whitish head patches, and spreading to the back, scapulars, and flanks, the latter being dark brown in both sexes. The bill in the young male now begins to show color, about as in the adult female, but not the swollen shape of the older male.

A complete postjuvenal molt takes place during the next summer, July, August, and September, at which the black plumage of the male is assumed, with the white eye patches and the pure white secondaries; the flanks are still dark brown in this and in all subsequent plumages that I have seen; the bill now becomes highly colored and approaches the adult bill in shape. The bird is now practically adult, at an age of 14 or 15 months, but the full development of the bill and highest stage of plumage will not be perfected for about a year more. The iris, which is brown in young birds, becomes white at this age. The female also assumes a practically adult plumage, at this molt, which is uniform dark brown; and she will be ready to breed the following spring.

Adult birds have an incomplete prenuptial molt in early spring, involving the head and body plumage and tail, and a complete postnuptial molt in late summer, at which time they become incapable of flight. There is no real eclipse plumage, but the appearance of one is created by the mixture of old, worn, faded feathers with fresh, new ones. There is also no marked seasonal change in any of the scoters.

Food.—The food of the white-winged scoter includes a varied bill of fare, differing greatly in the various localities which it visits. On the New England coast, where it is so abundant in the winter, it is strictly maritime and seems to feed mainly on small mussels and other small mollusks which it obtains by diving about the submerged ledges, often to a depth of 40 feet, tearing, with its powerful bill, the shellfish from the rocks to which they are firmly attached. I have seen the crop of one of these birds crammed full of mussels nearly an inch long, and have often wondered whether the tough shells were ground up in their muscular stomachs or chemically dissolved; probably both actions are necessary. Some of our fishermen have claimed that scoters are injurious to the shellfish interests on our coast, accusing them of feeding on young scallops and clams, but I doubt if they do much damage in this way; they certainly could not obtain many clams, which are usually buried in the sand. Sea clams, which are sometimes found on the surface of a sand flat, are probably more often taken by these birds. J. C. Cahoon (1889) recorded an instance where the clam was too big for the scoter, which was found floating on the water with a large sea clam firmly clasped on its bill; probably the weight of the clam had kept the bird's head under water until it was drowned. On inland lakes and ponds they live largely on crawfish, slugs, snails, and mussels. They have been known to eat, according to Dr. F. Henry Yorke (1889), small fishes, frogs, tadpoles, fish spawn, and the larvae of insects. On the western sloughs and marshes they evidently feed largely on vegetable food, such as flags, duckweed, pondweed, and pickerel weed. Doctor Yorke reports the following families of plants as identified among their foods: Lemnaceae, Naiadaceae, Selaginellaceae, Salviniaceae, Glatinaceae, Gentianaceae, Lentibulariaceae, Pontederiaceae, and Mayaceae; also the following genera: Iris, Myriophyllum, Callitriche, and Utricularia.

Behavior.—The flight of the white-winged scoter is heavy and apparently labored; it seems to experience considerable difficulty in lifting its heavy body from the surface of the water; except when facing a strong wind, it has to patter along the surface for some distance, using its feet to gain momentum. But, when well under way, it is much swifter than it seems, is strong, direct, and well sustained. Migrating flocks, in all sorts of irregular formations, fly high under favorable circumstances; but when flying against the wind or in stormy weather (northeast storms seem to be particularly favorable for the migration of the scoters) they fly close to the water and in rough weather they take advantage of the eddies between the waves. The flight is usually along the seacoast, following all the large indentations of the coast and crossing the smaller bays; but, where considerable distance is to be gained they often fly across capes or necks of land, usually all at about the same place. There is a regular crossing place on Cape Cod, Massachusetts, from Barnstable Harbor to Craigville beach, a distance of about 3 miles; this is one of the narrowest points on the cape and it saves them many miles of flight around the horn of the cape. Gunners take advantage of this confirmed habit and assemble there in large numbers to shoot at the passing flocks; when flying against a strong south wind here they usually fly low enough to shoot, but, if not, a loud shout from the gunners often brings them scurrying down to within range.

Edwin S. Bryant (1899) describes an interesting flight habit of this species, as follows:

This bird has some habits unlike those of other ducks. The most prominent habit is the morning flight. This does not occur so regularly as at first I supposed. But if a person is so fortunate as to be present when a great flight is in progress, he will witness what I consider to be a fascinating picture of bird life on the prairie.

Imagine if you can a body of water some 6 or 7 miles long and 2 miles wide where it leaves the main lake, extending northward, bounded on both sides by undulating prairie. Take for a background the steep hills on the far side of the lake, or the heavy timber of Grahams Islands—let the time be sunrise, with the dewdrop jewel accompaniment that the poets rave about. Fill the air with hundreds of scoters, circling and quartering after the manner of swallows, most of them fanning the weed tops in their flight. Flying by pairs, side by side, and in companies of pairs, they often circle about a person several times, within easy gunshot range; and if one is so disposed, he may shoot a pair with one discharge of the gun, so closely do they keep together. As would be supposed, the white wing patch is very conspicuously displayed as the birds glide around. In half an hour the performance is at an end.

In two minutes time the scene changes as if by magic. All the birds are making offshore together. With a glass I follow them. Their dark bodies stand out in contrast with the whitecaps, and the flash of a wing patch against a green wave is the last seen of them as they settle down far out in the lake. Later in the day they will swarm along shore or congregate on the numerous sandy points.

A. D. Henderson has sent me the following notes on the habits of this species in northern Alberta:

These birds are much esteemed by the halfbreed Indian population, and up to a few years ago residents in the northern part of the Province were allowed to kill them at any time for food by a special provision in the game act. They make the best wing shooting of all the ducks, as they seldom swerve from the gunner. They seem to be the most amorous of the ducks, readily decoying to a wounded female and chasing her on the water within easy range of a canoe in the breeding season. In spring they fly from daylight until about 9 o'clock, and it is then they are shot as they round the points on the lake or pass between narrows. Small flocks can be seen on the water, the males pursuing the females; then they will make a short flight and alighting again resume the sport with much splashing on the water. At this time they make a sound like tinkling ice, but whether this is made with their wings or voice I do not know. When flying during courtship their wings whistle like a goldeneye's, but much louder. They also utter a short croak while flying. Though hundreds of them breed here, I have never found a nest, but have heard of several being found, usually at quite a distance back from the large lakes and near smaller ones. It is said that Lac LaNoune takes it name from these ducks owing to the resemblance of their black and white coloring to the garb of the nuns.

J. M. Edson has sent me the following interesting notes:

This species is found at all seasons in the Puget Sound region, being particularly abundant during winter. Like *O. perspicillata*, this species has a habit of leaving the water and taking a daily flight off over the land, during the summer season. These flights are particularly noticeable in pleasant weather and in late afternoon. The birds rise in considerable flocks to an elevation of 200 or 300 feet, stringing out in line, or in converging lines, sometimes forming a V. The whistling of their wings can be heard for some distance. Often this sound is punctuated by the slapping together of interfering wings. They sometimes fly considerable distances inland before returning to the water. So far as known they do not nest in this region, although birds in the full adult plumage are frequently seen throughout the summer.

An instance of peculiar behavior of birds of this species came under my notice not long ago. Watching the sea birds from a bluff overlooking Bellingham Bay, on a calm evening in December (the 24th), my attention was attracted by unusual activity in a little group of white-winged scoters. They were about 50 yards from the beach. Ten of these birds were bunched together and actively swimming and plunging about within a circle of perhaps 10 or 12 feet in diameter. I was unable to distinguish the sexes with certainty, and have no knowledge to the effect that December is their courtship season. It looked like a game of tag of some sort. At the center of the group two birds would assume a pose as if billing and caressing each other, one with its head elevated, the other's depressed, the bills coming in contact. The pose would last only two or three seconds, till some other bird would approach one of them from behind, when the latter would suddenly turn upon it and chase it away, the pursued bird taking a circular course around the flock. Sometimes both the posing birds would be simultaneously approached, and each would turn upon his assailant. The other birds would hover close about, watching for a chance to tag the posers from behind. The two main actors would again come to the center and resume their pose, only to be promptly interrupted again with the same result. So far as I could observe, the same pair took the central part all the time. I watched them for perhaps half an hour, and the game was still in progress when I left. On February 11 I saw the same performance enacted at a greater distance from shore. There were about the same number of birds in the group and the play was as before. Although there were numerous other birds of the same species scattered about in the near vicinity, these paid no attention to the game.

Like the other scoters, this species is a strong swimmer and an expert diver. It dives to considerable depths for its food when necessary, though it prefers to feed at low tide when the mussel beds are nearer the surface. When wounded it can swim for such long distances under water that its pursuit is almost useless.

I have never heard any vocal sound from the white-winged scoter, but Dr. Charles W. Townsend has sent me the following notes from E. P. Richardson:

> In regard to the bell note of the white-winged scoter, I have only as much as this to say: On still nights in the fall, when we might be listening for black ducks, we would occasionally hear the rush of a flock of ducks overhead, with an occasional bell-like, low whistle, recurring at intervals in series of six or eight notes, entirely distinct from the rush of wings. The old gunner who took charge of our place at Eastham used to say that these birds were old white wings, and would add, "some call them bell coots." The sound is rather more slowly repeated than the usual wing beat of the black duck, but the impression that it always gave me was that it was produced by the wings and not by the voice of the fowl. At any rate, it is a very clear and distinct sound, a series of low, bell-like sounds which might occur repeated two or three times and then be lacking. These birds were crossing Cape Cod on their fall migration. I have not heard it for some years now, but it used to be a frequent experience. As I said, these birds were going over in the night and we could never, of course, identify them.

Game.—The time-honored sport of coot shooting has for generations been one of the most popular and important forms of wildfowl hunting on the New England coast. Next to the black duck, which undoubtedly stands first in the estimation of our sportsmen, there are probably more scoters killed on our coasts than any other of the Anatidae. Aside from the fact that the scoters are not of much value for the table, coot shooting has much to recommend it; it is a rough and rugged sport, testing the strength, endurance, and skill of an experienced boatman; the birds are strong fliers and hard to kill, requiring the best of marksmanship, under serious difficulties, and hard-shooting guns; during good flights game is almost always within sight, giving the sportsman much pleasant anticipation; and chances are frequently offered to show his skill at difficult and long shots. I was born and bred to be a coot shooter, inheriting the instinct from three generations ahead of me, and I only wish that I could impart to my readers a small fraction of the pleasure we have enjoyed in following this fascinating sport.

Rudely awakened at an unseemly hour, soon after midnight it seems, the party of gunners are given an early breakfast before starting out. It is dark as midnight as we grope our way down to the beach, heavily laden with paraphernalia, launch our boats in a sheltered cove among the rocks, and row out onto the ocean. The crisp October air is cool and fresh, as the light northeast wind comes in over the ledges, fragrant with the odors of kelp and rockweed. There is hardly light enough at first to see the line of boats, strung out straight offshore from the point, but soon we find our place in the line, anchor our several strings of wooden decoys, and then anchor our dory within easy gunshot of the nearest decoys, which if correctly placed are the smallest and most life-like; the largest decoys are merely to attract the birds from long distances. Perhaps before our decoys are set we have seen a few shadowy forms flitting past us in the gloom, or heard the whistle of their wings in the dark, the beginning of the morning flight; occasionally the flash of a gun is seen along the line and the day's sport has begun. As the gray of early dawn creeps upward from the sea we can clearly distinguish the long line of boats, perhaps a dozen or fifteen, anchored at regular intervals, a little less than two gunshots apart so that birds can not slip through the line, and extending for several miles offshore, an effective barrier to passing flocks. Every eye is turned northward, looking up the coast and straining to discover the minute specks in the distance, as the first flock appears several miles away. "Nor'ard," the warning signal is passed along the line, as some keen eye has made the longed-for discovery, and every gunner crouches in his boat to watch and wait and hope for a shot. Soon we can make them out, an irregular, wavering bunch of black specks, close to the water and well inshore. The boom of distant guns tells us that other gunners up the coast have seen them and perhaps taken their toll. On they come, now strung out in a long line headed straight for us, big black birds with flashing white wing patches, "bull white wings," as the males of this species are called; we shall surely get a shot. But no, they have seen us and swerved, flying along the line seaward; a shot from the next boat drops a single bird and they pass through the line beyond, dropping two more of their number. A bunch of young surf scoters, "gray coots," is headed for the next boat, and we try to attract their attention by imitating the whistling of their wings; they turn and swing in over our decoys, dropping their feet and preparing to alight; four barrels are fired in quick succession and three of them drop in the water. Two of them will die as they are lying on their backs with feet kicking the air, but the other has its head up and is swimming away. We throw over our anchor buoy and give chase, but cripples are hard to hit in the water and we have a long pull and plenty of shooting before we land him. Meantime we have missed a magnicent shot at a large flock of "skunk heads," surf scoters, which circled over our decoys and escaped through the gap, and on our return we find only one of our "dead" birds.

A temporary lull in the flight gives us a chance to rest and admire the beauty of the scene around us; the delicate blush of dawn deepens and brightens as the gorgeous hues of sunrise spread from the eastern horizon over the broad expanse of sky and sea, a rapidly changing play of colors until the sun itself appears over the water and bright daylight gilds the ocean. Bird life is not lacking in the scene; herring gulls are flying about on all sides, often coming near enough to tempt us to shoot at them, but never quite near enough to kill; they seem to know just how far a gun will shoot. Occasionally a black-backed or a few kittiwake gulls are seen. Loons are frequently passing, generally high in the air, with long outstretched necks, flying swiftly in a straight line, their bodies propelled by rapid wing strokes; they often fly within gunshot, but are tough and hard to kill. Large flocks of oldsquaws make interesting shooting, as they twist and turn and wheel in compact bunches; they are swift of wing and not easy to hit; their weird cries add a tinge of wildness to the scene. On rare occasions the sport is enlivened by a shot at a flock of brant, and our pulse runs high when we see a long line of big black birds with white bellies headed for our boat, flying close to the water; we are lucky if we get any for they are very shy.

The little "gray coots," the young of the American and the surf scoters, give the best shooting and are the best for the table; they decoy well, particularly when in small flocks, and are easily killed; a pair or a single bird will often circle about the decoys again and again, giving plenty of chances for long single shots. "Butter bills" and "skunk heads," the adults of these two species, decoy well in small flocks, but large flocks are usually wild and either pass the line high in the air or circle out around the end of it. Fifteen or twenty birds is considered a good day's sport, but as many as 135 birds have been killed in a day by two gunners in one boat, or over 90 by a single gunner. Although they are thus persecuted year after year throughout the whole length of their migration route, they do not seem to have diminished materially in numbers since the time of our earliest records, and vast numbers of them still migrate along our coast.

Fall.—The fall migration on the New England coast begins in September, a few early flocks sometimes appearing in August; the main flight is in October when, under favorable weather conditions, it is very heavy; before and during northeast storms large flocks of scoters are almost constantly in sight migrating southward; the flight is prolonged in lessening numbers during November, and by the end of that month they have reached their winter quarters.

Winter.—Their winter range extends from the Gulf of St. Lawrence southward along the Atlantic coast to South Carolina, and on the Pacific coast they winter from the Aleutian Islands to Lower California. White-winged scoters are particularly abundant in winter on the waters of Long Island and Vineyard Sounds, the center of their winter range, where they find an abundant food supply in the beds of shellfish which abound in this region. Great rafts of them may be seen bedded on the water way offshore where they sleep and rest. At certain stages of the tides, when the water is not too deep over their feeding grounds, they fly in regularly, day after day, to the same spot to feed on the mussel beds on the submerged ledges and sunken rocks along the shores and even in the harbors. Scallop fishermen are often guided by their movements in locating their quarry. Gunners soon learn to locate their feeding grounds and take advantage of their regular flights.

Winthrop Sprague Brooks (1915) described, under the subspecific name, *dixoni*, a subspecies of the white-winged scoter, the type of which was collected by Joseph Dixon at Humphrey Point, Alaska. He assumes that all of the white-winged scoters, which breed in Alaska and migrate down the Pacific coast, are referable to this subspecies, which he characterizes as "similar to *deglandi*, with the exception of the size and shape of bill, which in *dixoni* is shorter and broader in proportion to its length and more blunt at the tip, with the angles from its greatest width to the tip more abrupt." He says further: "On examining a large series of white-winged scoters from both sides of the continent there is no difficulty in separating Atlantic and Pacific birds by means of this character of the bill."

In order to establish a winter range for this subspecies I wrote Dr. Joseph Grinnell for his opinion on the status of California birds. He replied as follows:

> I know nothing about *Oidemia deglandi dixoni*. I have not used this name for any of the birds in this museum, because the authenticity of the alleged race has not been verified by anyone else. I have just looked at our birds, with Brooks's drawings of bills before me. I see every sort of variation from the narrow extreme to the broad extreme among birds taken in California. There are only three eastern birds here, and each of them finds a counterpart among California-taken specimens.

From this statement I should infer that the subspecies is untenable and that the characters ascribed to it are due to individual variation. The study of more specimens from the supposed breeding range of *dixoni* might establish a local breeding race, which mingles in its winter range with the commoner form. A careful study of a large series from many localities is necessary to settle the question.

Dr. H. C. Oberholser and I have recently made a careful study of the large series of white-winged scoters in the collections of the United States National Museum and the Biological Survey, containing birds from many different parts of North America, and find that the characters on which *dixoni* are supposed to be based can be matched in many birds from the Atlantic coast, and that they are no more prevalent in birds from the Pacific coast or the interior than elsewhere. We therefore came to the conclusion that these characters represent merely individual variation and should not be regarded as establishing a subspecies.

DISTRIBUTION

Breeding range.—Northern North America. East to the Labrador coast (Hamilton Inlet and Nain), occasionally in Newfoundland (Gaff Topsail). South to the north shore of the Gulf of St. Lawrence, southern Manitoba (Shoal Lake), central North Dakota (Devils and Stump Lakes), and northeastern Washington (east of the Cascade Mountains). West to northwestern British Columbia (Stikine River) and sparingly to northwestern Alaska (Kotzebue Sound). Seen in summer and perhaps breeding in the Aleutian Islands as far west as Tanaga Island. North to the barren grounds of northern Alaska and Canada.

Winter range.—Mainly on the seacoasts. On the Atlantic coast from the Gulf of St. Lawrence southward to South Carolina and rarely to Florida. On the Pacific coast from the Commander, Pribilof, and Aleutian Islands southward to Lower California (San Quintin Bay). In the interior on the Great Lakes and irregularly or casually as far west and south as southern British Columbia (Okanogan Lake), Colorado (9 records), and Louisiana.

Spring migration.—Northward along the Atlantic coast to Labrador and northwestward from the coast to the interior; also northward in the interior and along the Pacific coast. Early dates of arrival: Ontario, Toronto, April 13; Ohio, Lorain, April 27; Kentucky, Bowling Green, April 6; Minnesota, April 5; Manitoba, Aweme, April 15; Alberta, Alix, May 10; Mackenzie, Fort Simpson, May 18. Late dates of departure: Connecticut, May 15; Massachusetts, May 25; Ontario, Ottawa, May 4, and Toronto, May 26; Pennsylvania, Pittsburgh, May 13; Ohio, Lorain, May 3. The main flight up the Pacific coast and across British Columbia to the interior is in May.

Fall migration.—Apparently a reversal of spring routes. Early dates of arrival: Massachusetts, September 6; Maryland, Baltimore, September 12; Ontario, Beamsville, October 8; Minnesota, Heron Lake, October 11; Colorado, Loveland, October 11; Idaho, Coeur d'Alene, October 22; Utah, Bear River, October 8. Late dates of departure: Mackenzie, Nahanni River, October 14; Ontario, Beamsville, November 26; Idaho, Coeur d'Alene, December 1.

Egg dates.—North Dakota: Thirteen records, June 18 to August 10; seven records, June 29 to July 19. Arctic Canada: Nine records, June 14 to July 10. Alberta, Saskatchewan, and Manitoba: Seven records, June 21 to July 6.

MELANITTA PERSPICILLATA (Linnaeus)

SURF SCOTER

HABITS

This is probably the most abundant and certainly the most widely distributed of the three American species of scoters. It is widely and well known on the Atlantic and Pacific coasts, and in some of the more northern localities it is exceedingly abundant. The enormous flights of scoters, or "coots," as they are called, which pour along our coasts in the spring and fall are made up mainly of surf scoters and white-winged scoters; every gunner knows them, and most of the residents along the New England coasts have tasted the delights of coot stew.

Spring.—The abundance of the surf scoter on the spring migration is well illustrated by the following quotation from Mr. Dresser, given by J. G. Millais (1913), based on observations made at Lepreaux Lighthouse, in the Bay of Fundy:

> On my arrival there on April 25 myriads of ducks were flying past, among which surf scoters were more numerous than any other species. They followed the line of the coast at a short distance from the shore, and in passing the point generally steered close in or flew over the end of the point itself. On the 26th I spent the day among the rocks, and I never recollect seeing waterfowl in such countless numbers as I did on that day, all wending their way northward. Velvet, common, and especially surf scoters were the most numerous; but there were also many eiders, brent geese, long-tailed ducks, with a few harlequins, great northern divers, and some others. The surf scoters flew in large, compact flocks, from 8 to 10 deep. I estimated the length of the flocks by watching them as they passed certain points, the distance between which was known to me, and I found that one compact flock was at least half a mile in length, a second reaching from one point to another, distant nearly a mile and a quarter. I made several telling shots amongst them, knocking over 8 at one discharge and 6 and 4 at a double shot, though I was only using a light 15-bore gun. I found them, however, very hard to recover, for during the time the dog was retrieving them one or two were sure to come to and paddle off, and the sea was too rough to go out in a boat to pick up the cripples. The males proved to be far more numerous than the females, of which sex I only killed 3 during the whole day.

George H. Mackay (1891) writes:

> In the spring mating begins before the northward migration commences, as I have taken eggs from females, between the 15th and 25th of April, which varied in size from a cherry stone to a robin's egg. During this period the duck when flying is always closely followed by the drake, and wherever she goes he follows; if she is shot, he continues to return to the spot until also killed. I have often on firing at a flock shot out a female; the moment she commences to fall she is followed by her mate; he remains with her, or flies off a short distance, only to return again and again until killed, regardless of previous shots fired at him. I have never seen any such devotion on the part of the female; she always uses the utmost speed in flying away from the spot, and never returns to it.

Courtship.—W. Leon Dawson (1909) thus describes the courtship.

I have seen a surf scoter courtship in mid-April. Five males are devoting themselves to one female. They chase each other about viciously, but no harm seems to come of their threats; and they crowd around the female as to force a decision. She in turn chases them off with lowered head and outstretched neck and great show of displeasure. Now and then one flees in pretended fright and with great commotion, only to settle down at a dozen yards and come sidling back. If she will deign a moment's attention, the flattered gallant dips his head and scoots lightly under the surface of the water, showering himself repeatedly with his fluttering wings. One suitor swims about dizzily, half submerged, while another rises from the water repeatedly, apparently to show the fair one how little assistance he requires from his feet in starting, a challenge some of his corpulent rivals dare not accept, I ween. I have watched them thus for half an hour, off and on, and the villains still pursue her.

Charles E. Alford (1920) describes another interesting performance, as follows:

I once watched 8 male surf scoters wooing one female, and a most absurd spectacle it was. Immediately the female dived, down went all her admirers in pursuit. Then after a lapse of about 40 seconds the males would reappear one by one, the female, who was always the last to rise to the surface, being invariably accompanied by one male; but whether it was the same male on each occasion I was unable to distinguish. For a few seconds pandemonium would reign, the rejected suitors splashing through the water and pecking at their rivals in the most vicious manner, whilst the object of their desire floated serenely in their midst, apparently well pleased that she should be the object of so much commotion. Then she would dive again, and so the performance continued for over an hour, when they drifted out of sight.

He writes again (1921): "When displaying, the male surf scoter swims rapidly to and fro, keeping head and neck erect, and at intervals dipping its beak into the water. Should several males be present, the female swims from one to the other, bowing her head, or darting occasionally at some undesirable suitor." Maj. Allan Brooks (1920) says that he has "seen them vigorously courting in central British Columbia, well along in June; three or four males whirling about a female on the water like whirling beetles, and uttering a curious low, liquid note, like water dropping in a cavern." Mr. Charles L. Whittle writes to me that he has seen active courtship in the fall, October 5, which he describes as follows:

The males would face the females and bow rapidly and repeatedly, even to the extent of emersing their heads, thereby spraying themselves with water, the females watching the operation with interest. Another pretty and characteristic maneuver on the part of the males was to fly away suddenly about 75 feet, their wings being raised over their backs till the tips nearly touched as they alighted on the water, and then to swim back to their mates with great velocity, only to repeat their bowing. The males chased each other away from their respective mates by lowering their heads and swimming fiercely at their offending neighbor. That the females were also parties to these courtship performances is shown by the fact that they also would similarly attack the male, paired to the other female, if he approached too near.

Nesting.—In spite of the abundance of this species over a wide breeding range very few naturalists have ever found its nest, and remarkably little has ever been published regarding its breeding habits. The reason for this is that the breeding grounds are usually in such inaccessible places in the marshy interior that few explorers have ever visited them; moreover, the nests are probably so widely scattered and so well hidden that few have been found. The following from Dr. E. W. Nelson (1887) shows how abundantly the surf scoter must breed in northern Alaska and yet he never found a nest. He says:

On August 23, 1878, I visited Stewart Island, about 10 miles to the seaward of St. Michael. As I neared the island in my kyak I found the water literally black with the males of this species, which were united in an enormous flock, forming a continuous band around the outer end of the island for a distance of about 10 miles in length and from one-half to three-fourths of a mile in width. As the boat approached them those nearest began to rise heavily, by aid of wings and feet, from the glassy surface of the gently undulating but calm water. The first to rise communicated the alarm to those beyond, until as far as could be seen the water was covered with flapping wings, and the air filled with a roar like that of a cataract. The rapid vibrations produced in the air by tens of thousands of wings could be plainly felt. In all my northern experience among the waterfowl which flock there in summer I never saw any approach to the number of large birds gathered here in one flock, nor shall I soon forget the grand effect produced by this enormous body of birds as they took wing and swept out to sea in a great black cloud and settled again a mile or so away.

MacFarlane (1891) found a number of nests of the surf scoter in the Anderson River region which he said were much like those of the white-winged scoter, "the only difference noted being that generally less hay and feathers was observed in the composition of its nest, while only one contained as many as 8 eggs, the usual number being from 5 to 7." Of the white-winged scoter's nests he said:

These were always depressions in the ground, lined with down, feathers, and dry grasses, and placed contiguous to ponds or sheets of fresh water, frequently amid clumps of small spruce or dwarf willow, and fairly well concealed from view.

In a letter to Professor Baird, dated July 16, 1864, he writes:

The surf duck is numerous, but as its nest is usually placed at a considerable distance from open water, and always well concealed underneath the low-spreading branches of a pine or spruce tree, we never get many of its eggs. The female never gets off the nest until very closely approached, and then invariably (so far as I had an opportunity of judging) makes off to the nearest lake, where it will remain for hours, and thus exhaust the patience of the finder, who is, when traveling, at least, obliged to secure the eggs without their parent.

Audubon (1840) gives an interesting account of the finding of a surf scoter's nest in southern Labrador, which is about the only detailed account we have of the nesting habits of this common species. He writes:

For more than a week after we had anchored in the lovely harbor of Little Macatina, I had been anxiously searching for the nest of this species, but in vain; the millions that sped along the shores had no regard to my wishes. At length I found that a few pairs had remained in the neighborhood, and one morning while in the company of Captain Emery, searching for the nests of the red-breasted merganser, over a vast oozy and treacherous fresh-water marsh, I suddenly started a female surf duck from her treasure. We were then about 5 miles distant from our harbor, from which our party had come in two boats, and fully 5½ miles from the waters of the Gulf of St. Lawrence. The marsh was about 3 miles in length and so unsafe that more than once we both feared as we were crossing it that we might never reach its margin. The nest was snugly placed amid the tall leaves of a bunch of grass and raised fully 4 inches above its roots. It was entirely composed of withered and rotten weeds, the former being circularly arranged over the latter, producing a well-rounded cavity 6 inches in diameter by 2½ in depth. The borders of this inner cup were lined with the down of the bird, in the same manner as the eider duck's nest, and in it lay 5 eggs, the smallest number I have ever found in any duck's nest. They were $2\frac{5}{16}$ inches in length by $1\frac{5}{8}$ in their greatest breadth; more equally rounded at both ends than usual; the shell perfectly smooth and of a uniform pale yellowish or cream color.

We saw no signs of breeding surf scoters in southern Labrador in 1909, and apparently the few which bred there in Audubon's time have long since ceased to breed there regularly. On the northeast coast of Labrador, however, or rather a few miles inland, they probably still breed regularly and abundantly. We saw large numbers of males in the inner harbors and in the mouths of rivers at a number of places all along the coast in July and August, which suggested that probably the females were incubating sets of eggs or tending broods of young on the inland ponds or marshes. We hunted for nests in many suitable places, but never succeeded in finding one. Samuel Anderson, an intelligent observer and collector of birds at Hopedale, told me that surf scoters breed about the inland ponds and lakes, making their nests in the grass or under bushes close to the edge of the water. There is a Labrador set of 7 eggs in the collection of Herbert Massey, of Didsbury, England, for which he has kindly given me the data; it was taken by R. S. Duncan on Akpatok Island on June 11, 1903, and the female was shot for identification.

Eggs.—The surf scoter evidently lays from 5 to 9 eggs, usually about 7. The eggs are, I think, usually recognizable by their shape, size, and color. They vary in shape from ovate to elliptical oval and are often quite pointed. The shell is smooth but not at all glossy.

The color is a very pale "cartridge buff," or a pinkish or buffy white. Mr. Millais (1913) describes the eggs in the Massey collection as "rather pointed in shape, creamy in color." The measurements of 33 eggs, in various collections, average 61.6 by 43 millimeters; the eggs showing the four extremes measure **67.5** by 43, 59 by **45**, **58** by 41, and 59 by **40.5** millimeters.

Plumages.—Strangely enough there does not seem to be a single specimen of the downy young surf scoter in any American or European collection, except two half-grown young in the Museum of Comparative Zoology, in Cambridge, Massachusetts, collected by Francis Harper on Athabasca Lake, on July 28, 1920. Although as large as teal, these birds are still wholly downy, with no trace of appearing plumage. The smaller, a female, has the crown, down to and including the eyes, a deep glossy "clove brown" in color; the color of the black varies from "olive brown" anteriorly to "clove brown" on the rump; the sides of the head and throat are grayish white, mottled with "clove brown"; the entire neck is pale "clove brown"; the colors of the upper parts shade off gradually into paler sides and a whitish belly. In younger birds these colors would probably be darker, brighter, and more contrasted, as they are in other species.

In the juvenal plumage the sexes are alike. The crown is very dark, blackish brown, conspicuously darker that the rest of the plumage; the upper parts are dark brown and the lower parts lighter brown and mottled; there is a whitish loral space and a smaller whitish auricular space; the tail feathers are square tipped; and there is no trace of the white nuchal patch. During the first winter, beginning sometimes as early as October but often not until February, the sexes differentiate by the growth of new black feathers in the male and brown feathers in the female; this growth begins on the head, scapulars, and flanks, whence it spreads, before spring, until it includes all the fore part of the body and much of the back, leaving only the juvenal wings, part of the back, and the central under parts, which fade out almost to white; the tail is molted during the winter and the new feathers are pointed at the tip. The white nuchal patch is acquired by the young male before spring, but not the frontal patch; the bill assumes its brilliant coloring and increases in size, but it does not reach its full perfection for at least another year.

A complete postnuptial molt takes place in August, September, or even later in young birds, which produces a plumage which is practically adult. The male acquires the white frontal patch and the female the white nuchal patch at this molt and the bills become more mature, but full perfection is probably not attained for another year.

Young birds probably breed the following spring and at the next postnuptial molt become fully adult, when 27 or 28 months old.

Adults have a partial prenuptial molt, involving mainly the head and flanks, in March and April, and a complete molt in August. There is no true eclipse plumage and no marked seasonal change. I have a highly plumaged adult male in my collection, collected October 4, in which the white nuchal patch is merely indicated by a narrow, broken outline and the frontal patch by a short row of small white feathers.

Food.—The food and feeding habits of the surf scoter are practically the same as those of the other scoters and other diving sea ducks. Their food consists almost entirely of various small mollusks, such as mussels, sea clams, scallops, and small razor clams. The large beds of the common black mussel which are so numerous and so extensive in the tidal passages of our bays and harbors or on outlying shoals are their favorite feeding grounds. Large flocks, often immense rafts, of scoters spend the winter within easy reach of such beds, which they visit daily at certain stages of the tide; although they can dive to considerable depths to obtain food if necessary, they evidently prefer to feed at moderate or shallow depths and choose the most favorable times to visit the beds which can be most easily reached. Their crops are crammed full of the small shellfish, which are gradually ground up with the help of small stones in their powerful stomachs and the soft parts are digested. A small amount of vegetable matter, such as eelgrass and algae, is often taken in with the other food, perhaps only incidentally. Dr. F. Henry Yorke (1899) says that, on the lakes of the interior, "it feeds on shellfish, especially mussels, crayfish, and fish spawn; besides a few bulbs of aquatic plants."

Behavior.—The flight of the surf scoter is not quite so heavy as that of the white-winged scoter; it is a smaller, lighter, and livelier bird on the wing, but it so closely resembles the American scoter in flight that the two can not be distinguished at any great distance. It rises heavily from the surface of the water and experiences considerable difficulty in doing so unless there is some wind, which it must face in order to rise. This necessity of rising against the wind is well understood by gunners, who take advantage of it to approach a flock of bedded birds from the windward, forcing the birds to rise toward the boat and thus come a little nearer. When once under way the flight is strong, swift, and well sustained. In calm weather or in light winds migrating birds fly high, but in windy or stormy weather they plod along close to the waves. They often fly in large flocks or irregular bunches without any attempt at regular formation, following the coast line, as a rule, but sometimes passing over capes or points to make short cuts. Mr. Mackay (1891) writes:

I have noticed during the spring migration northward in April that frequently the larger flocks of the surf scoter are led by an old drake. That the selection of such a leader is a wise precaution has frequently been brought to my notice, for on first perceiving such a flock coming toward me in the distance they would be flying close to the water; as they neared the line of boats, although still a considerable distance away, the old drake would become suspicious and commence to rise higher and higher, the flock following him, until the line of boats is passed, when the flock again descends to the water. When over the boats shots are frequently fired up at them, but so well has the distance been calculated that it is seldom a bird is shot from the flock.

As a diver the surf scoter is fully equal to the other sea ducks, depending on its diving powers in its daily pursuit of food and to escape from its enemies in emergencies. It dives with an awkward splash, but very quickly and effectively, opening its wings as it goes under, and using them in its subaqueous flight. It can remain under for a long time and swim for a long distance without coming up; it is useless to attempt to chase a slightly wounded bird. Mrs. Florence Merriam Bailey (1916) has graphically described the ability of this species to dive through the breaking surf, as follows:

It was a pretty sight when, under a gray sky, the beautiful long green rolls of surf rose and combed over and the surf scoters came in from the green swells behind to feed in front of the surf and do skillful diving stunts to escape being pounded by the white waterfalls. As the green wall ridged up over their heads they would sit unmoved, but just as the white line of foam began to appear along the crest they would dive, staying under till the surf had broken and the water was level again. When diving through the green rollers near the shore the black bodies of the scoters, paddling feet and all, showed as plainly as beetles in yellow amber.

I have never heard the surf scoter utter a sound; and Mr. Mackay (1891) says: "My experiences show that all the scoters are unusually silent and seem to depend entirely on their sight in discovering their companions. I have rarely heard the surf scoter make any sound, and then only a low, guttural croak, like the clucking of a hen; they are said to utter a low whistle." Doctor Nelson (1887) says: "In the mating season they have a low, clear whistle for a call note, and may be readily decoyed within gunshot by imitating it from a blind."

Fall.—Referring to the fall migration, Dr. Charles W. Townsend (1905) writes:

Although scoters fly most in stormy weather and are often found quietly feeding on calm days, still they sometimes go south in great numbers even in pleasant weather. This flight is greatest in the early morning, but may be continued all day. At times flock succeeds flock as far as the eye can see off the beach at Ipswich. Occasionally four or five exclusive ones go along together, but usually the flocks are much larger, up to five or six hundred. These sweep along at times in one long line close to the water. Anon they press together in a compact and solid square. Again they spread out into a long line abreast or form a V, and at all times they rush along with irresistible energy. On reaching the angle at Annisquam where Cape Ann juts out boldly, the birds are often at a loss what to do. Sometimes they fly first one way and then another, rising higher and higher all the time, and then strike out toward the end of the cape, over which they resume their southerly course at a considerable height. Another flock will turn at the angle without pausing and skirt the shore around the cape. Again, a flock will pause and fly high at the angle, and then along the coast, soon to descend to the original height above the water and round the end of the cape. All these are methods commonly adopted. Occasionally a flock will get discouraged on reaching the solid barrier of the cape, will turn back and drop into the water to talk it over. All this shows the dislike of the scoter to fly over the land.

As a result of many years of observation, Mr. Mackay (1891) says:

The old birds of the surf scoter appear about the middle of September, with a very large movement about the 20th, according to the weather, the young birds making their appearance the last of September or first of October. I have known a considerable flight to occur on the last day of September, the wind all day being very fresh from the southwest, which deflected them toward the land; such an early movement is, however, unusual. An easterly storm about the middle of August is likely to bring them along, the wind from this direction being particularly favorable for migration; if, on the other hand, the weather is mild and warm, it is not usual to see them so early.

From this time on they continue to pass along the coast until near the end of December, the main flight coming between the 8th and 20th of October, depending upon the weather, when the migration appears to be at an end. During such migration they are estimated to fly at a rate of about 100 miles an hour, but this rate is also governed by the weather. The greater part of these scoters pass around Cape Cod, as I have never heard of, nor seen, any of the immense bodies of "bedded" fowl north or east of it as occur south and west of the cape; probably because they are unable to find either the security or profusion of food north of it that they can obtain in the waters to the south. They therefore congregate here in large numbers.

Winter.—In the waters lying south of Cape Cod, Massachusetts, in the vicinity of Nantucket, Muskeget Island, and Marthas Vineyard, vast numbers of scoters spend the winter. Mr. Mackay (1891) writes:

Most of these places being inaccessible to ordinary sportsmen, the birds can live undisturbed during the late autumn, winter, and spring months, undoubtedly returning year after year to these same waters, which appear to have become their winter home.

Where there are large ponds adjacent to the coast, separated from the ocean by a strip of beach, all three of the scoters will at times frequent them to feed, and will collect in considerable numbers if the supply of food is abundant; in which case they are very unwilling to leave such ponds, and, although much harassed by being shot at and driven out, continue to return until many are killed. An instance of this kind occurred the 1st of November, 1890, when some 400 scoters collected in the Hummuck Pond on Nantucket Island; they were composed entirely of the young of the surf and white-winged scoters, only one American (a female) being obtained out of about 50 birds shot in one day (November 3) by a friend and myself. On March 18, 1875, I saw on a return shooting trip from the island of Muskeget to Nantucket a body of scoters, comprising the three varieties, which my three companions and myself estimated to contain 25,000 birds.

DISTRIBUTION

Breeding range.—Northern North America. East to the Atlantic coast of Labrador and probably Newfoundland. South nearly or quite to the Gulf of St. Lawrence, to James Bay (both sides), northern Manitoba (Churchill), northern Saskatchewan and Alberta (Athabasca Lake), perhaps northern British Columbia, and to southern Alaska (Sitka). West to the Bering Sea coast of Alaska (Yukon delta). North to northern Alaska (Kotzebue Sound), the barren grounds of Canada and northern Labrador. Said to have bred in northeastern Siberia (Tschuktschen Peninsula) and in Greenland (Disco Island).

Winter range.—Mainly on the sea coasts. On the Atlantic coast from the Bay of Fundy southward to Florida (St. Lucie, Jupiter, etc.), most abundantly from Massachusetts to New Jersey. On the Pacific coast from the Aleutian Islands southward to Lower California (San Quintin Bay). It winters commonly on the Great Lakes and more sparingly westward to southern British Columbia (Okanogan Lake) and southward rarely to Louisiana (New Orleans).

Spring migration.—Early dates of arrival: New Brunswick, April 10; Central Alberta, McMurray, May 14; Alaska, Kowak River, May 22. Late dates of departure: Louisiana, New Orleans, March 20; Georgia, Cumberland Island, May 6; North Carolina, Pea Island, May 15; Rhode Island, May 21; Massachusetts, May 9.

Fall migration.—Early dates of arrival: Massachusetts, September 4; Rhode Island, September 1; South Carolina, Mount Pleasant, October 24; Minnesota, Jackson County, October 1; Idaho, Fernan Lake, October 9; Colorado, Barr Lake, October 22; Utah, Bear River, October 24.

Casual records.—Three records for Bermuda (January 8, 1849, October 7, 1854 and November 17, 1874). Said to have occurred in Jamaica. There are numerous records for Great Britain and France, three for Finland and several others for western Europe; these may come from a Siberian breeding range.

Egg dates.—Arctic Canada: Twelve records, June 19 to July 8; six records, June 25 to July 1.

ERISMATURA JAMAICENSIS (Gmelin)

RUDDY DUCK

HABITS

This curious little duck is in a class by itself, differing in several peculiarities from any other North American duck. It is widely scattered over the most extensive breeding range of any of our ducks, from far north to far south and from our eastern to our western coasts. Its molts and plumages are unique, involving a complete seasonal change from the gaudy nuptial to the dull and somber autumn dress; even the seasonal changes in the oldsquaw are less striking. But its eggs furnish the greatest surprise of all; for, although this is one of our smallest ducks, it lays eggs which are about as large as those of the great blue heron or the wild turkey. In its appearance and behavior it is also unique and exceedingly interesting. One must see it on its breeding grounds, in all its glory, to appreciate what a striking picture is the male ruddy duck. In the midst of a sea of tall, waving flags a quiet, sheltered pool reflects on its glassy surface the dark green of its surroundings, an appropriate setting for the little gem of bird life that floats gently on its surface, his back glowing with the rich, red brown of his nuptial attire, offset by the pure white of his cheeks, his black crown, and above all his wonderful bill of the brightest, living, glowing sky blue. He knows he is handsome as he glides smoothly along, without a ripple, his saucy sprigtail held erect or even pointed forward till it nearly meets his upturned head; he seems to strut like a miniature turkey gobbler.

Courtship.—His mate knows that he is handsome, too, as she shyly watches him from her retreat among the flags, where perhaps she is already building her basketlike nest. As she swims out to meet him his courtship display becomes more ardent; he approaches her with his head stretched up to the full extent of his short neck and his eyes gleaming under two swollen protuberances above them like the eyes of a frog; with his chest puffed out like a pouter pigeon, he bows and nods, slapping his broad, blue bill against his ruddy breast; its tip striking the water and making a soft, clucking sound. Should a rival male appear upon the scene, he rushes toward him, they clash in an angry struggle, and disappear beneath the surface in desperate combat, until the vanquished one skulks away and leaves the victor to strut and display his charms with more pride than ever. Since the above was written, Dr. Alexander Wetmore (1920) has published an accurate description of what is apparently the same performance, but rather than repeat it here, I would refer the reader to it.

Mrs. Florence Merriam Bailey (1919) describes it more briefly, thus:

> When I arrived only two pairs were in evidence, the puffy little drakes looking very cocky and belligerent, suggesting pouter doves with their air of importance and the curious muscular efforts by which they produced their strange notes. When I first saw one perform, not knowing about his tracheal air sac, I thought he might be picking at his breast or have something stuck in his throat and be choking. With quick nods of the head that jerked the chin in, he pumped up and down, till finally a harsh guttural cluck was emitted from his smooth, blue bill. Often in doing chin exercises the little drakes pumped up a labored *ip-ip-ip-ip-u-cluck; cluck*, producing it with such effort that the vertical tail pressed forward over the back, as if to help in the expulsion, afterwards springing erect again.

Nesting.—In the deep-water sloughs of North Dakota we found the ruddy ducks nesting in abundance; the ideal conditions found here are to be found in many places throughout the west, where the nesting habits of the species are probably similar. In these large sloughs there are extensive tracts of tall reeds, bullrushes, or flags, often higher than a man's head and growing so thickly that nothing can be seen through them at a little distance. In these excellent hiding places the ruddy duck conceals its nest, and so well is this done that even after the nest has once been located it is extremely difficult to find it again. The nests are basketlike structures, well made of the reeds, bullrushes, or flags, closely interwoven; the material always matches the surroundings of the nest, so sometimes the nest is made of the dry stalks only and sometimes partially or wholly of the green material, producing a very pretty effect. The nest is built up some 7 or 8 inches above the level of the water, which is often more than knee deep, and attached firmly to the growing reeds; a sloping pile of reeds is usually added as a stairway leading to the nest, down which the duck can quickly slide into the water on the approach of danger; and the growing reeds above are often arched over the nest in such a way as nearly to conceal it. There is no lining in the nest except a few finer bits of reeds and flags; and what little down is found there may be more accidental than an intentional lining. From such a well-concealed nest the departure of the duck could never be seen; she simply slides into the water and slinks away like a grebe. The female is particularly shy during the breeding season and seldom shows herself near the nest.

The man who found the first ruddy duck's nest must have been surprised and puzzled, for he would never suppose that such large eggs could belong to such a small duck. W. H. Collins (1881) mistook the first eggs of this species that he found at St. Clair Flats for brant's eggs, because the ruddy ducks kept out of sight and some brant happened to be flying about the marsh. But the next season, when no brant were to be seen, he succeeded in identifying the eggs by a careful study of the feathers in the nest, the parents keeping out of sight, as usual. He did finally succeed in seeing a female ruddy leave her nest and swim away under water to the nearest clump of rushes. According to Rev. J. H. Langille (1884), the nesting habits of this duck are somewhat different in this vicinity from what they are in North Dakota. He says:

The nest, built some time in June, is placed in the sedges or marsh grass over the water, and may contain as many as 10 eggs, remarkably large for the size of the bird, oval or slightly ovate, the finely granulated shell being almost pure white, tinged with the slightest shade of grayish blue. The nest may be quite well built of fine colored grasses, circularly laid, or simply a mere matting together of the tops of the green marsh grass, with a slight addition of some dry, flexible material. I found one nest on a hollow side of a floating log. It consisted of a few dried grasses and rushes laid in a loose circle. Indeed, the bird inclines to build a very slight nest.

Robert B. Rockwell (1911) has found the ruddy duck nesting in still more open situations in Colorado. On May 31, 1907, he found a fine set of 10 eggs in an excavation in the side of a large muskrat house, without any downy lining whatever, and only a few inches above the water level. On June 8 this nest contained 11 eggs, 2 of which were canvasback's or redhead's; there was also a new nest of the canvasback, containing 8 fresh eggs, on the other side of the same muskrat house and only 4 feet away; and, moreover, a new ruddy duck's nest, containing 3 fresh eggs, was found on top of the house and about midway between the two nests. "This was a mere unlined depression in the litter composing the house, entirely without concealment of any kind, and the great snowy white eggs could be seen from a distance of many yards." Three ducks' nests on one muskrat house is certainly a remarkable record.

The ruddy duck has been known to use an abandoned nest of the American coot, which sometimes is not much unlike its own. Doctor Wetmore tells me that in the Bear River marshes in Utah the old nests of the redheads are commonly appropriated by the ruddy ducks. It also lays its eggs in other duck's nests and even in grebe's nests. At Crane Lake, Saskatchewan, I flushed a female ruddy duck from a clump of bulrushes in the midst of a large colony of western grebes; a careful search through the clump revealed only grebes' nests, but one of the nests held 2 eggs of the western grebe and 1 egg of the ruddy duck. I have found ruddy ducks' eggs in the nests of the redhead and the canvasback, and others have mentioned the same thing; the other two species often lay in the ruddy ducks' nests also, so that it is sometimes difficult to decide which was the original owner of the nest.

Eggs.—The ruddy duck is said to lay as many as 19 or 20 eggs, but such large sets are not common; the numbers usually run from 6 to 9 or 10. The eggs are often deposited in two layers and with the largest numbers in three layers; it is obviously impossible for so small a duck to cover any large number of such large eggs. The indications are that in the more southern portions of its range two broods may be raised in a season, which seems to be very much prolonged. William G. Smith says in his notes that he has taken young birds in the down as late as October 16 in Colorado. The eggs are distinctive and could hardly be mistaken for anything else. They vary somewhat in shape from short ovate to elongate ovate, or from oval to elliptical oval. The shell is thick and decidedly rough and granular, much more so than any other duck's egg. When first laid the eggs are pure, dull white or creamy white, but they become more or less stained during incubation. The measurements of 80 eggs, in the United States National Museum and the writer's collections, average 62.3 by 45.6 millimeters; the eggs showing the four extremes measure **67.6** by 44.5, 66.5 by **48**, **59.4** by 45.4, and 61.3 by **42.6** millimeters.

Young.—The period of incubation seems to be unknown, but it is probably not far from 30 days. It is apparently performed by the female alone, although the male does not desert the female at this season, and, contrary to the rule among ducks, he remains with the young family and helps care for them until they are fully grown. Dr. Alexander Wetmore writes to me as follows:

> The male ruddy ducks in most instances remain with the females after the young hatch, and it is a common sight to see a male, with tail erect and breast and throat puffed out, swimming at the head of a brood of newly hatched young in a compact flock, while the female follows behind. When such families are approached the adults submerge quietly and disappear with no demonstration whatever, while the young, left to their own devices, make off as rapidly as possible, still maintaining their close formation. Only when seriously threatened do they dive and then scatter. As they grow older the young birds become more independent, and usually when half grown are found separated from their parents. Occasionally, however, well-grown young are found with the female. Young as well as adults are more or less helpless on land, resembling grebes in this respect. Young birds half grown were able to waddle a few steps, but fell on the breast almost at once and then usually progressed by shoving along in a prostrate position with both feet stroking together. These half-grown birds were sullen and ferocious, and none that I had became tame at all. They invariably snapped and bit at my fingers when handled, and with open mouths resented every approach. When first hatched, the feet of these birds are truly enormous in proportion to the size of the body and form a certain index to the future activities of the ducklings.

According to Maj. Allan Brooks (1903), the "young when first hatched are, as might be expected, very large, and dive for their food, unlike all other young ducks, which take their food from the surface for several weeks."

G. S. Miller, jr. (1891), published the following observations on the behavior of young ruddy ducks on Cape Cod, Massachusetts:

On August 11 I found four young, accompanied by the female parent, on a large shallow pond which lies between the towns of Truro and Provincetown. At the approach of my boat the old bird left her young and joined five other adults which were resting upon the water half a mile away; the young ones, however, were too young to fly, and so attempted to escape by swimming and diving to the shelter of a cat-tail island near which they happened to be when surprised. Two of them reached this place of safety, but the others were secured after a troublesome chase. They were very expert divers, remaining beneath the surface for a considerable length of time, and on appearing again exposing the upper part of the head only, and that for but a few seconds. As the water just here happened to be filled with pond weed (*Potamogeton pectinatus* and *P. perfoliatus*), it was not difficult to trace the motions of the birds when beneath the surface by the commotion which they made in passing through the thick masses of vegetation. The flock of old birds contained at least two adult males, which were very conspicuous among their dull-colored companions. They were all very shy, so that it was impossible to approach to within less than 100 yards of them. The adults, as well as the two remaining young, were seen afterwards on several visits to the pond.

Plumages.—The downy young, when first hatched is a large, fat, awkward, and helpless looking creature, covered with long coarse down, which on the upper parts is mixed with long hair-like filaments, longest and coarsest on the rump and thighs. The upper parts are "drab" or "hair brown," deepening to "Prout's brown" or "mummy brown" on the crown and rump, with two whitish rump patches, one above each thigh; the brown of the head extends below the eyes to the lores and auriculars, a broad band of grayish white separating this from a poorly defined malar stripe of "drab"; the under parts are mostly grayish white, shading into the darker colors on the sides and into an indistinct collar of "drab" on the lower neck. The colors fade out paler with increasing age. The young bird is almost fully grown before the juvenal plumage is complete; it comes in first on the flanks, scapulars and head; the down is replaced last on the center of the belly, back, and rump. In this plumage the upper parts are dark brown, "clove brown" or "bone brown" on the back and "blackish brown" on the crown; the feathers of the mantle are indistinctly barred, tipped, or sprinkled with fine dots of pale buffy shades; and the crown feathers are tipped with brownish buff. The flank feathers are more distinctly barred with dusky and grayish buff; the breast feathers are dusky, broadly tipped with buff and the rest of the plumage is more or less mottled with dusky, grayish, and buffy tints. There is no clear white on the side of the head which is mottled with dusky, the mottling forming a more or less distinct malar strip. The sexes are alike in this plumage, except that the female is decidedly smaller. This plumage is worn without much change until the spring molt begins. This molt is nearly complete, involving everything but the wings, and produces the decided seasonal change peculiar to this species.

Mr. A. J. van Rossem has sent me some notes on the molts and plumages of this unique duck, based on extensive studies, from which I quote as follows:

> The juvenal plumage is retained until about January or February, when it is replaced (including the tail) by a plumage closely resembling that of the winter adults. The male, at least, about the middle or end of May then assumes a red plumage in general resembling the midsummer adult, except that the reds are darker and apt to be obscured by an admixture of darker (similar to the winter) feathers. With the taking on of this first red plumage, the tail is again molted. It is molted again in the fall, at the time of the transition into winter plumage. Thus two years are required to attain the brilliant red plumage of the fully adult male.

The ruddy duck is one of very few species which have a strictly nuptial plumage and two extensive molts. The prenuptial molt in April and May produces the well-known nuptial plumage of the male, involving practically all of the contour plumage and the tail, and characterized by the brownish black crown, the white cheeks, the sky-blue bill, and the "chestnut" back. The nuptial plumage of the female is not so striking; it is much like that of the first winter, but the cheeks, chin, and throat become purer white.

There is no eclipse plumage. The summer molt, occurring from August to October, is complete, producing an adult winter plumage much like that of the first winter, except that the cheeks, chin, and throat are pure white, including the lores and nearly up to the eyes; the sexes are much alike in this plumage, but the male is decidedly larger, and many of the mottled feathers of the mantle and flanks are more or less washed with chestnut. Adults can always be distinguished from young birds by the white cheeks and throats.

Food.—Being decidedly a diving duck, the ruddy duck obtains most of its food on the bottom and subsists very largely on a vegetable diet, hence its flesh is usually well flavored. While living on the inland ponds, marshes, and streams, it feeds on the seeds, roots, and stems of grasses and the bulbs and leaves of aquatic plants, such as flags, teal moss, wild rice, pond lilies, duckweed, and wild rye. Dr. F. Henry Yorke (1899) says it also eats small fishes, slugs, snails, mussels, larvae, fish spawn, worms, and creeping insects. Prof. W. B. Barrows (1912) "once took from the crop and stomach of a single ruddy duck, at Middletown, Connecticut, 22,000 seeds of a species of pondweed (*Naias*), which at that time was growing in great abundance in the city reservoir, where the bird was shot." Dr. J. C. Phillips (1911) found in the stomachs of ruddy ducks, shot in Massachusetts, "seeds of bur reed, pondweed, bulrush, and *Naias*, and buds, etc., of wild celery," also "chironomid and hydrophilid larvae."

In the Currituck Sound region of North Carolina and Virginia I have found them feeding almost exclusively on the seeds of the fox-tail grass. Nuttall (1834) mentions "seeds and husks of the *Ruppia maritima*," which is apparently the same thing. Audubon (1840) says: "When on salt marshes they eat small univalve shells, fiddlers, and young crabs, and on the seacoast they devour fry of various sorts. Along with their food they swallow great quantities of sand or gravel."

Behavior.—In its flight, swimming, and diving habits the ruddy duck more closely resembles the grebes than does any other American duck. Its small, rounded wings are hardly sufficient to raise its chunky little body off the water, except with the aid of its large, powerful feet, pattering along the surface for several yards. But, when well under way, it makes good progress in flight, though it flies usually close to the water and seldom rises to any great height in the air, even when migrating. It has a peculiar, uneven, jerky gait in flight by which it can be easily recognized at a long distance, and it usually flies in good-sized or large flocks. Audubon (1840) says:

They alight on the water more heavily than most others that are not equally flattened and short in the body, but they move on that element with ease and grace, swimming deeply immersed, and procuring their food altogether by diving, at which they are extremely expert. They are generally disposed to keep under the lee of shores on all occasions. When swimming without suspicion of danger they carry the tail elevated almost perpendicularly and float lightly on the water; but as soon as they are alarmed, they immediately sink deeper, in the manner of the anhinga, grebes, and cormorants, sometimes going out of sight without leaving a ripple on the water. On small ponds they often dive and conceal themselves among the grass along the shore, rather than attempt to escape by flying, to accomplish which with certainty they would require a large open space. I saw this very often when on the plantation of General Hernandez in east Florida. If wounded, they dived and hid in the grass, but, as the ponds there were shallow, and had the bottom rather firm, I often waded out and pursued them. Then it was that I saw the curious manner in which they used their tail when swimming, employing it now as a rudder, and again with a vertical motion; the wings being also slightly opened, and brought into action as well as the feet.

Walter H. Rich (1907) writes:

The wings are small in proportion to their chunky little bodies, and their flight at the outset is heavy and labored, but once fairly going they fly fast, their wings making considerable noise from their rapid motion. With all these drawbacks the ruddy is wonderfully quick, either in the air or on the water. He is quite capable of taking care of himself once he gets it into his head that harm is intended. He can get under water with a celerity that falls little short of the marvelous. One of his tricks has always been a mystery to me: He will sink himself completely beneath the surface without diving—simply settles down like a sinking craft and beats a retreat under water, where he is as much at home as any duck of them all. I have seen black ducks, when they thought themselves undiscovered and their wit said it was dangerous to fly, sink themselves so that only the head showed above water, and have seen shell drakes settle down in the same style until only their heads were visible and so go darting and zigzagging away when they had flown in and settled among a bunch of decoys before discovering the cheat, but I have never seen any of these go completely below the surface without an attempt at diving as does the ruddy.

Audubon (1840) says: "Their notes are uttered in a rather low tone and very closely resemble those of the female mallard." Rev. J. H. Langille (1884) observes: "The ruddy duck is nearly noiseless, occasionally uttering a weak squeak." Doctor Wetmore tells me that the female is entirely silent and that the only note heard from the male is the courtship call, *tick tick tickity quo-ack.*

Fall.—On its migrations the ruddy duck follows the courses of the streams and the lakes, flying low and in large flocks, often close to water and below the level of the banks of the streams. The flights are made mainly early in the morning or during the dusk of evening, perhaps even during the night; they seem to appear suddenly in the ponds and small lakes and disappear as mysteriously; they are seldom seen coming or going. The flocks are made up largely of the dull-colored young birds, and even the old males have acquired their somber autumn dress. They are said to be unsuspicious and easily approached by gunners, but my experience has shown that they are well able to take care of themselves. When in a large flock on an open lake they are particularly difficult to approach, for they will fly long before the gunner can come within gunshot; I have chased them for hours in this way and seen them go spattering off close to the surface with a great whirring of little wings, only to drop into the water again at no great distance, without checking their speed, sliding along the surface and making the spray fly; only when cornered in some narrow bay and forced to fly past the boat do they give the gunner a chance for a shot. Even when suddenly surprised they can escape by diving in remarkably quick time and, swimming under the water for a long distance, come up at some unexpected place; often they seem to have vanished entirely until a careful search reveals one crawling out on a grassy bank to hide or skulking somewhere in the reeds. To chase a wounded bird is almost hopeless. When swimming under water the wings are closed and both feet work simultaneously. William G. Smith states in his notes that while hunting in a boat, where the water was clear, he has "often observed" a wounded ruddy duck "dive down, grasp a weed," and "remain in this position for 20 minutes"; but he does not say whether the duck was alive or not at the end of this remarkable performance.

Game.—Ruddy ducks resort in large numbers, late in the fall, to Back Bay, Virginia, where they are known as "boobies," and furnish good sport for the numerous duck clubs located in that famous resort for sportsmen. Here they spend the winter in the broad expanse of shallow fresh and brackish water bays and estuaries, with the hosts of other wild fowl that frequent that favored region, growing fat and tender on the abundance of foxtail grass, wild celery, and other duck foods. Their feeding grounds are mainly in the shallower, more protected parts of the bays and near the shores, where they are most intimately associated with the American coots which gather there in immense rafts. Large flocks of these sprightly little ducks are frequently seen flying back and forth and they are popular with the sportsman, as they are lively on the wing, decoy readily under proper conditions, and are excellent table birds when fattened on clean vegetable food. They are usually shot from the batteries, such as are used for canvasbacks, but, as they are a little shy about coming to the large rafts of canvasback decoys that are used for the larger ducks, better results are obtained by "tying out" the battery with a smaller number of "booby" decoys. Under favorable circumstances it does not take long for a gunner to secure his legal limit of 35 ducks a day. Another method of shooting them, which is often very succesful, is for a number of boats to surround a flock of birds or drive them into some small bay, where they are eventually forced to fly out past the boats, as they do not like to fly over the land.

DISTRIBUTION

Breeding range.—Mainly in the sloughs and marshes of central and western North America. East to southern Manitoba (Shoal Lake), west central Minnesota (Becker County), southeastern Wisconsin (Lakes Koshkonong and Pewaukee), and southeastern Michigan (St. Clair Flats). South to northern Illinois (Lake and Putnam Counties), northern Iowa (Hancock County), south central Texas (Bexar County), northern New Mexico (Lake Burford), central Arizona (Mogollon Mountains), and northern Lower California (latitude 31° N.). West to southern and central California (San Diego, Los Angeles, Monterey, and Siskiyou Counties), central Oregon (Klamath and Malheur Lakes), northwestern Washington (Seattle and Tacoma), and central British Columbia (Cariboo District). North throughout much of Alberta (Buffalo Lake and Belvedere), probably to Great Slave Lake (Fort Resolution) and to northern Manitoba (York factory).

Outlying, and probably casual, breeding stations have been recorded as far east as Ungava (Richmond Gulf), southeastern Maine

(Washington County), eastern Massachusetts (Cape Cod), southern Rhode Island (Seaconnet Point), and central New York (Seneca River). Extreme southern breeding colonies have been found in southern Lower California (Santiago), the Valley of Mexico, at the Lake of Duenas, Guatemala, and in the West Indies (Cuba, Porto Rico, the Grenadines, Carriacou, etc.), many of which are probably permanent colonies.

Winter range.—The northern portions of the breeding range are vacated in winter. It winters abundantly on the Atlantic coast as far north as Chesapeake Bay and more rarely north to Long Island and Massachusetts; south to Florida, the Bahamas, and the West Indies (Cuba, Porto Rico, Jamaica, Martinique, Grenada, Barbados, etc.). On the Pacific coast, from southern British Columbia (Boundary Bay) southward to Lower California, Guatemala, and Costa Rica (Irazu). In the interior north to central Arizona (Pecks Lake), southern Illinois, and western Pennsylvania (Erie).

Spring migration.—Early dates of arrival: Massachusetts, March 20; Utah, Bear River, March 30; Minnesota, Heron Lake, April 3; Ohio, Oberlin, April 7; Manitoba, southern, April 26; Alberta, Edmonton, May 1. Average dates of arrival: Pennsylvania, Erie, April 16; Ohio, Oberlin, April 15; Nebraska, April 7; Minnesota, Heron Lake, April 10; Wyoming, Cheyenne, April 21; Manitoba, southern, May 5.

Late dates of departure: Kentucky, Bowling Green, April 18; Lower California, Colnet, April 8.

Fall migration.—Early dates of arrival: Virginia, Potomac River, August 20 (average September 30); Massachusetts, Pembroke, September 5; West Indies, Barbados, September 13.

Casual records.—Accidental in Bermuda (November 24, 1846) and Alaska (Kupreanof Island, August 15, 1916). Rare straggler in Nova Scotia and New Brunswick.

Egg dates.—California: Thirty records, April 26 to August 11; fifteen records, May 22 to June 10. North Dakota: Twelve records, June 8 to July 19; six records, June 13 to July 9. Colorado: Nine records, May 31 to August 6. Porto Rico: December to March.

NOMONYX DOMINICUS (Linnaeus)

MASKED DUCK

HABITS

This curious little duck resembles the ruddy duck in many ways and has been placed by some writers in the same genus with it. It is a tropical species, with its center of abundance somewhere in eastern South America, found frequently if not regularly in some of the West Indies, and with at least five well-established records in five widely scattered localities within the United States.

Nesting.—A. H. Holland (1892) found it breeding at Estancia Espartella, in Argentine Republic, where he speaks of it as—

Rare, living singly or in pairs in the small lagoons, either open or containing rushes. It is next to impossible to flush this peculiar duck, as it takes after the grebes and invariably dives when disturbed, so that I have never seen it on the wing. When swimming it holds its stiff tail spread out and erected, inclined somewhat toward its head, and as it swims very low in the water the duck is only visible by its head, tail, and the top of its back. It builds amongst the rushes early in November, making a nest of green rushes with scarcely any lining, being a very flat construction. The eggs are three in number and white in color, very rough and very round.

Eggs.—There is an egg of this species in the R. M. Barnes collection, in Lacon, Illinois, which I have examined. It was taken from the oviduct of a bird in Yucatan on November 17, 1904. It is ovate in shape and broadly rounded at both ends; the shell is rough and granulated; and the color is a dirty white. It looks very much like a ruddy duck's egg. It measures 63 by 45.8 millimeters.

Behavior.—According to Baird, Brewer, and Ridgway (1884):

Léotaud mentions this duck as being one of the birds of Trinidad, where it is by no means rare. While to a certain extent it seems to be migratory some are always present on that island. It is social in its habits and seems more disposed than any other duck to keep to the water. Its flight is rapid, but is not so well sustained as that of most of the other kinds. When it is on the land it keeps in an upright position, its tail resting on the ground. Its movements on dry land are embarrassed by its claws, which are placed so far back as to disturb its equilibrium. Its flesh is excellent, and is held in high esteem in that island.

In a paper on the birds of Jamaica, W. E. D. Scott (1891) says:

In the ponds about Priestmans River I met with this species on two occasions, and from native hunters learned that it was not at all uncommon, especially early in the fall. At Priestmans River, 9th February, 1891, I took an adult male, No. 11000, of *Nomonyx dominicus*. The bird was in a small and very shallow pond, and did not attempt to fly away upon being approached, but tried to hide in some thin grass growing where an old stump of a tree projected from the water, and remained so motionless as almost to escape notice, though not more than 20 feet away. It was killed with a light load of dust shot.

These little ducks do not seem at all rare on the island, and have much the habits of the grebes, frequenting small fresh-water ponds and depending rather on hiding in the grass or diving than on flight to escape pursuit. They are said by the native gunners to breed at various points on the island.

T. M. Savage English (1916) writes:

Nomonyx dominicus seems to be more or less abundant throughout the year, on the secluded ponds of salt water which are frequent among the tall black mangrove (*Avicennia*) woods in the north of Grand Cayman; it most probably breeds somewhere near them—very possibly among the dense thickets of red mangrove (*Rhizophora*), by which they are mostly surrounded. Anyone who has ever been among red mangroves will appreciate the difficulty of finding the nest of a diving bird among them—except by a fortunate chance, which never came to the writer.

Most of the resident birds of Grand Cayman are remarkably fearless of man, very much as robins are in Europe, but these ducks are more wary, and when their pond is approached generally make their first appearance in the middle of it, having dived at the sight or sound of the intruder and, if near the shore, found their way under water to what they think is a safer place. When at rest they float very much as most waterfowl do, the water line being in about its usual place, but when swimming they are almost always deeply submerged, and if approaching or receding from the observer, seem to have a relatively enormous "beam." Of course this effect may be only due to the very low elevation of the bird's back above the water. Their method of diving is interesting. It has the appearance of being done without the movement of a muscle, just as if the bird were a leaking vessel which was going down on an even keel. This downward progress is often interrupted, when just the head, the neck, and the upper part of the upstanding tail are showing above the surface, or a little later, when only the head and part of the neck, which is habitually kept stiffly upright (as is the tail), are visible. In either of these positions the bird seems able to rest as well as to swim at some speed.

Nomonyx dominicus has at least two calls, one of them very like the clucking of a hen to her chickens and the other more reminiscent of a short note from a motor horn.

H. B. Conover writes to me, of the habits of this species in Venezuela, as follows:

We first saw the masked duck in a small pond on the edge of the savannah about 60 miles south of Maracaibo. Here one day I ran into a pair of males in the hen plumage. They were sitting a short way offshore and allowed us to walk up within 15 or 20 yards. The native with me shot at them, killing one, and the other bird jumped. This pair were probably stragglers; I was around there for a week and never saw any more. At Lagunillas in May these birds were abundant in the same marsh as the tree ducks. They sat around in open patches of water among the aquatic plants in small lots of about 5 to 15. They seemed to stay almost entirely in these open patches of water and rarely, if ever, were seen to alight among the floating aquatic plants. They would start to flush at about 75 yards and would rarely let one get within 50 yards of them in a boat. As a general rule, when the flock was approached, they would go off one or two at a time, not rising in a body. They would again alight within 200 or 300 yards. They rose fairly easily from the water, which was a great surprise to me, as they got off the surface very quickly and were not anywhere nearly as clumsy as the ruddy duck. A few birds were seen at this time which showed the red plumage, but the greater majority were in the brown stage. I tried my best to get an adult male, but the birds were so wild I was unsuccessful.

DISTRIBUTION

Breeding range.—In the West Indies (Cuba, Jamaica, Haiti, Porto Rico, St. Croix, Barbados, and Trinidad). Mainly in eastern South America, in Argentina, eastern and central Brazil, Guiana, Venezuela, etc.

Has been taken in Chile (Concepcion, June and September), Peru (Eten, October 11, 1899), Bolivia (Tatarenda and Lake Titacaca), Ecuador (Sarayacu and Peripa), Panama, Guatemala, Mexico (Orizaba, Jalapa, Matamoras, and Esquinapa), and southern Texas (Brownsville, July 22, 1891), and it may breed at some or all of these places.

Winter range.—Includes the breeding range. The records for western South America and Central America may represent winter wanderings.

Casual records.—Has wandered widely in North America to Maryland (Elkton, September 8, 1905), Massachusetts (Malden, August 27, 1889), Vermont (Albury Springs, September 26, 1857), and Wisconsin (near Newville, November, 1870). Some of these may be escapes from captivity.

CHEN HYPERBOREA HYPERBOREA (Pallas)

SNOW GOOSE

HABITS

As fully explained under the next subspecies, a careful study of the available specimens of birds and eggs, from various portions of the breeding range and the winter range of this species, has demonstrated that, while the greater snow goose (*nivalis*) occupies a limited breeding range in northern Greenland and adjacent lands and a narrow winter range on the Atlantic coast, the lesser snow goose (*hyperborea*) is a much more abundant bird of much wider distribution. It breeds along the entire Arctic coast of this continent and on the islands north of it, from Alaska to Baffin Land. Its winter range extends from the Atlantic to the Pacific, but it is very rare east of the Mississippi Valley and much more abundant from there westward. It is especially abundant in winter in California, Texas, and Mexico.

Spring.—The breeding grounds of the snow goose are so far north that we know very little about them in their summer home. They are known to us mainly as winter residents or as migrants. The lesser snow goose seems to have two main lines of flight in the spring, one from the Gulf coast directly northward through the Mississippi Valley and the Athabaska-Mackenzie region, or Hudson Bay route, to the Arctic coast, and the other from California northward, by an overland route west of the mountains, to northern Alaska and then eastward to the mouth of the Mackenzie River or beyond it. The Alaska route is not well known, and it may be that many, perhaps a majority, of the birds pass northeastward across the mountains to the Mackenzie Valley long before they reach northern Alaska. Illustrating these two lines of flight we have the following statements by E. A. Preble and Dr. E. W. Nelson; Mr. Preble (1908) says:

The valleys of the Athabaska and the Mackenzie lie in the path of migration of great numbers of snow geese of both the eastern and western forms. The rivers themselves, however, are seldom followed by the birds, except for short distances, since their general courses trend somewhat toward the west, while the lines of flight of the geese are usually nearly due north and south. Flocks of snow geese, leaving in spring the marshes at the delta of the Peace and Athabaska, a favorite stopping place, strike nearly due northward over the rocky hills, probably not again alighting until several hundred miles nearer their breeding grounds. Thus they press onward, close on the heels of retreating winter, feeding, when suitable open water is denied them, on the various berries which have remained on the stems through the winter.

Pursuing the course of the river northward, the next favorite goose ground is the delta of the Slave, where great numbers stop both spring and fall for rest and food. The low country about the outlet of Great Slave Lake is also a favorite resort. Leaving this point the geese in spring take a general northerly course, which suggests that their breeding grounds are north of the east end of Great Bear Lake.

Doctor Nelson's (1887) remarks would seem to indicate that only a small portion of the birds come as far north as St. Michael and Point Barrow before they turn eastward. He writes:

The handsome lesser snow goose is uncommon on the coast of Norton Sound and about the Yukon mouth. It arrives in spring from the 5th to the 15th of May, according to the season, and after remaining a very short time passes on to its more northern summer haunts. In the vicinity of Nulato, on the Yukon, Dall found them arriving about May 9, on their way up the Yukon; "they only stop to feed and rest on the marshes during the dusky twilight of the night, and are off with the early light of an Arctic spring."

According to Murdoch they are occasionally seen at Point Barrow in spring. This is all seen of these geese in spring throughout Alaska, except perhaps on the extreme northern border, for south of this none breed, and none are found after about May 25. They are far less numerous in spring than in fall along the coast of Bering Sea, and their spring migration is over so quickly that they are rarely killed at that season. Doctor Adams, while at St. Michael in 1851, noted the arrival of these birds from the south in spring and their departure to the north in fall, agreeing with my own observations, as noted elsewhere.

Nesting.—Although there are quite a number of sets of eggs of the snow goose in collections, the information we have regarding its nesting habits is scanty enough. MacFarlane (1891) apparently never found the nest of this bird himself, for he says:

The Esquimaux assured us that large numbers of "white waves" annually breed on the shores and islands of Esquimaux Lake and Liverpool Bay, but strange to say, we never observed any in the Barren Grounds proper or on the shores of Franklin Bay. The Esquimaux brought in to Fort Anderson about 100 eggs, which they claimed to have discovered among the marshy flats and sandy islets on the coast of the former, as well as from similar localities on and in the vicinity of the lake of that (Esquimaux) name.

There is a set of 7 eggs in the collection of Herbert Massey, Esq., taken for Bishop J. O. Stringer, on an island in the center of the mouth of the Mackenzie River on June 20, 1896; the nest is described as a depression in the ground, lined with a beautiful lot of gray down; it was collected by an Eskimo, but the bird was shot and the head, wings, and feet were sent with the eggs. I have a set of 5 eggs in my collection from the same missionary, taken on Garry Island in the mouth of the Mackenzie River on June 10, 1912.

Eggs.—The eggs of the snow goose vary in shape from ovate to elliptical ovate. The shell is thick and smoothly granulated, with a slight gloss on incubated specimens. The color is dull white or creamy white. They are usually much nest stained. The measurements of 103 eggs in various collections average 78.6 by 52.3 millimeters; the eggs showing the four extremes measure **88** by 55.5, 79.6 by **57.2**, **63.2** by 42.4, and 67.8 by **41.8** millimeters.

Plumages.—In the small downy young snow goose, recently hatched, the color of the head shades from "olive buff" above to "pale olive buff" below, suffused with "colonial buff" or pale yellow on the throat, forehead, and cheeks; the down on the back is quite glossy and appears "hair brown," "light drab," or "light grayish olive" in different lights; the under parts are "pale olive buff," suffused on the breast and sides with pale yellow shades.

I have seen no specimens showing the change into the juvenile plumage. In this plumage in the fall the head and neck is mottled with brownish gray or dusky, faintly below, more heavily and thickly above; the mantle and wing coverts, and early in the season the breast are washed or finely sprinkled with grayish; the scapulars, tertials, and secondaries are heavily sprinkled and clouded with grayish; the primaries are more grayish black and not so extensively black tipped as in the adult.

During the first winter and spring much progress is made toward maturity by wear and molt. The dusky markings gradually disappear, much of the contour plumage is molted, as well as the wing coverts and tail, until by summer there is little left of the immature plumage, except a small amount of grayish mottling on the head and the juvenal wings. At the complete molt that summer young birds become practically indistinguishable from adults, when 14 or 15 months old.

Food.—The food of the snow goose is largely vegetable, in fact almost wholly so, during the greater part of its sojourn in its winter home. In the spring this consists largely of winter wheat and other sprouting grains and grasses; and in the fall the stubble fields are favorite feeding grounds, where large flocks are known to congregate regularly. According to Swainson and Richardson (1831) it "feeds on rushes, insects, and in autumn on berries, particularly those of the *empetrum nigrum*." Doctor Coues (1874) gives the best account of its feeding habits, as follows:

Various kinds of ordinary grass form a large part of this bird's food, at least during their winter residence in the United States. They gather it precisely as tame geese are wont to do. Flocks alight upon a meadow or plain, and pass over the ground in broken array, cropping to either side as they go, with the peculiar tweak of the bill and quick jerk of the neck familiar to all who have watched the barnyard birds when similarly engaged. The short, turfy grasses appear to be highly relished; and this explains the frequent presence of the birds in fields at a distance from water. They also eat the bulbous roots and soft succulent culms of aquatic plants, and in securing these the tooth-like processes of the bill are brought into special service. Wilson again says that, when thus feeding upon reeds, "they tear them up like hogs;" a questionable comparison, however, for the birds *pull* up the plants instead of *pushing* or "rooting" them up. The geese, I think, also feed largely upon aquatic insects, small mollusks, and marine invertebrates of various kinds; for they are often observed on mud flats and rocky places by the seaside, where there is no vegetation whatever; and it is probable that when they pass over meadows they do not spare the grasshoppers. Audubon relates that in Louisiana he has often seen the geese feeding in wheat fields, where they plucked up the young plants entire.

Behavior.—Dr. D. G. Elliot (1898) says that the snow geese fly—

very high in a long, extended curved line, not nearly so angular as the V-shaped ranks of the Canada and other geese. With their snowy forms moving steadily along in the calm air, the outstretched wings tipped with black, glowing in the sun's rays with the faint blush of the rose, they present a most beautiful sight. Usually they fly silently with hardly a perceptible movement of the pinions, high above

"* * * the landscape lying so far below
With its towns and rivers and desert places,
And the splendor of light above, and the glow
Of the limitless blue ethereal spaces."

Occasionally, however, a solitary note like a softened "*honk*" is borne from out the sky to the ear of the watcher beneath. Should they perceive a place that attracts them, they begin to lower, at first gradually, sailing along on motionless wings until near the desired spot, and then descend rapidly in zigzag lines until the ground or water is almost reached, when with a few quick flaps they gently alight.

Vernon Bailey (1902) writes:

They are oftenest seen on the wing high overhead in long diagonal lines or V-shaped flocks, flying rapidly and uttering a chorus of shrill falsetto cries.

Illustrating the sociability of the snow goose, in its relation to other species, W. Leon Dawson (1909) says:

Snow geese dispense shrill falsetto cries as they fly about in companies of their own kind, or else mingle sociably with other species. Doctor Newberry says he has often seen a triangle of geese flying steadily, high overhead, "composed of individuals of three species (*Chen hyperborea*, *Branta canadensis hutchinsii*, and *Anser albifrons gambelli*), each plainly distinguishable by its plumage, but each holding its place in the geometrical figure as though it was composed of entirely homogeneous material, perhaps an equal number of the darker speecies, with three, four or more snow-white geese flying together somewhere in the converging lines."

At Moses Lake and again on the Columbia River I have seen a single snow goose attach itself to a company of resident Canadas—in each case through several days' observation—appearing now alone and now in company with the larger birds. A specimen taken May 9, 1907, at Wallula was with three Canada geese (one pair and a presumed "auntie"), and these were very reluctant to leave their fallen companion.

Fall.—On the fall migration, when the vast hordes of snow geese begin to wing their way southward from their Arctic summer homes, we begin to realize the astounding abundance of the species. George Barnston (1862), of the Hudson's Bay Co., writes:

The snow goose, although it plays a less conspicuous part in the interior of the country, where it seldom alights, except along the margin of the larger lakes and streams, becomes, from its consolidated numbers, the first and greatest object of sport after the flocks alight in James Bay. The havoc spread throughout their ranks increases as the season advances and their crowds thicken, and even the Indian becomes fatigued with the trade of killing. In the fall of the year, when the flocks of young "wewais," or "wavies," as they are called, are numerous and on the wing between the low-tide mark and the marshes or are following the line of coast southerly, it is no uncommon occurrence for a good shot, between sunrise and sunset, to send to his lodge about 100 head of game.

These "wavies," or white geese, form the staple article of food as rations to the men in James Bay and are the latest in leaving the coast for southern climes, an event which takes place toward the end of the month of September, although some weak broods and wounded birds linger behind until the first or second week in October. They are deliberate and judicious in their preparation for their great flight southward and make their arrangements in a very businesslike manner. Leaving off feeding in the swamps for a day or more, they keep out with the retreating ebb tide, retiring, unwillingly as it were, by steps at its flow, continually occupied in adjusting their feathers, smoothing and dressing them with their fatty oil, as athletes might for the ring or race. After this necessary preparation the flocks are ready to take advantage of the first north or northwest wind that blows, and when that sets in in less than 24 hours the coast that has been covered patchlike by their whitened squadrons and widely resonant with their petulant and incessant calls is silent as the grave—a deserted, barren, and frozen shore.

J. R. Forster (1772), the naturalist, who sailed with Captain Cook in this region, says:

The Indians have a peculiar method of killing all these species of geese, and likewise swans. As these birds fly regularly along the marshes, the Indians range themselves in a line across the marsh, from the wood to high-water mark, about musket shot from each other, so as to be sure of intercepting any geese which fly that way. Each person conceals himself, by putting round him some brushwood; they likewise make artificial geese of sticks and mud, placing them at a short distance from themselves, in order to decoy the real geese within shot; thus prepared they sit down, and keep a good lookout; and as soon as the flock appears they all lie down, imitating the call or note of geese, which these birds no sooner hear, and perceive the decoys, than they go straight down toward them; then the Indians rise on their knees, and discharge one, two, or three guns each, killing two or even three geese at each shot, for they are very expert. Mr. Graham says he has seen a row of Indians, by calling, round a flock of geese, keep them hovering among them, till every one of the geese was killed. Every species of geese has its peculiar note or call, which must gradually increase the difficulty of calling them.

Dr. George Bird Grinnell (1901) draws a pretty picture of migrating snow geese, as follows:

The spectacle of a flock of these white geese flying is a very beautiful one. Sometimes they perform remarkable evolutions on the wing, and if seen at a distance look like so many snowflakes being whirled hither and thither by the wind. Scarcely less beautiful is the sight which may often be seen in the Rocky Mountain region during the migration. As one rides along under the warm October sun he may have his attention attracted by sweet, faint, distant sounds, interrupted at first, and then gradually coming nearer and clearer, yet still only a murmur; the rider hears it from above, before, behind, and all around, faintly sweet and musically discordant, always softened by distance, like the sound of far-off harps, of sweet bells jangled, of the distant baying of mellow-voiced hounds. Looking up into the sky above him he sees the serene blue far on high flecked with tiny white moving shapes, which seem like snowflakes drifting lazily across the azure sky; and down to earth, falling, falling, falling, come the musical cries of the little wavies that are journeying toward the southland.

Winter.—Doctor Coues (1874) refers to the abundance of snow geese in the winter resorts in California, as follows:

On the Pacific coast itself, particularly that of California, the birds are probably more abundant in winter than anywhere else. Upon their arrival in October, they are generally lean and poorly flavored, doubtless with the fatigue of a long journey; but they find abundance of suitable food and soon recuperate. At San Pedro, in southern California, in November, I saw them every day, and in all sorts of situations—some on the grassy plain, others among the reeds of a little stream or the marshy borders of the bay, others on the bare mud flats or the beach itself. Being much harassed they had grown exceedingly wary and were suspicious of an approach nearer than several hundred yards. Yet with all their sagacity and watchfulness—traits for which their tribe has been celebrated ever since the original and classic flock saved Rome, as it is said—they are sometimes outwitted by very shallow stratagem—the same that I mentioned in speaking of the speckle-bellies. It is strange, too, that the noise and general appearance of a carriage should not be enough to frighten them, but such is the case. I have driven in a buggy along the open beach directly into a flock of snow geese that stood staring agape, "grinning" the while, till they were almost under the horse's hoofs; the laziest flock of tame geese that were ever almost run over in a country by-road were in no less hurry to get out of the way. Advantage is often taken of this ignorance to shoot them from a buggy; and, though they have not yet learned that anything is to be dreaded when the rattling affair approaches, yet no doubt experience will prove a good teacher, and its acquirements be transmitted until they become inherent. A wild goose of any species is a good example of wariness in birds, as distinguished from timidity. A timid bird is frightened at any unusual or unexpected appearance, particularly if it be accompanied by noise, while a wary one only flies from what it has learned to distrust or fear through its acquired perceptions or inherited instincts. Doctor Heermann's notice of this species gives an idea of the immense numbers of the birds in some localities, besides relating a novel method of hunting them. He says they "often cover so densely with their masses the plains in the vicinity of the marshes as to give the ground the appearance of being clothed in snow. Easily approached on horseback, the natives sometimes near them in this manner, then suddenly putting spurs to their animals, gallop into the flock, striking to the right and left with short clubs, and trampling them beneath their horses' feet. I have known a native to procure 17 birds in a single charge of this kind through a flock covering several acres."

Walter E. Bryant (1890), in comparing their status then with past conditions in California, writes:

There has not, so far as I am aware, been a very marked decrease in the number of geese which annually visit California, but the area over which they now feed is considerably less than in 1850. In the fall of that year, my father, while going from San Francisco to San Jose, met with acres of white and gray geese near San Bruno. They were feeding near the roadside, indifferent to the presence of all persons, and in order to see how close he could approach he walked directly toward them. When within 5 or 6 yards of the nearest ones they stretched up their necks and walked away like domestic geese; by making demonstration with his arms they were frightened and took wing, flying but a short distance. They seemed to have no idea that they would be harmed, and feared man no more than they did the cattle in the fields. The tameness of the wild geese was more remarkable than of any other birds, but it must be understood that in those days they were but little hunted and probably none had ever heard the report of a gun and few had seen men. This seems the most plausible accounting for the stupid tameness of the geese, 40 years ago. What the wild goose is to-day on the open plains of the large interior valleys of California those who have hunted them know. By 1853 the geese had become wilder and usually flew before one could get within shotgun range, if on foot, but in an open buggy or upon horseback there was no difficulty. There was a very marked contrast between the stupidly tame geese after their arrival in the fall and the same more watchful and shy birds before the departure in spring of the years 1852 and 1853. This is an important fact, showing not only the change in the instinct occurring within three years, but the more remarkable change, or it may be called the revival of the instinct of fear, which was effected within a few months; to this point I will refer again.

The following quotations from Grinnell, Bryant, and Storer (1918) will give a fair idea of present conditions in California:

There has been a more conspicuous decrease in the numbers of geese than in any other game birds in the State. Many observers testify that there is only 1 goose now for each 100 that visited the State 20 years ago, and some persons aver that in certain localities there is not more than 1 to every 1,000 which formerly occurred here. Not only have these birds been slaughtered for the market, but gangs of men have been paid to destroy them where they were feeding in grain fields. Until 1915 they were afforded no protection whatever, and as a natural result their ranks have been so often decimated that, comparatively speaking, only a remnant now remains.

In former years, when passing through the Sacramento or San Joaquin Valleys by train, great flocks of white geese, in company with other, dark-colored species, were often to be seen sitting on the grain fields or pasture lands almost within gunshot of the cars. The days are past and gone when a man has to drive geese from his grain field. In many places where formerly the ground was so covered with white geese as to look snowclad, not a single goose is now to be observed feeding and but few flying overhead. In spite of the extreme shyness and watchfulness of these geese, the ingenuity of the hunter and the increased efficiency of firearms has so far overbalanced the natural protection thus afforded that the birds are now actually threatened with extinction. Unless the protection now furnished proves adequate in the very near future, this State, which at one time appeared to have an inexhaustible supply of geese, will have entirely lost this valuable game resource.

That snow geese are abundant in winter in certain parts of Texas is well illustrated by the following note made by Herbert W. Brandt in Kleberg County:

On March 23, 1919, we went up to the Laureless Ranch headquarters and got Mr. Cody, the foreman, to go for a ride with us. He showed us a new road and took us to Laguna Larga, a great marshy tract 6 miles long in the plains. The water is not deep and grass grows up through it all over, and there are a few small patches of tules or cat-tails, but it all dries up if the summer is dry. As we approached it looked as if it was covered with snow, but it proved to be thousands upon thousands of snow geese and other wild geese. Here is their winter home, coming into the great pastures at night to feed on the abundant grass. Last year for the first time known a couple of large flocks remained the entire summer. It was the most wonderful sight in bird life I ever saw, and it will never be forgotten, as cloud after cloud of white and black birds took to wing and then settled down in a distant part of the marsh.

Mr. Kleberg told us that the geese we saw were just a few left from the great winter flocks, most of them having now departed for the northland. He has seen 500 acres of solid geese, he said, just one snow bank. He hunts them by taking his big Packard car and runs toward them on the prairies at 60 miles an hour. The wind is always blowing here and the geese must rise and fly against it; as they are overtaken they work the pump shotguns on them.

DISTRIBUTION

Breeding range.—Arctic America, along the coast from northern Alaska (Point Barrow) eastward to Hudson Bay (Southampton Island) and Baffin Land, and on Arctic lands and islands north of North America. Has been seen in summer on the Arctic coast of northeastern Siberia (Tchuktchen Peninsula), where it probably breeds. It may breed farther west in Siberia.

Winter range.—Includes the whole of temperate North America from the Atlantic to the Pacific, rare or straggling, mainly as a migrant, on the Atlantic coast, uncommon east of the Mississippi

Valley and most abundant in California, Mexico, Texas, and Louisiana. East rarely to Rhode Island (Narragansett Bay, January 10, 1909), more frequently to the coasts of Virginia and North Carolina and probably only casually to the West Indies. South regularly to the Gulf coasts of Alabama, Louisiana, and Texas, and to central Mexico (Tamaulipas and Jalisco). West to the Pacific coast States. North to southern British Columbia (Vancouver), Nevada, Utah, southern Colorado (San Luis Valley), southern Illinois, and sparingly to the coast of Virginia. South on the Asiatic coast to Japan.

Spring migration.—Mainly northward in the interior and northwestward or northeastward from the coasts to inland valleys. Early dates of arrival: Iowa, Sac County, March 28; Montana, Teton County, April 9; Manitoba, Aweme, April 5; Mackenzie, Fort Simpson, May 2, and Fort Anderson, May 20; Banks Land, Mercy Bay, 74° N., May 31; Alaska, St. Michael, May 5, Nulato, May 9, Kowak River, May 23, and Cape Prince of Wales, May 31. Late dates of departure: Southern Texas, Rio Grande River, March 29, and San Angelo, April 16; California, Gridley, May 1; Utah, Bear River, May 5; Montana, Teton County, April 23; Manitoba, Shoal Lake, April 30; Mackenzie, Fort Simpson, May 25.

Fall migration.—A reversal of spring routes, with more eastward wanderings (practically all New England records are in fall). Early dates of arrival: Maine, Cape Elizabeth, October 2; Massachusetts, Essex County, October 7; New York, Shinnecock Bay, October 8; Mackenzie, Providence, August 30; Alberta, Buffalo Lake, September 26; Manitoba, Aweme, September 24; Arkansas, Helena, October 19; Louisiana, Cameron County, October 7; Texas, San Angelo, October 1; Montana, Terry, September 12; Utah, Bear River, September 3; California, Stockton, September 28; Alaska, Wainwright, September 6, St. Michael, September 1, and Taku River, September 17. Late dates of departure: Banks Land, September 7; Alaska, St. Michael, October 10; Alberta, Buffalo Lake, October 26; Manitoba, October 31; Montana, Teton County, November 24; Massachusetts, Ipswich, December 7.

Casual records.—Snow geese, probably of this form, have wandered on migrations to Labrador (Independent Harbor, October 1, 1914), Florida (Wakulla County, October 30, 1916, and November 23, 1918, and Key West), Bermuda (October 19, 1848), South Carolina (Mount Pleasant, October 16, 1916), the Bahama Islands, Cuba, Jamaica, and Porto Rico. Said to have occurred in the Hawaiian Islands. It has been recorded in Iceland, Norway, Holland, Great Britain, Germany, Hungary, etc. The numerous European records suggest the probability of a more extended breeding range in Palaearctic regions than is now known, with a westward migration.

Egg dates.—Arctic Canada: Nineteen records, June 9 to July 6; ten records, June 15 to 23.

CHEN HYPERBOREA NIVALIS (J. R. Forster)

GREATER SNOW GOOSE

HABITS

When I first began to study the distribution of this large subspecies I was skeptical as to its status, for it did not seem to have any well-defined breeding range or winter range, and it looked very much as if the large birds could be nothing more than extra large individuals. My confusion was due to the fact that I did not know where to draw the line between the greater and the lesser snow geese. I found that my friend Frederic H. Kennard had been studying the same problem for some time, and while he had collected considerable data, which he placed at my disposal, he was waiting for additional data before publishing it.

For the purposes of this life history it will suffice to give merely the general conclusion I have arrived at and a brief statement of the steps which led to them. A collection of the measurements of over 250 birds from various parts of the country, when tabulated according to size, shows very clearly that an extra large subspecies, now called *nivalis*, occupies a very narrow winter range on the Atlantic coast, which it reaches by a decidedly eastern migration route from its breeding range in northern Greenland. All of the largest birds come from extreme eastern localities; I have seen only one bird from the interior that I should call *nivalis;* that is an immature bird in the United States National Museum, labeled Hudson Bay, which, if it came from there, is probably a straggler. All of the birds from Atlantic coast States and Provinces are *nivalis*, except a few very small ones which are doubtless stragglers from the westward and are referable to *hyperborea*. The average measurements of all the birds from Atlantic coast points, including the small birds referred to above, are decidedly larger than the average measurements of birds from the interior or from the Pacific coast States.

The average measurements of all the birds from the interior, from Hudson Bay to Texas, agree very closely with the average measurements of a series of birds from California. This shows conclusively that the birds of the interior are unquestionably referable to the smaller form, *hyperborea*, and that the larger birds from that region, which have been called *nivalis*, are merely large specimens of *hyper-*

borea, which can be nearly, if not quite, matched with birds from California. The measurements of the greater snow goose do not well illustrate its real superiority in size; it is a much heavier bird than its western relative, with a much more stocky build, thicker neck, and larger head. It is generally recognizable at a glance, in the flesh.

The breeding range of the greater snow goose must be determined largely by elimination, though it is clearly indicated by two specimens from northern Greenland; these are the only Greenland specimens of snow geese that we have; they are both typical *nivalis;* and one was the parent of a downy young. The average measurements of 26 birds from the Arctic coasts of Alaska, Canada, and Baffin Land agree very closely with those from California and the interior, and none of them are any larger than the largest birds from these localities. A study of the average measurements of 20 sets of eggs from Arctic America, collected at various points from Point Barrow to Baffin Land, shows no correlation of size with locality; the largest 2 sets came from Cape Bathurst and Franklin Bay; and the smallest 2 came from Mackenzie Bay and Point Barrow. Judging from the evidence shown in the measurements of both birds and eggs, it seems fair to assume that *nivalis* does not breed anywhere on the Arctic coast from Alaska to Baffin Land and that all the breeding birds of that region are referable to *hyperborea.* This leaves for *nivalis* a known breeding range in northern Greenland, which probably extends into Ellesmere Land, Grinnell Land, and Grant Land.

Spring.—Although we have very little data on the subject, what evidence we have seems to indicate that the greater snow geese, which spend the winter on the Atlantic coast, migrate overland across New England to the Gulf of St. Lawrence and then across the Labrador Peninsula to their Arctic summer home. William Brewster (1909) published a letter from M. Abbott Frazar giving an account of a large flock of snow geese which he saw migrating at Townsend, Massachusetts, on April 13, 1908. There were at least 75 birds in the flock. Although the subspecies is in doubt, the chances are that these were greater snow geese. The following note, published by Harrison F. Lewis (1921), throws some light on the subject:

Most recent writers on the waterfowl of northeastern North America speak of the greater snow goose (*Chen hyperboreus nivalis* [Forst.]) as a rare bird in that area and appear to pay little or no attention to the fact that Mr. C. E. Dionne, on pages 109–110 of his book, "Les Oiseaux de la Province de Quebec" (1906), states of this subspecies that it "is very common and often occurs in considerable flocks in spring and fall in certain places on our shores, notably at St. Joachim, where I have seen flocks of three or four thousand individuals, on the Island of Orleans, and as far as the Sea-Wolves' Batture." The three points mentioned by Mr. Dionne are within sight of one another. In their vicinity probably all the greater snow geese in existence in a wild state gather each spring and autumn. From the independent statements of various careful observers, I should conclude that their number is now about five or six thousand. When I visited St. Joachim on March 31, 1921, I saw about 2,000 greater snow geese there and was told that the maximum number would be present about 10 days later. They are well protected by a resident warden maintained by the Cap Tourmente Fish and Game Club.

Kumlien (1879) "saw a few specimens in early spring and late autumn" at Cumberland Sound, where it was apparently "rare and migratory."

W. Elmer Ekblaw has sent me the following notes on the arrival of these geese on their breeding grounds in northern Greenland:

June is almost gone when the first snow geese arrive in northwest Greenland. The land is almost bare of snow, the inland lakelets are open, and rushing streams are flush to the brim with clear, cold water. Spring is at its height when the snow geese come. The first notice of their arrival is a high-pitched *honk-honk*, almost resembling the call of the domestic guinea fowl, that rings out clear and sharp from the swales and valleys of the inland slopes. The birds fly low and swift, their gleaming white plumage dazzlingly conspicuous against the dark-brown hills. When they fly near, the black tips of their wings are easily recognizable. They stalk regally about the lakelets and along the streams like the snow king's soldiers, stately and dignified. They are mated when they arrive in the North, and though they stay in flocks most of the time, they pair as soon as they alight, either on land or in water. Wherever they appear they grace the landscape.

On July, 2, 1914, I watched a flock of 10 at close range while they fed in a small shallow pool in which *Pleuropogon* and *Hippuris* grew abundantly. There were 5 pairs in the flock, and though they did not separate far, the pairs kept somewhat to themselves as they floated idly about in the pool or marched about the shore. They apparently found food on the bottom of the pool, because they dipped under much like canvasbacks feeding on wild celery. I watched them for at least an hour, delighted with their grace and quiet beauty. Their calm behavior contrasted strongly with the wild antics of the oldsquaws in near-by pools.

Nesting.—A long time ago Dr. Witmer Stone (1895) published a list of birds collected by the Peary expedition of 1891 and 1892 in northern Greenland, in which were included an "adult female in worn plumage and one young gosling entirely in down" of the greater snow goose. These birds were collected by Langdon Gibson, the ornithologist of the expedition, in the vicinity of McCormick Bay, latitude 77° 40′ north. The measurements of this bird clearly indicate that it is a typical specimen of *Chen hyperborea nivalis* and this constitutes the only definite breeding record we have for this subspecies. Recently Mr. Gibson (1922) has published his notes on this collection, from which I quote as follows:

It was my good fortune to record, for the first time, the breeding of this species in north Greenland. A family was found in Five Glacier Valley on

July 11, 1892. The male disputed my advance with head lowered and much hissing, quite after the fashion of the barnyard goose, and before I was aware of the existence of goslings I shot the female. Then I took two of the goslings, that were about 2 weeks old, leaving the gander to rear the remaining six. The birds were on the nest at the time of capture. The nest itself was well lined with grasses and placed near a pile of broken stone beside a marshy spot some acres in extent and about 100 yards from a shallow pond.

On August 21, when again passing through the valley, I was happy to see the male proudly marching at the head of his family of six at least 10 miles from the nest. As he had a broken wing and his family then had every indication of being able to shift for themselves, I reluctantly, and in the interest of science, dispatched him.

A brief note in the report of the Greeley (1888) expedition to Grinnell Land indicates that this goose probably breeds on this and other lands west of northern Greenland, as a pair was seen June 12, 1882, near Fort Conger, latitude 82° north, and another June 13, 1882, on the shore of Sun Bay. The snow geese found breeding in northern Greenland by the Crocker Land Expedition were undoubtedly greater snow geese, but unfortunately no specimens of birds or eggs were preserved. Mr. Ekblaw's notes state:

The geese nested in the grassy swales and flats along the lake-dotted flood plain of the streams which empty into North Star Bay. The nests are placed in depressions among the tussocks so that the brooding birds are not readily detected; built up somewhat with mud and grass and dead vegetation and lined with white feathers and down, they are much better constructed than are the nests of the eider and the oldsquaw.

The first eggs are laid soon after July 1. A full clutch is 6 or 7 eggs. In about four weeks they hatch. The mothers and the young frequent the larger inland lakes until the young are able to walk and swim and dive fairly well, and then they take to the open sea. In late August or early September the fall molting season comes on. The geese then repair to the most remote and isolated lakes to be safe and free from disturbance while their wing feathers are renewed. At this time they are relatively helpless and the Eskimo find them easy prey. By mid-September the molting season is over and the geese leave at once.

Eggs.—Apparently there are no eggs of the greater snow goose in collections. All the eggs in collections came from regions where this subspecies is not known to breed and are almost certainly referable to the smaller race.

Plumages.—The downy young, referred to above, is described by Dr. D. G. Elliot (1898) as follows:

Lores, dusky. Two black stripes from bill, passing above and beneath the eye. Top of head, dark olive brown. Sides of head, neck, and entire under parts, light yellow. Upper parts, dark olive brown. Bill, black; nail, yellowish white.

The subsequent molts and plumages are apparently the same as those of the lesser snow goose.

Food.—The food and the feeding habits of the greater snow goose are very much like those of its western relative. It has less opportunity, in the Eastern States, to feed in grain and stubble fields, as such cultivated areas are scarcer and more restricted. Perhaps for this reason these geese seem to be more often seen on the seacoast marshes and beaches in the East than they are in the West. Mr. Elisha J. Lewis (1885) the veteran sportsman, writes:

Snow geese are numerous on the coast of Jersey and in the Delaware Bay. They frequent the marshes and reedy shores to feed upon the roots of various marine plants—more particularly that called sea cabbage. Their bills being very strong and well supplied with powerful teeth, they pull up with great facility the roots of sedge and all other plants.

Harold H. Bailey tells me that on the coast of Virginia they come into the hollows on the sandy beaches to pull up the beach grass and other scanty sand-dune plants to feed on the roots; they do not come into the fresh-water bays with the Canada geese to feed on the fox-tail grass.

Behavior.—Audubon (1840) writes:

The flight of this species is strong and steady, and its migrations over the United States are performed at a considerable elevation, by regular flappings of the wings, and a disposition into lines similar to that of other geese. It walks well, and with rather elevated steps; but on land its appearance is not so graceful as that of our common Canada goose. Whilst with us they are much more silent than any other of our species, rarely emitting any cries unless when pursued on being wounded. They swim buoyantly, and when pressed, with speed. When attacked by the white-headed eagle, or any other rapacious bird, they dive well for a short space. At the least appearance of danger, when they are on land, they at once come close together, shake their heads and necks, move off in a contrary direction, very soon take to wing, and fly to a considerable distance, but often return after a time.

Winter.—Regarding its winter habits Doctor Elliot (1898) says:

On the northern portion of the Atlantic coast the snow goose can not be said to be common, and in many parts is seldom seen. Small flocks are occasionally met with on the waters of Long Island, but the species becomes more abundant on the shores of New Jersey and the coasts of Virginia and North Carolina, where, in the latter State in the vicinity of Cape Hatteras and along the beaches and inlets of Albemarle Sound, it sometimes congregates in great multitudes. Occasionally flocks of considerable size may be seen on the inner beach of Currituck Sound where the water is brackish, but the birds do not remain any length of time in such situations. They present a beautiful sight as they stand in long lines upon the beach, their pure, immaculate plumage shining like snow in the sun, against the black mud of the marshes or the dingy hues of the shore. It is very difficult to approach them at such times, as they are exceedingly watchful and wary, but occasionally a few may leave the main body and, if flying by, will draw perhaps sufficiently near to geese decoys or live geese tied out in front of a blind to afford an opportunity for a shot. The chances are better, however, for the sportsman when these geese are moving in small flocks of six or seven, as they are then more apt to come near the shore looking for favorable feeding places or spots on the beach to sand themselves.

H. H. Bailey tells me that these geese are not now as common on the Virginia coast as formerly, that they do not come until cold weather in midwinter, and that they spend most of their time in Chesapeake Bay or on the ocean, resorting to the hollows among the sand dunes of the outer beaches when these are partially covered with snow and ice.

DISTRIBUTION

Breeding range.—Positively known to breed only in northern Greenland (McCormick Bay, latitude 77° 40′ north, and North Star Bay). Probably breeds also in Ellesmere Land, Grinnell Land (Fort Conger), and Grant Land.

Winter range.—Mainly, if not wholly, on the Atlantic coasts of Maryland, Virginia, and North Carolina, from Chesapeake Bay to Core Sound. Probably all the birds from farther south are referable to *hyperborea.*

Spring migration.—Directly north, overland across New England and the Labrador Peninsula. Early dates of arrival: New York, Shelter Island, April 3; Massachusetts, Townsend, April 13; Maine, Scarborough, April 4; Quebec, St. Joachim, March 31, and Hatley, April 6; Greenland, Etah, June 10. Late dates of departure: North Carolina, Currituck Sound, March 6; Maine, Georgetown, April 25, and Lubec, April 30.

Fall migration.—A reversal of spring route. Early dates of arrival: Quebec, St. Lawrence River, October 12; Massachusetts, Westfield, November 24; Connecticut, Portland, November 20; North Carolina, Currituck Sound, December 11.

Casual records.—An immature bird in the United States National Museum is labeled Hudson Bay, with no further data. Probably all records for Bermuda and the West Indies are referable to *hyperborea*, as that is the wider-ranging form, but some records may refer to *nivalis.*

CHEN CAERULESCENS (Linnaeus)

BLUE GOOSE

HABITS

The blue goose is one of the few North American birds which we know only as a migrant and a winter resident, and within the narrowest limits. It has generally been regarded as a rare species, but it is really astonishingly abundant within the narrow confines of its winter home on the coast of Louisiana. Its apparent rarity is due to the fact that on its migrations to and from this favorite resort its seldom straggles far from its direct route to and from its unknown breeding range. To find the breeding resorts of the blue goose is one of the most alluring of the unsolved problems in American ornithology. It is really surprising that such a large and conspicuous species, which is numerically so abundant, can disappear so completely during the breeding season.

Spring.—Numerous records from various observers indicate a heavy spring migration northward through the Mississippi Valley and over the Great Lakes to James Bay and Hudson Bay, but beyond there the species vanishes completely. No one knows where the blue goose goes to spend the summer and none of the numerous Arctic explorers have ever found its breeding grounds.

The blue goose migrates generally in flocks by itself and usually the old white-headed birds are in separate flocks from the young birds; but occasionally one or more dark-blue geese may be seen leading a flock of pure white-snow geese, which makes a striking picture. The main flight in the spring seems to pass up the east side of James Bay.

Owen Griffith says, in a letter published by Mr. W. E. Saunders (1917):

> About 3 miles north of Fort George Post there is a big bay (salt water) with lots of mud and grass at low tide, and in the spring almost every flock of wavies and other geese feed in this bay on their way north; the Indians never hunt them on their arrival in this bay, but gather on a long hill on the other side and then shoot at the birds as they are going off; they generally get up in small flocks, and as they have to rise considerably to clear the hill they can be seen getting up some time before they get to the hill, and then everyone runs along a path and tries to get right under where the flock is going to pass; of course, if three or four flocks get up at the same time, there is shooting on different parts of the hill and the hunters are apt to spoil one another. The Indians say that once these birds leave this bay that they do not feed again till they get far north (Hudson Straits or Baffin Land) in fact a wavey's nest is a great rarity. Strange to say they do not feed in this bay in the fall.

Dr. Donald B. MacMillan, who spent the spring and part of the summer of 1921 at Bowdoin Harbor in southern Baffin Land, says that the blue geese and snow geese migrate from Cape Wolstenholme across to southern Baffin Land. He was told by the natives of that country that both of these geese breed in immense numbers in the marshy lands near some lakes in the interior, a region too difficult to reach and too remote from where his ship was frozen in until August.

Nesting.—There seems to be no authentic record of the finding of a nest of the blue goose and, so far as I know, the nest and eggs in a wild state are unknown to science. All that has been published on the breeding habits or the probable breeding range of the species seems to be based on speculation or hearsay. George Barnston (1862), one of the best authorities on the geese of the Hudson Bay region, says that:

According to Indian report a great breeding ground for the blue wavy is the country lying in the interior of the northeast point of Labrador, Cape Dudley Digges. Extensive swamps and impassible bogs prevail there, and the geese incubate on the more solid and the driest tufts dispersed over the morass, safe from the approach of man. * * * In May it frequents only James Bay and the Eastmain of Labrador, and it is probably the case that its hatching ground is on the northwest extremity of that peninsula and the opposite and scarcely known coast of Hudson Straits.

Eggs.—Hon. R. M. Barnes has kept blue geese on his estate at Lacon, Illinois, for a number of years and has succeeded in raising them to maturity in confinement. He says in a letter to Mr. Frederic H. Kennard that the eggs of the blue goose are quite different from those of the snow goose. He describes the eggs of the snow goose as "more elongated and of a slightly yellowish color," whereas the eggs of the blue goose "are pyriform, of thicker diameter, shorter in proportion to their length, and have a very slight bluish cast," the eggs of both appearing white at first glance; moreover, the eggs of the blue goose have "more minute pit holes and apparently most of these pit holes have a very small deep black center, which can only be disclosed by a microscope." An egg in Mr. Kennard's collection is pure white and very finely granulated. It measures 78 by 51 millimeters.

The measurements of four eggs, laid by one of Mr. Barnes's birds, are 81 by 54.2, 84.6 by 56, 81.2 by 55.8, and 81 by 60.2 millimeters. The nest was lined with "the purest of white down."

Plumages.—Mr. Barnes says that the downy young of the blue goose "is of a deep smoky or slaty bluish color." F. E. Blaauw, who has also raised this species in captivity, describes it as "olive green, darkest on the upper side and yellowish on the belly," with "a little white spot under the chin."

In the fresh juvenile plumage of the first fall, October, the chin is white, the entire head and neck are uniform bluish gray, the back very dark gray, with brownish edgings, and the under parts dull gray, almost whitish on the belly; the wings show a dull reflection of the adult color pattern, the lesser coverts are more or less edged or tipped with brownish, the greater wing coverts are plain pale gray, the primaries and secondaries are duller and browner black than in adults, and the tertials and scapulars are either plain dusky brown or are less conspicuously patterned than in adult birds.

During the first winter there is a nearly continuous molt, with a gradual advance toward maturity. White feathers appear in the head and neck early in the winter, and by spring these have become nearly all white in some birds, but there is generally more or less black on the crown and hind neck. The bluish-gray feathers of the adult plumage invade the lower neck and breast, much new plumage comes in on the back, and new scapulars are partially or wholly acquired. The tail is molted in the spring, beginning sometimes as early as the last of February, but not the wings. Before the birds go north in the spring many of the first-year birds are practically indistinguishable from adult birds except for the immature wings.

What takes place during the following summer we can only guess at, as summer specimens are lacking, but apparently a complete summer molt produces the second winter plumage, which is practically adult. The head and neck become wholly white, or nearly so. The wing is practically adult, with the pure gray lesser coverts and the pale gray primary coverts; the greater coverts have black centers, shading off into silvery gray, and broad, white edges; the scapulars and tertials have the adult color pattern; and the primaries and secondaries are deep black. The under parts are mainly bluish gray, and the rump and upper tail coverts are clear pale gray.

Subsequent molts produce similar plumages, and probably third-year and older birds show greater perfection of plumage and more brilliant color patterns. Many adult birds show more or less white on the under parts, in strong contrast with the bluish gray, varying from a small spot to a large area covering nearly all of the belly; this may be the result of crossing with the snow goose or it may be a character which develops and increases with age. Some observers believe that the blue goose is a dark-color phase of the snow goose. Mr. Blaauw, who has bred both species for some 22 years, has come to this conclusion. On the other hand, Mr. Barnes, who has also bred both, has come to the opposite conclusion; in addition to the difference in the eggs and young, he says that the "build of the two birds is very different, and their physical appearance is very distinct. The call notes are not very similar." It seems to me that they are too unlike in many ways to be color phases of one species, and I can find no conclusive evidence to prove that they are. They are very closely related; so that, like the mallard and black duck, they can interbreed and raise fertile hybrids, which they probably frequently do.

Food.—The feeding habits of the blue goose have been well described by W. L. McAtee (1910), as follows:

In the Mississippi Delta the blue geese rest by day on mud flats bordering the Gulf. At the time of my visit (January 29 to February 4, 1910) these were entirely destitute of vegetation, a condition to which the geese had reduced them by their voracious feeding. Every summer these flats are covered by a dense growth of "cut grass" (the local name for *Zizaniopsis mili-*

acea), "goose grass" (*Scirpus robustus*), "oyster grass" (*Spartina glabra*), "Johnson grass" (*Panicum repens*), and cat-tails or "flag grass" (*Typha augustifolia*), and every fall are denuded by the blue geese, or brant, as they are called in the Delta. The birds feed principally upon the roots of these plants, but the tops of all are eaten at times, if not regularly. Each goose works out a rounded hole in the mud, devouring all of the roots discovered, and these holes are enlarged until they almost touch before the birds move on. They maintain themselves in irregular rows while feeding, much after the manner of certain caterpillars on leaves, and make almost as clean a sweep of the area passed over.

In the Belle Isle region the method of feeding is the same except that the birds feed by day, but the places frequented are what are locally known as "burns"; that is, areas of marsh burned over so that new green food will sooner be available for the cattle. These pastures, for the most part, are barely above water level, so that the holes dug by the geese immediately fill with water. Continued feeding in one area produces shallow, grass-tufted ponds, where formerly there was unbroken pasture. Some of these ponds are resorted to for roosting places, in which case the action of the birds' feet further deepens them, and veritable lakes are produced, which the building-up influence of vegetation can not obliterate for generations, and never, in fact, while the geese continue to use them.

The numbers of the blue geese are so great that these effects are not local but general. At Chenjere-au-Tigre, one proprietor formerly hired from two to four men at a dollar a day, furnishing them board, horses, guns, and ammunition, and keeping them on the move constantly in the daytime to drive the geese away. The attempt was unsuccessful, however, and fully 2,000 acres of pasture were abandoned. Other proprietors had similar experience and suffered loss of the use of hundreds of acres.

The stomachs and crops of the birds in my collection were sent to the Biological Survey for examination by Mr. McAtee, who reported that the contents consisted entirely of the stems of spikerush (*Eleocharis*), of which those in the crops were whole and those in the gizzards finely ground.

Mr. Hersey was told, while collecting blue geese for me in Louisiana, that they also feed on the duck potato, one of the principal duck foods in that vicinity. In his notes on their feeding habits he states:

In reviewing my experience with the blue geese it seems that *normally* they begin to feed about 2 p. m. and continue to do so until dark. They then fly to their roosting ground, where they spend the night. Some time before daylight the flocks again begin to feed, and do so until about 9 or 10 o'clock. They then rest until the afternoon, usually without leaving the feeding ground.

While feeding, small parties are continually flying into the air and moving to a new spot on the outskirts of the flock. If they see anyone approaching at such times, they at once warn their companions and the whole flock takes wing with great clamor.

Behavior.—O. J. Murie has sent me some interesting notes on this species in which he says:

The blue geese are apparently not as prone to fly in the V formation as the Canada geese. The flocks are often broken in a mixture of V's, bars, curves, and irregular lines. Perhaps this is due to the immense numbers in the flocks noted on James Bay. When a flock of "waveys" is passing, the Indian hunter will imitate their call by a single, high-pitched "*guop*"—very different from the double "*au-unk*" in the case of Canada geese. As the blue geese approach with answering calls, an accompanying undertone is heard, a conversational "*ga-ga-ga-ga-ga-ga*," with an occasional clear "whistle." The whistling note is the call of young birds. The Indian makes use of all these sounds, employing the "cackling" notes and an occasional whistle when the birds are near enough to hear it.

On one occasion, while lying in a blind, I heard a peculiar, startled "squawk." Looking up I saw a single blue goose pursued by a duck hawk. The goose ducked and swerved here and there in his flight, with the duck hawk swooping after. The chase continued some distance down the marsh, when finally the duck hawk turned aside and gave it up.

Fall.—Mr. Murie says of the fall flight:

The extensive salt-water marshes around the south and part of the west shore of James Bay furnish an excellent feeding place for shore birds and various ducks and geese, including thousands of blue geese. I was told that blue geese are seen as early as August. By September at least they begin to arrive in James Bay, and during this month and most of October they congregate in immense flocks, principally in Hannah Bay, at the extreme south end of James Bay. Here the Indians go for their annual goose hunting. A blind of willows is placed at a favorite feeding spot, often beside a small streamlet cutting its way by several channels over the mud flat. For decoys, lumps of mud or sod are turned up with a wooden spade. In the top of each lump is thrust a small stick or twig, at the end of which is fastened a piece of folded paper, or, better yet, a small bundle of white quills from a snow goose is stuck in to represent the white head. These crude decoys are very realistic at a distance and prove effective.

The blue geese feed on the open tide flats, while the Canada geese are often found in the swamps or open muskegs well within the margin of the forest. The birds become extremely fat, sometimes bursting open in the fall to the ground when shot. According to native information they do not feed at all the last few days before they begin their flight farther south. In 1914 the blue geese were seen leaving for the south, up the Moose River, on November 1. The following autumn they went south October 21 and 22. In each case snow was falling, with a north or northeast wind.

Winter.—Few people, who have not seen them, appreciate the astonishing abundance of blue geese in the narrow confines of their winter home on the coast of Louisiana.

Mr. Hersey states in his notes:

I am told that before going north most of the flocks congregate in the vicinity of Great and Little Constance Lakes on the Gulf coast west of Vermilion Bay. These flocks are said to be enormous, but the estimates I heard of their numbers were too vague to be of use. One game warden, a very conservative man, told me, however, that he once saw a spot 5 miles long and 1 mile wide (approximately), covered with blue geese, all standing as close together as they could get. Three men fired 5 shots into this flock and picked up 84 dead birds.

Mr. McAtee (1910 and 1911) found these geese exceedingly abundant in a very restricted area on the Louisiana coast. He writes:

The center of abundance of the species is a narrow strip extending along the coast of Louisiana from the Delta of the Mississippi to a short distance west of Vermilion Bay. To the eastward the bird is known only as a straggler, and to the west it diminishes gradually in numbers, being scarce on the extreme western coast of Louisiana and rare on the Texas coast. * * * Being so localized in their winter range, it might seem that the blue geese are in danger of extermination. But they are so wary and so few hunters molest them that at present there is no appreciable reduction in their numbers by man. The same is true, I feel sure, of the winter colonies of snow geese and swans on Currituck Sound, North Carolina. So long as conditions remain the same, the birds being very wary, and having little market value, there is no incentive to kill them, nothing occurring during their stay in the United States will materially lessen their numbers, nor even interfere with the increase of these fine birds. However, if they should become an object of pursuit, it is equally true that they would diminish very rapidly.

DISTRIBUTION

Breeding range.—Recently reported as breeding in large numbers in the interior of southern Baffin Land. Breeding range otherwise unknown.

Winter range.—Mainly in a very restricted area on the coast of Louisiana; from the mouth of the Mississippi River to Vermilion Bay, decreasing very rapidly in abundance eastward and more gradually westward to the coast of Texas (Rockport, Corpus Christi, and Brownsville). Has been recorded in winter as far north as Nebraska, southern Illinois, and Ohio (New Bremen, January 17, 1916), but probably only casually.

Spring migration.—Northward through the Mississippi Valley up the east coasts of James Bay and Hudson Bay. Early dates of arrival: New York, Amagansett, March 21; Rhode Island, Westerly, March 16; Illinois, Lacon, March 23; Iowa, March 28; Manitoba, Aweme, April 9; Ontario, Kingsville, April 6. Late dates of departure: New York, Miller Place, April 28; Manitoba, Shoal Lake, May 29.

Fall migration.—Southward across the eastern United States; more easterly than in the spring. Early dates of arrival: Ontario, Ottawa, October 11; Maine, Umbagog Lake, October 2; Massachusetts, Gloucester, October 20; Rhode Island, Charlestown Beach, October 16; Manitoba, Shoal Lake, October 1; Illinois, Gary, October 21; Louisiana, November 1. Late dates of departure: James Bay, Moose River, November 1; Ontario, Thames River, November 16; Maine, Little Spoon Island, November 13; Rhode Island, Dyers Island, November 9; New York, Amityville, November 22; Manitoba, Aweme, October 24.

Casual records.—Rare in Atlantic coast States, but records are too numerous to be regarded as casuals. Has been recorded in North and South Carolina, the Bahama Islands, and Cuba. There are two good records for California (Stockton, February 1, 1892, and Gridley, December 15, 1910).

EXANTHEMOPS ROSSII (Cassin)

ROSS GOOSE

HABITS

The smallest and the rarest of the geese which regularly visit the United States is this pretty little white goose, hardly larger than our largest ducks, a winter visitor from farthest north, which comes to spend a few winter months in the genial climate of California.

Spring.—Whither it goes when it wings its long flight northeastward across the Rocky Mountains in the early spring no one knows, probably to remote and unexplored lands in the Arctic regions. At certain places it is abundant at times, as the following account by Robert S. Williams (1886), of Great Falls, Montana, will illustrate; he writes:

On the 17th of April, 1885, after several days of stormy weather, with wind from the northwest, accompanied at times by heavy fog and rain, there appeared on a bar in the Missouri River at this place a large flock of Ross's snow geese. In the afternoon of the same day, procuring a boat, we rowed toward the flock, which presented a rather remarkable sight, consisting as it did of several thousand individuals squatting closely together along the edge of the bar. Here and there birds were constantly standing up and flapping their wings, then settling down again, all the while a confused gabble, half gooselike, half ducklike, arising from the whole flock. We approached to within a hundred yards or so, when the geese lightly arose to a considerable height and flew off over the prairie, where they soon alighted and began to feed on the short, green grass. While flying, often two or three birds would dart off from the main flock, and, one behind the other, swing around in great curves, quite after the manner of the little chimney swift in the East. Apparently these same birds remained about till the 26th of April, long after the storm was over, but they became broken up into several smaller flocks some time before leaving. Some five or six specimens were shot during their stay.

Mr. Roderick MacFarlane (1891) never succeeded in finding its breeding grounds or learning anything definite about where it goes in summer; he says:

A male bird of this species was shot at Fort Anderson on 25th May, 1865, where it is by far the least abundant of the genus during the spring migration. The Esquimaux assured us that it did not breed in Liverpool Bay, and it may therefore do so, along with the great bulk of the two larger species, on the extensive islands to the northwest of the American continent. At Fort Chipewyan, Athabasca, however, it is the last of the geese to arrive in spring, but among the first to return in the autumn.

Nesting.—Absolutely nothing seems to be known about its breeding habits in a wild state. Probably nothing will be known until some of the vast unexplored areas in the Arctic regions are better known. But these regions are so inaccessible that their exploration would involve more time, greater expense, and more enthusiasm than even the valuable results to be attained are likely to warrant. Therefore this and several other similar problems are likely to remain for a long time unsolved.

For all that we know about the nesting habits of the Ross goose, we are indebted to F. E. Blaauw (1903) who has succeeded in breeding this species in captivity on his place at Gooilust in Holland. He writes:

At a meeting of the British Ornithologists' Club on March 20, 1901, I exhibited an egg of the rare Ross's snow goose (*Chen rossi*) laid in captivity by a solitary female kept by me at Gooilust. A year later, through the courtesy of Doctor Heck, of Berlin, I received a second specimen of this species, which fortunately proved, as I hoped it would, to be a male. The birds soon paired. and in the beginning of May, 1902, the female made a nest under a bush in her inclosure. The nest was, as is usual with geese, a small depression in the soil, lined with dry grass and grass roots.

Toward the end of the month the female began to lay, and on the 30th, when the full complement of 5 eggs had been deposited, she began to sit, having in the meantime abundantly lined her nest with down from her own breast. The two birds had always been of a very retiring disposition, but after the female had laid her eggs the male, who nearly always kept watch close by the nest, became quite aggressive. He would fearlessly attack anybody that approached.

Eggs.—There is an egg in the collection of Adolph Nehrkorn, probably one of Mr. Blaauw's eggs, which is described as white and which measures 74 by 47 millimeters.

Young.—Mr. Blaauw's bird had incubated for only 21 days when he was surprised to find the eggs hatched. "All the 5 eggs had hatched, and the little birds were still in the nest when I noticed them, forming a most charming group, ever watched as they were by their anxious parents." Another season, when 3 eggs were set under a hen, the period of incubation proved to be 24 days.

Plumages.—Mr. Blaauw (1903) describes the downy young as follows:

The chicks are of a yellowish gray, darker on the upper side and lighter below, and have, what makes them most conspicuously beautiful, bright canary-yellow heads, with the most delicate grayish sheen over them, caused by the extremity of the longer down hairs being of that color. The bill is black, with a flesh-colored tip. A little spot in front of each eye is also blackish. The legs are olive green. The down is wonderfully full and heavy, and it seems almost incredible how such large birds can have come out of such small eggs. Three of the chicks were as described above, but two of them had the part white which in the others was yellow. All that I can add is that, as usual with chicks, the intensity of the coloration gradually diminished as they got older, and in particular the brightness of the yellow of the head and the depth of the black in front of the eyes slowly diminished, so that even when a week old the delicate glory of it had largely disappeared.

The young birds described above all died in the downy stage, but another season he raised one young bird, in which he describes (1905) the development of the plumage as follows:

I am now able to give a complete account of the development of Ross's goose, *Chen rossi.* This season my female laid three eggs, and, as in previous years, she had proved to be a bad mother, I took the eggs away and put them under a common hen. The period of incubation was 24 days this time, and the eggs were hatched on the 10th of July. All the three eggs were hatched, but unfortunately the hen in some way or other killed two of the chicks the same day that they were born. The third escaped this fate and was tenderly cared for by its foster mother. I have described in detail the color of the down in a previous letter (*Ibis,* 1903, p. 245), so that it will suffice to say that the chick was a fluffy object with gray down and a bright canary-yellow head.

The little bird grew very rapidly, and when 2 weeks old was about the size of a Japanese bantam hen. The bill was still black at this stage, with a pink tip (the nail), and the legs were greenish. When 3 weeks old the feathers began to appear on the shoulders, the flanks, the tail, and the wings. When 4 weeks old the bird was about the size of a small hen. The body was almost entirely feathered, but the head and neck were still in down. The legs were bluish and the bill was getting lighter in color. When 5 weeks old the whole body was feathered, and when 6 weeks old even the flight feathers were of their full length. The first plumage may be described as follows: General color, white. A brownish-gray spot on the occiput, which runs down along the back of the neck. The base of the neck and the mantle brownish gray, forming a crescent of that color, of which the points are turned forward on each side of the base of the neck. The smaller wing coverts are of the palest brownish gray, with a dark spot at the tip of each feather. The flanks are gray, the large flight feathers black. The first five secondaries have a dark spot in the center; those that follow are white, with only a very slight sprinkling of brownish; the three innermost have dark centers, and the white edges are finely spotted with gray. The tail is white, with only a suspicion of a grayish tint on the middle feathers. The legs are greenish gray with pink shining through. The bill is pinkish, the lores are blackish gray, which color extends over and behind the eyes. When 10 weeks old the bird began to molt, and the gray feathers of the juvenile dress were rapidly replaced by white ones. Also the large tail feathers were molted, the central rectrices being dropped first. The legs now began to turn pink in earnest, and the bill assumed its double coloration of a greenish base and a pink tip.

From the above account one would infer that the Ross goose acquires its fully adult plumage at its first prenuptial molt, when about 10 months old. As this molt probably involves the tail, all the contour plumage, and the wing coverts, it would leave only the secondaries and tertials to be replaced at a complete postnuptial molt the following summer.

Winter.—The principal winter home of the Ross goose within our limits seems to be in the central valleys of California, where it associates with the snow goose in the stubble fields and is often quite common. It seems to be tamer than other species of geese which visit that region; hence many are shot for the market and quite a number have found their way into scientific collections. Often the wing-tipped birds are kept in captivity and become easily domesticated; I have seen some interesting photographs illustrating the tameness of such captured birds.

DISTRIBUTION

Breeding range.—Entirely unknown, probably on some unexplored Arctic lands.

Winter range.—The main winter range is in California, in the interior valleys (Sacramento and San Joaquin) and nearer the coast farther south (Ventura and Orange Counties). A few may winter occasionally in neighboring States or in Mexico, but probably only casually.

Spring migration.—Northeastward to the Athabasca-Mackenzie region and beyond into Arctic regions. Early dates of arrival: Montana, Lewistown, March 14; Oregon, Camp Harney, April 12; Alaska, Wrangell, April 15; Mackenzie, Fort Anderson, May 25. Average dates of arrival in Montana are April 7 and 8 and of the departure April 24. Later dates of departure: California, Merced County, April 2; Montana, Teton County, May 8; Alberta, Athabasca River, June 4; Arctic coast, Kent Peninsula, June 2.

Fall migration.—A reversal of the spring route. Early dates of arrival: Great Slave Lake, September 1; Alberta, Buffalo Lake, September 6; Montana, Columbia Falls, October 10; California, Stockton, October 6; Utah, Bear River, October 22. Late dates of departure: Alberta, Buffalo Lake, October 10 and Athabasca Lake, October 18; Montana, Columbia Falls, October 28.

Casual records.—Outside of regular migration, it has occurred in Manitoba (Winnipeg, September 20, 1902), Louisiana (Little Vermilion Bay, February 23, 1910), Arizona (Fort Verde, October 24, 1887), Mexico (Bustillos Lake, Chihuahua) and British Columbia (Comox, January, 1894, and Lumby, May, 1920).

ANSER ALBIFRONS ALBIFRONS (Scopoli)

WHITE-FRONTED GOOSE

HABITS

Two forms of the white-fronted goose have long stood on our Check List unchallenged—a smaller European form (*albifrons*) and a larger American form (*gambelli*). The status of the European form as an American bird was based on a somewhat doubtful record for eastern Greenland, where it was supposed to occur only as a straggler. It was supposed to be entirely replaced in North

America by the larger form, where all the American geese of this species were called *gambelli*. This was not a very satisfactory arrangement, for the two forms were so much alike that it was very difficult to distinguish them, and some European writers refused to recognize them. Recently Messrs. Swarth and Bryant (1917) have demonstrated that there are probably two subspecies of white-fronted geese which spend the winter in California between which there is a striking difference in size, and there are some other differences. They have also shown that Hartlaub's name, *gambelli*, belongs to the larger and the rarer of the two and that all of the smaller white-fronted geese, which are far commoner, should be called *albifrons*.

The white-fronted goose has a wide distribution; in the eastern half of this continent it is everywhere rare, but on migrations and in its western winter range it is locally abundant; in much of its breeding range in the far northwest it is one of the commonest and best known of the geese.

Spring.—In writing of this species at Fort Klamath, Oregon, Dr. J. C. Merrill (1888) says:

Very common in April, the main flight occurring between the 20th and 30th, and many flocks stopping to feed in the grassy meadows bordering the marsh. The upper part of the valley is inclosed on the west and north by the main divide of the Cascade Mountains and on the east by a spur from the same range, all averaging a height of over 6,500 feet. On stormy days, if the wind was not blowing from the south, geese flying low up the valley had great difficulty in rising sufficiently to cross the abrupt divide, and most of them would return to the marsh and its vicinity to wait for a more favorable opportunity. At such times geese of this and the next species gathered by thousands and afforded great sport. The immense numbers of these birds that migrate through western Oregon can not be appreciated until one has seen their spring flight, which, I am informed, extends in width from the coast inland about 250 or 300 miles. About 50 of this species were seen at the marsh on May 23 and 20, on May 27 and June 3, after which none were observed; their remaining so late excited general remark among the settlers.

Nesting.—Of the arrival and nesting of this goose at St. Michael, Alaska, Dr. E. W. Nelson (1887) writes:

When the white-fronted goose first arrives in the north the lakes are but just beginning to open and the ground is still largely covered with snow. The last year's heath berries afford them sustenance, in common with most of the other wild fowl at this season. As the season advances they become more numerous and noisy. Their loud call notes and the cries of the males are heard everywhere.

The mating season is quickly ended, however, and on May 27, 1879, I found their eggs at the Yukon mouth. From this date on until the middle of June fresh eggs may be found, but very soon after this latter date the downy young begin to appear. These geese choose for a nesting site the grassy border of a small lakelet, a knoll grown over with moss and grass, or even a flat, sparingly covered with grass. Along the Yukon, Dall found them breeding gregariously, depositing their eggs in a hollow scooped out in the sand. At the Yukon mouth and St. Michael they were found breeding in scattered pairs over the flat country. Every one of the nests examined by me in these places had a slight lining of grass or moss, gathered by the parent, and upon this the first egg was laid; as the complement of eggs is approached the female always plucks down and feathers from her breast until the eggs rest in a soft warm bed, when incubation commences.

John Murdoch (1885) says that at Point Barrow—

the eggs are always laid in the black, muddy tundra, often on top of a slight knoll. The nest is lined with tundra moss and down. The number of eggs in a brood appears subject to considerable variation, as we found sets of 4, 6, and 7, all well advanced in incubation. The last-laid egg is generally in the middle of the nest and may be recognized by its white shell unless incubation is far advanced, the other eggs being stained and soiled by the birds coming on and off the nest.

Roderick MacFarlane (1891) writes:

A considerable number of nests of this "gray wavy" was discovered in the vicinity of fresh-water lakes in timber tracts, as well as along the Lower Anderson River to the sea. Some were taken on the Arctic coast, and several also on islands and islets in Franklin Bay. In all, about 100 nests were secured. The nest, which was always a mere shallow cavity in the ground, in every observed and reported instance had more or less of a lining of hay, feathers, and down, while the maximum number of eggs in no case exceeded 7. On the 5th of July, 1864, on our return trip from Franklin Bay we observed 30 molting ganders of this species on a small lake in the Barrens. Our party divided, and by loud shooting and throwing stones at them they were driven to land, where 27 of them were run down and captured. Their flesh proved excellent eating; it is seldom, indeed, that I have come across a gray wavy that was not in good condition in the far North.

A nest of the white-fronted goose in the writer's collection was taken near Point Barrow, Alaska, on June 27, 1916. It consists of a mass of pale gray and white down, thoroughly mixed with breast feathers of the goose, bits of dry, coarse grasses, lichens, mosses, dead leaves of the dwarf willow, and other rubbish found on the tundra; it is quite different in appearance from the nests of other geese. It contained four eggs advanced in incubation.

Eggs.—The white-fronted goose lays from 4 to 7 eggs, usually 5 or 6. These vary in shape from elliptical oval or elliptical ovate to elongate ovate. The color varies from "light buff" to creamy white or pale pinkish white. I have never seen any tinge of greenish in the eggs of this goose. The eggs often become very much stained with buffy or reddish brown stains, such as "cinnamon buff" or "ochraceous buff," which rub off or scratch off in irregular patches, exposing the original color; there are often several degrees of color in the same set of eggs, the freshest egg being quite clean and the oldest eggs decidedly dark colored. The measurements of 109 eggs, in various collections, average 79 by 52.5 millimeters; the eggs showing the four extremes measure **89.6** by 48, 82 by **58**, **70** by 52, and 77.2 by **46.7** millimeters.

Young.—The period of incubation does not seem to be positively known, but probably it is about 28 days. The male does not desert the female during the process, and both sexes help in caring for and protecting the young. Mr. Hersey, while collecting for me on the Yukon delta, encountered a family of these birds, about which he wrote in his notes for June 21, 1914, as follows:

On the edge of a little pond on the tundra about 5 miles back from the mouth of the river I found a pair of these geese and a brood of five young. The birds had been resting under a clump of dwarf willows, and on my approach the old birds came out into the open and attempted to lead the young away over the open tundra. The young, although not more than a day or two old, could run as fast as a man could travel over the rough ground. I had to remove my coat before I could overtake them. They did not scatter, but ran straight ahead, keeping close together, one of the parents running by their side and guiding them and the other flying along above them and not more than 3 feet above the ground. The young kept up a faint calling, and the old birds occasionally gave a low note of encouragement.

Doctor Nelson (1887) says:

The young are pretty little objects and are guarded with the greatest care by the parents, the male and female joining in conducting their young from place to place and in defending them from danger. The last of June, in 1877, I made an excursion to Stewart Island, near St. Michael, and while crossing a flat came across a pair of these geese lying prone upon the ground in a grassy spot, with necks stretched out in front and their young crouching prettily all about them. Very frequently during my visits to the haunts of these birds the parents were seen leading their young away through the grass, all crouching and trying to make themselves as inconspicuous as possible. At Kotzebue Sound, during the *Corwin's* visit in July, 1881, old and young were very common on the creeks and flats at the head of Eschscholtz Bay.

John Koren collected for me in northeastern Siberia a strange family party consisting of a female spectacled eider and two downy young white-fronted geese.

Plumages.—The downy young white-fronted goose is a beautiful creature, thickly covered with long, soft down and brightly colored. It somewhat resembles the young Canada goose, but the upper parts are a trifle duller in color, and the bill is brown with a light-colored nail, instead of all black as in the Canada. The colors of the upper parts, including the central crown, back, wings, rump, and flanks, vary from "buffy olive" to "ecru olive," darkest on the crown and rump and palest on the upper back, with a yellowish sheen; there is a faint loral and postocular stripe of olive; on the remainder of the head and neck the colors shade from "olive ocher" on the forehead, cheeks, and neck to "colonial buff" on the throat; the colors on the under parts shade from "mustard yellow" on the breast to "citron yellow" on the belly. The colors become duller and browner with increasing age; large downy young are "olive brown" above and grayish or "deep olive buff" below.

I have seen no specimens showing the development of the juvenal plumage. This is much like the adult plumage, except that the "white front" is lacking and there are no black spots on the under parts, which are mottled with whitish and gray; the upper parts are duller colored with lighter edgings; the tail feathers are more pointed and narrower and the wing coverts are narrower than in adults. This plumage is worn without much change during the first winter and spring; but more or less white appears in the "white front," and sometimes a few black spots appear in the breast. The tail is molted in the spring, and during the next summer a complete postnuptial molt produces a plumage which is practically adult.

Food.—Lucien M. Turner (1886) says, of the food of this species: "It inhabits the fresh-water lagoons, and is essentially a vegetarian. The only animal food found in their crops was aquatic larvae and insects. I am not aware that it eats shellfish at any season of the year. The young grass shoots found in the margins of the ponds form its principal food." Doctor Nelson (1887) says: "During August and September the geese and many other wild fowl in the north feed upon the abundant berries of that region and become very fat and tender." In the interior valleys of California, where it spends the winter, and on its migrations through agricultural districts, it feeds in the grain fields on fallen grain in the fall and on the tender shoots of growing grain in the spring. In some places where these geese were formerly abundant they did so much damage to the young crops that the farmers hired men to drive them away.

Audubon (1840) says:

In feeding they immerse their necks, like other species; but during continued rains they visit the cornfields and large savannahs. While in Kentucky they feed on the beech nuts and acorns that drop along the margins of their favorite ponds. In the fields they pick up the grains of maize left by the squirrels and racoons, and nibble the young blades of grass. In their gizzards I have never found fishes nor water lizards, but often broken shells of different kinds of snails.

Behavior.—The flight of the white-fronted goose is similar to that of the Canada goose, for which it might easily be mistaken at a distance. It flies in V-shaped flocks, led by an old gander, and often very high in the air. Its flight has been well described by Neltje Blanchan (1898) as follows:

A long clanging cackle, ***wah, wah, wah, wah,*** rapidly repeated, rings out of the late autumn sky, and looking up, we see a long, orderly line of laughing geese that have been feeding since daybreak in the stubble of harvested grain fields, heading a direct course for the open water of some lake. With heads thrust far forward, these flying projectiles go through space with enviable ease of motion. Because they are large and fly high, they appear to move slowly; whereas the truth is that all geese, when once fairly launched, fly rapidly, which becomes evident enough when they whiz by us at close range. It is only when rising against the wind and making a start that their flight is actually slow and difficult. When migrating, they often trail across the clouds like dots, so high do they go—sometimes a thousand feet or more, it is said—as if they spurned the earth. But as a matter of fact they spend a great part of their lives on land; far more than any of the ducks.

On reaching a point above the water when returning from the feeding grounds the long defile closes up into a mass. The geese now break ranks, and each for itself goes wheeling about, cackling constantly, as they sail on stiff, set wings; or, diving, tumbling, turning somersaults downward, and catching themselves before they strike the water, form an orderly array again, and fly silently, close along the surface quite a distance before finally settling down upon it softly to rest.

The peculiar laughing cry of this bird has given it the name of "laughing goose." Its cries are said to be loud and harsh, sounding like the syllable *wah* rapidly repeated; the note is easily imitated by striking the mouth with the hand while rapidly uttering the above sound.

The following, taken from Mr. MacFarlane's unpublished notes, illustrates the methods employed by the natives to capture these and other geese during the flightless molting season:

On 12th July we observed about 30 geese (*Anser gambelli*) on the edge of a small lake (in the water) in the Barren Grounds; they were all ganders, and molting. On our approach they went sailing (swimming) across the lake, which was about 2 miles in extent. Our party then divided—half taking one side, and half the other side of the lake—and by the time we reached the spot where the geese had quitted the water, they had all concealed themselves as well as the scant grass and low tangled willows in the vicinity would admit. After we discovered their whereabouts there was some sport and a lively chase after them, and we soon succeeded in securing 27 out of the 30—the remaining 3 having escaped beyond our reach, although followed for some distance into the water. They were all in good condition, in fact, gray wavies are always fat and excellent eating, while it is but seldom in spring and never in summer that a really good Canada goose is met with. The Indians inform me that when they observe a flock of swans or geese on a lake, during the molting season, they at once make a fire on the shore, and they state that this course on their part never fails to drive the geese, etc., on land, where most of them easily fall a prey to the hunter. If they were only wise enough to remain in the water at a proper distance they would be safe enough.

Fall.—The white-fronted goose does not start to migrate until driven south by cold weather. Doctor Nelson (1887) says:

All through September, old and young, which have been on the wing since August, gather in larger flocks, and as the sharp frosts toward the end of September warn them of approaching winter, commence moving south.

The marshes resound with their cries, and after some days of chattering, flying back and forth, and a general bustle, they suddenly start off in considerable flocks, and a few laggards which remain get away by the 7th or 8th of October.

There is a southward migration through the Mississippi Valley to the Gulf coast of Louisiana and Texas, but the main flight trends more to the southwestward to the principal winter home of the species in California. Doctor Coues (1874) writes:

The "speckle-bellies," as they are called in California, associate freely at all times with both the snow and Hutchins geese, and appear to have the same general habits, as well as to subsist upon the same kinds of food. Their flesh is equally good for the table. As is the case with other species, they are often hunted, in regions where they have become too wild to be otherwise successfully approached, by means of bullocks trained for the purpose. Though they may have learned to distrust the approach of a horse, and to make off with commendable discretion from what they have found to be a dangerous companion of that animal, they have not yet come to the same view with respect to horned cattle, and great numbers are slaughtered annually by taking advantage of their ignorance. The bullock is taught to feed quietly along toward a flock, the gunner meanwhile keeping himself screened from the birds' view by the body of the animal until within range. Though I have not myself witnessed this method of hunting, I should judge the gunners killed a great many geese, since they talk of its "raining geese" after a double discharge of the tremendous guns they are in the habit of using. Man's ingenuity overreaches any bird's sagacity, no doubt, yet the very fact that the geese, which would fly from a horse, do not yet fear an ox, argues for them powers of discrimination that command our admiration.

Winter.—Dr. L. C. Sanford (1903) has given us a good account of hunting white-fronted and snow geese in their winter haunts, as follows:

The large bodies of water that are found at rare intervals in northern Mexico are the resort through the winter of countless numbers of geese; not the Canada goose of the East and Middle West, but the snow goose and the white-fronted goose. In early October the hordes arrive, announcing their coming with discordant clamor. They choose as a resting place the shallow alkali waters, and as a feeding ground the neighboring corn stubble, if such there be. A short distance from Minaca is one of these lakes, some 20 miles in length. In the Mexican summer rains replenish the scanty water supply left over from the spring, and October finds it a paradise for waterfowl. Shut in by the rolling hills of the mesa, yellow with wavy grass, its blue surface reflects a bluer sky. All around, as far as the eye can reach, are herds of cattle, for some 6 miles away is a ranch; and at this spot one fall recently we stopped. Early in the morning a breakfast of tortillas and coffee was served, and before it was finished a Mexican boy appeared with the horses. Guns were slipped into the saddle cases. Our attendant found room for most of our ammunition in his saddlebag, and we started for the lake. It was a ride of about 6 miles over an open country, but the horses were fast, and in less than half an hour we looked down from a knoll on the sheet of water some 2 miles away. Along the farther shore was a bank of white, shining in the light of sunrise—a solid bank of snow geese. Scattered over its surface everywhere were flocks of ducks and geese, black masses of them. We hurried on, passing through herd after herd of cattle, which increased in numbers as the water was approached. A coyote stopped to take a fleeting glance from the top of a hill opposite, then disappeared. A jack rabbit scurried from in front. A familiar cry overhead caused us to look up. It came from a flock of sandhill cranes, far out of reach, which were sailing on toward their feeding ground in the stubble. We reached the edge of the lake, and hundreds of ducks rose as the horses neared them, mostly shovelers and teal, but mallard, widgeon, and pintail were all there. The geese were across the lake, thousands in one band. Every now and then a white line jointed the resting birds, and at the approach of a flock their discordant cries could be heard a mile away. How to get a shot seemed more or less of a problem, owing to lack of cover. Finally we noticed a few bunches of rushes extending well out into the lake, the only possible chance to hide. We waded out and took a position in the farthest clump. The Mexican led off the horses and started on a tour to the farther shore. It was a long way off, almost 4 miles, but there was plenty to watch. Every few minutes flocks of ducks would pass over us in range, but we let them go. Gulls circled around, crying at the unusual sight of two men with guns. We looked over at the geese. At times cattle seemed almost among them; yet the white assembly did not move, and we only heard them when a flock was about to alight to those on the ground. The horses were getting closer, and finally a part of the body started, to settle down a little farther on. But presently a tumultuous clamor, and the entire company was in motion. Line after line separated and led out into the lake. Some followed the opposite shore; an immense flock led toward our clump, and we crouched in the water. On they came, scarcely a hundred yards off. But geese are uncertain, even in Mexico, and for some reason best known to themselves they turned when just out of range and led toward the shore beyond us. In a few minutes they were reassembled and the immediate prospect of a shot gone. The Mexican, with his string of horses, continued down the opposite side, evidently after birds we could not see. Ducks were around us all the time, and flocks drifted by within easy range, unmolested. Before long we heard the familiar cry and looked to see a mass of white heading for the flock on the shore; our blind was right in their line, and they came on, low down over the water, nearer and nearer; finally, 50 or more seemed directly over us, so close we could see their red bills and legs. This was the chance; back to back we raked them, four barrels; 3 birds fell on one side, 2 on the other. The reports started all of the wild fowl in the country. In a few minutes part of the first flock came over us from the opposite direction, and 2 dropped. A flock of geese swung in range over the dead birds, and we killed 2 more. For an hour the shots were frequent, but the birds became wiser every minute and kept to the middle of the lake or else came over the blind out of range. We picked up 18, a dozen white, the rest white-fronted—all one Mexican could pack on a horse.

DISTRIBUTION

Breeding range.—Nearly circumpolar. On the barren grounds and Arctic coasts of North America, east at least as far as the Anderson River and Beechey Lake, in the district of Mackenzie, and west as far as the Yukon Valley (Fort Yukon, Lake Minchumina and the Yukon delta). On the west coast of Greenland, mainly between 66° and 72° N. In Iceland, Lapland, Nova Zembla, Kola, Kolguev, and along the Arctic coast of Siberia to Bering Straits. The only gap in the circumpolar breeding range seems to be between the district of Mackenzie and Greenland; this may prove to be the breeding range of the large tule goose, now called *gambelli*.

Winter range.—In North America, mainly in western United States and Mexico. East to the Mississippi Valley, rare east of that, and hardly more than casual on the Atlantic coast. South to the coast of Louisiana and Texas and to central western Mexico (Jalisco and Cape San Lucas). West to the Pacific coasts of Mexico and United States. North to southern British Columbia, southern Illinois, and perhaps Ohio. In the Eastern Hemisphere, south to Japan, China, India, the Caspian, Black, and Mediterranean Seas and northern Africa.

Spring migration.—Early dates of arrival: Manitoba, Aweme, April 5; Mackenzie, Fort Simpson, May 11, Fort McMurray, May 15, and Fort Anderson, May 16; Coronation Gulf, May 31. Alaska dates of arrival: Forrester Island, April 24, St. Michael, April 25, Kuskokwim River, April 29, Kowak River, May 10, Wainwright, May 27, Point Barrow, May 16. Late dates of departure: California, Stockton, May 2; Washington, Grays Harbor, May 5; Manitoba, Shoal Lake, May 26; Alaska, Kuiu Island, May 6; Oregon, Fort Klamath, June 3.

Fall migration.—Early dates of arrival: Manitoba, Aweme, September 7 (average October 2); Alberta, Red Deer River, September 12; Alaska, Sitka, September 29; British Columbia, Porcher Island, September 6; Washington, Tacoma, October 1; Colorado, Brighton, October 1; Utah, Bear River, October 10; California, Stockton, September 7. Late dates of departure: Mackenzie, Great Bear Lake, October 9; Alaska, St. Michael, October 8, and Craig, November 8; Manitoba, Aweme, November 1.

Casual records.—Has wandered east to Labrador (Hopedale, May, 1900), Massachusetts (Essex County, October 5, 1888, Plymouth, November 26, 1897, and Ipswich, August, 1907), North Carolina (Currituck Sound, January 1897), and Cuba. Said to have occurred in the Hawaiian Islands.

Egg dates.—Arctic Canada: Seventeen records, June 2 to July 10; nine records, June 24 to July 6. Alaska: Twelve records, May 23 to July 25; six records, June 5 to 24. Greenland: Five records June 4 to July 26.

ANSER ALBIFRONS GAMBELLI Hartlaub

TULE GOOSE

HABITS

The above scientific name has been in use for many years to designate the North American race of the white-fronted goose, which was understood to be slightly larger and to have a decidedly larger bill than the European bird. But the characters were not sufficiently well marked and not constant enough to fully satisfy all the European writers, many of whom questioned the validity of the race. Recently, however, Messrs. Swarth and Bryant (1917) have established the fact "that two well-defined subspecies of *Anser albifrons* occur in California during the winter months, instead of the single race heretofore recognized." Tule goose is the common name they have proposed for the newly discovered larger race, and they seem to have shown that Hartlaub's name, *gambelli*, refers to the larger rather than to the smaller and commoner race. In addition to a marked difference in size between the two races, comparable to that existing between the Canada and the Hutchins geese, "the larger birds are of a browner tint, and the smaller ones more gray. This is especially noticeable in the heads and necks. In some individuals of the larger race the head is extremely dark brown, almost black." Also the larger bird is said to have the "naked skin at edge of eyelid, yellow or orange," whereas in the smaller bird it is "grayish brown."

At present the tule goose is known only from its limited winter range in California. Its center of abundance seems to be in Butte and Sutter Basins in the Sacramento Valley, but there are persistent rumors among hunters that it occurs also in the Los Baños region in the San Joaquin Valley and at Maine Prairie in Solano County. I have examined and measured perhaps half a dozen large specimens of white-fronted geese in eastern collections, taken at widely scattered localities in the Mississippi Valley, Hudson Bay, and even on the Atlantic coast, that measured well up within the range of measurements of the tule goose, but they have not been compared with typical large birds from California, nor have they been examined by anyone who is familiar with the characteristics of the tule goose; they may be stragglers of the larger race or they may be only extra large individuals of the smaller race.

Nesting.—In my attempt to establish a breeding range for the larger race I find nothing but negative evidence. There are very few specimens of breeding birds in American collections, and all of those that I have seen are referable to the smaller race. I have collected the measurements of 109 eggs, taken in various localities in northeastern Siberia, Alaska, northern Canada, and Greenland, and they show no correlation of size with locality; the largest two sets came from Greenland and Siberia and the smallest two from Point Barrow and Greenland; average measurements from one locality are not materially different from those from another locality. Moreover, the average measurements of the 109 eggs are very close to the average measurements of 81 eggs of European birds; and the extremes in our series are inclusive. From the above we can infer only that the breeding range of the larger race has never been found and that none of its eggs are in existence. The breeding range of the species, *Anser albifrons*, is circumpolar, except for a decided gap between Greenland and the district of Mackenzie. Somewhere in this gap, or in the Arctic regions north of it, may be the breeding grounds of the big tule goose. An interesting parallel is seen in the case of the Ross goose, which also is found in a restricted winter range in California; the breeding grounds of both are entirely unknown; perhaps some day both may be found breeding somewhere in the vast unexplored regions of the Arctic Archipelago.

Behavior.—Messrs. Swarth and Bryant (1917) have referred to certain characteristic habits of the tule goose, as follows:

> It is said that the two kinds flock separately, for the most part; and that the larger race is never seen in such big flocks as is customary with the other, but is most frequently noted singly or in pairs. Also that while the smaller variety is a common frequenter of grain fields and uplands generally the larger one is preeminently a denizen of open water or of ponds and sloughs surrounded by tules and willows. The predilection of the latter species for such localities has given rise to the local names by which it is known, "tule goose," or "timber goose," as contrasted with the upland-frequenting "speckle-belly."

The notes of the tule goose are said to be "coarser and harsher" than those of the smaller bird.

DISTRIBUTION

Breeding range.—Unknown. It may fill in the gap in the known breeding range of *albifrons*, between the district of Mackenzie and Greenland, where much far northern land is unexplored.

Winter range.—Mainly in California (Sacramento Valley). It may also occur in other central valleys of California and perhaps rarely elsewhere.

ANSER FABALIS (Latham)

BEAN GOOSE

HABITS

Because this common European species has been recorded as a straggler in northern Greenland it has been included in our American list. It was named the bean goose because of its well-established habit of arriving in England with great regularity during bean-harvesting time in October; beans were very extensively cultivated in certain districts, to which these geese resorted in large numbers to feed on the remains of the harvest. John Cordeaux (1898) says on this subject:

We learn from Arthur Young's Agricultural Survey (1798) that the small country towns and villages in the middle-marsh and sea-marsh districts of Lincolnshire were surrounded by vast open fields, arable lands, cow and horse pastures, and furze; on strong land the rotation was fallows, wheat, beans, and again fallows. The area under beans in the low country was enormous, the wheat stubbles being plowed once, and the beans sown broadcast in the spring and never cleaned. These were harvested late in the autumn, usually got with much loss from the jaws of winter. These were the days of the gray goose, which our observant forefathers called the bean goose (*Anser segetum*), coming in great flocks in the later autumn to feast on the shelled beans in the open fields; and this continued till the change in cultivation and general inclosure banished them from their ancient haunts.

Most of the old wild-fowl shooters, who have long since gone over to the majority, used to assert that these autumn flights fed regularly in the bean fields as long as the old system of agriculture continued—a system in which quite one-third of the cultivated land was under that crop.

Nesting.—Rev. F. C. R. Jourdain writes to me that in Nova Zembla the nests of the bean goose are found on grassy tussocks on low ground, and that in Lapland it breeds in the partly wooded marshes where a few birch trees grow, nesting generally on the top of a grassy hummock.

Witherby's Handbook (1920) says that its main breeding haunts are in more wooded districts than those of most geese; that it nests "on islets in rivers or swamps, sheltered by rank vegetation and sometimes by willows or other bushes," and that the nest is "composed of down mixed with grass, moss, etc."

Eggs.—Mr. Jourdain says of the eggs:

They are large as compared with other geese, bulky in appearance, creamy white when first laid, but rapidly becoming nest stained with yellowish, which becomes more pronounced as incubation advances.

The set usually consists of 4 or 5 eggs, occasionally 6. The measurements of 51 eggs, as given in Witherby's Handbook (1920), average 84.2 by 55.6 millimeters; the eggs showing the four extremes measure **91** by 57.2, 84 by **59**, and **74.5** by **53.3** millimeters.

Food.—Mr. Cordeaux (1898) says:

The bean goose is very partial to all sorts of grain, and, in this respect, differs from the gray lag, whose chief food is grass. A local name is "corn-goose," in France "harvest-goose," and in Transylvania it is known as the "growing-grain goose"; it will, however, eat grass and clover as readily as its congeners when the stubbles are exhausted.

Behavior.—Seebohm (1901) describes the flocking habits of old and young birds in Siberia, as follows:

I then skirted the margin of a long, narrow inlet, exactly like the dried-up bed of a river, running some miles into the tundra, bending round almost behind the inland sea. I had not gone more than a mile when I heard the cackle of geese; a bend of the river bed gave me an opportunity of stalking them, and when I came within sight I beheld an extraordinary and interesting scene. At least 100 old geese, and quite as many young ones, perhaps even twice or thrice that number, were marching like a regiment of soldiers. The vanguard, consisting of old birds, was halfway across the stream, the rear, composed principally of goslings, was running down the steep bank toward the water's edge as fast as their young legs could carry them. Both banks of the river, where the geese had doubtless been feeding, were strewn with feathers, and in five minutes I picked up a handful of quills. The flock was evidently migrating to the interior of the tundra, molting as it went along.

Yarrell (1871) quotes Sir Ralph Payne-Gallwey as saying that in Ireland—

It is by far the commonest species, and may be seen in enormous "gaggles" for six months of every year. It is essentially an inland feeder on bogs and meadows, but will fly to the mud banks and slob of the tide at dusk to pass the night. These geese frequent every bog and marsh in Ireland which afford food and security from molestation. They are always found inland in large numbers save in frost, when they fly down to the meadows and soft green reclaimed lands that lie near the tide. A small proportion will, in the wildest weather, frequent the mud banks to feed and rest. They usually quit their inland haunts at dusk, disliking to remain on land by night when dogs, men, or cattle may disturb them, and accordingly fly to the estuaries to rest and feed. At first dawn they again wing inland and pass the day in open, unapproachable ground.

DISTRIBUTION

Breeding range.—Northern Palaearctic region. East in northern Siberia to the Taimyr Peninsula. South to about 64° N. in Siberia, Russia, Kola, Finland, and Scandinavia. Also on Kolguev and Nova Zembla. Replaced in eastern Siberia by closely allied forms.

Winter range.—Europe and western Asia. South to northern China, casually northern India, Persia, the Mediterranean Sea, and rarely to northern Africa. West to Great Britain.

Casual record.—A specimen in the Zoological Museum of Copenhagen is said to have come from northern Greenland.

Egg dates.—Northern Europe: Eight records, April 29 to June 20. Novaya Zemlia: One record, July 9.

ANSER BRACHYRHYNCHUS Baillon

PINK-FOOTED GOOSE

HABITS

An accidental occurrence of this Old World species in eastern Greenland had long been the slim excuse we had for including the pink-footed goose in the list of American birds, until recently, September 25, 1924, one was taken in Essex County, Massachusetts.

This and the bean goose resemble each other so closely in general appearance and habits that much confusion has arisen as to the distribution and comparative abundance of the two species. It has even been suggested that they may be only subspecies or varieties, as the characters on which the description of the pink-footed goose was based are not very constant. It seems to be now conceded, however, that they are distinct species and that the latter is now the commonest of the gray geese in Great Britain.

Nesting.—Rev. F. C. R. Jourdain has kindly sent me the following notes on his experiences with the pink-footed goose in Spitsbergen:

Although it is probable that this species breeds in Iceland, there is as yet no definite proof, and the only certainly known breeding place is Spitsbergen, so that perhaps the following notes may have some interest.

The pink-footed goose is still a fairly common bird along the west coast of Spitsbergen. Here it has only two enemies—man and the Arctic fox. In former years the Arctic fox was the more dangerous foe, and the habits of the goose have been gradually evolved to contend with this wily little enemy, while the only men to be feared were the few trappers and sealers who robbed the nests occasionally in the spring and shot the molting birds in the summer. Now the foxes have been greatly thinned down, but the little sealing sloops, no longer dependent on their sails, but filled with noisy little oil engines, penetrate everywhere, so that the birds are badly harried. Still the eggs of the geese can not be as readily collected as those of the eider and in consequence have less value, and to discover the isolated nests on the tundra and shingle banks in marketable numbers would be a hopeless task. But during the season of molt the goose has only its speed of foot to trust to, and no doubt large numbers are killed from time to time.

While the brent have found security from the foxes by breeding on the islets round the coast and the barnacle has attained the same end by nesting on steep cliffs, the pink foot, which is a much larger and stronger bird, has to a certain extent managed to hold its own on the open tundra, though it is much more usual to find the nest in somewhat similar sites to those used by the barnacle. I think it is quite possible for a couple of pink-footed geese to keep a prowling fox at bay, though a single bird might have a very unpleasant time, and probably a fair proportion of nests come to grief in this way every year. Like the other species which breed here, the male pink foot is an excellent father and stands by his mate during the incubation period. The first nest we met with was about 10 miles up a wide valley running into Ice Fjord. Here on a slightly raised mossy ridge, which gave a wide view over the snow-sodden flats, we put up a pair of pink-footed geese from the nest, which contained 2 eggs, highly incubated on June 26. This was a curiously small clutch, and yet there is no reason to suppose that the birds had been already robbed. Subsequently we found another nest with 2 incubated eggs on a grassy cliff, and this, too, was in a locality which had not been disturbed. Koenig, who examined a very large number of nests of this species, only met with full sets of 2 on two occasions and considers 4 as the normal number, while sets of 3 and 5 occur commonly. He also met with an instance of 7 eggs in one nest and believed them to be the product of one female, but in another case where 9 eggs were found, the stages of incubation proved that two females had laid together. Curiously enough we never met with more than 4 eggs or young, but the number of eggs taken in 1921 was not large.

Like the brent and barnacle the pink-footed shows a decided tendency to sociability in the breeding season, though many nests are also quite isolated.

Along the west side of Prince Charles Foreland, where the mountains rise steeply from the sea and are almost perpetually wrapped in drifting mists and fogs, there are low grassy slopes and bluffs which lie at the foot of the main range. Here many pink-footed geese breed in hollows on the green ledges, sometimes two or three pairs nesting not far from one another, and nesting hollows which have been used in previous years are quite plentiful; sometimes four or five being visible close together. Most of these nests can be reached by a scrambling climb from below, but there may be a sheer drop of 15 or 20 feet below the nest. In one case we saw four newly hatched goslings in a nest, and the most enterprising of them scrambled over the side of the nest for about a foot. The others, however, did not follow, and it was only by repeated efforts that the youngster was able, after several attempts, to regain its place in the nest. By next day every one of the young birds had disappeared completely. On a subsequent occasion we saw an old goose at the foot of a cliff, attending closely to something on the ground below. Little by little it descended the slope, guiding and helping with its bill something which looked like a downy young bird. One of our party coming from another direction shot the old goose, and going to the spot discovered the young which was evidently being conducted to the shore. The empty nest was on the cliff exactly above where we had first seen the bird, and I have no doubt that it had fallen from the top without suffering any injury, for we took the bird back with us to the ship and attempted to rear it. On another occasion we put up two geese from a vast expanse of shingle at the mouth of one of the valleys, and going to the spot found three addled eggs and a newly hatched gosling. The little bird had already left the nest and made off for us with wonderful speed over the shingle. It was not the least use replacing it, for immediately it was released it set off again in pursuit of the nearest of our party. On returning later with cameras, the young bird was not to be found.

One nest was in a curious position, quite close to the sea, on some sloping ground, sheltered by big bowlders which had fallen from the rocks above.

By July 7 many of these geese had already shed their primaries and were unable to fly in North Spitsbergen. Great numbers of their feathers lay strewn along the shore, while the birds made off at top speed directly they sighted us.

Eggs.—Witherby's Handbook (1920) describes the eggs of the pink-footed goose as "dull whitish." The measurements of 292 eggs average 78.2 by 52.3 millimeters; the eggs showing the four extremes measure **88** by 52.6, 82.6 by **56.7**, **69** by 51, and 77 by **47.4** millimeters, according to the same authority.

Food.—John Cordeaux (1898) says:

Geese, on reaching their feeding grounds, whirl in wide circles over the selected spot and, when satisfied that all is safe, sweep suddenly downwards with considerable velocity, and commence feeding at once on alighting.

When, through the depth of snow on the high wolds, food is not to be got, geese entirely change their habits, loafing about on the coast and sand banks during the day, and in the evening flying and dropping anywhere in the low country where they can get green food; the snow seldom lies long in coast districts, and there are always places which the winds have left bare, and the ground is more or less uncovered. I have often seen their paddlings and droppings in pasture, corn, and turnip fields, near the coast. If the neighborhood is quiet and retired, they come inland just as readily in the daytime as at night

Geese feed very greedily anywhere at the break up of a snowstorm, and they are then least difficult to approach, being too much engrossed in eating to heed slight indications of disturbance or interruption. The pink-footed geese, when associated with other species on a feeding ground, keep apart and are not inclined to be sociable. In the day they are visible on a hillside at a very considerable distance and, if a yellow stubble, look like a blue cloud on the land. They are also very conspicuous objects on the sands of the coast, lining the tide edge in long extended line, like a regiment on parade.

In the dusk of evening or at night geese are not so wide-awake as in the day, or they do not see so well, and I have sometimes walked into a flock to our mutual astonishment.

Behavior.—The same writer says of the habits of this species:

The habits of the pink-footed goose so closely resemble those of the bean goose that much which has been written of the one will hold good of the other. They arrive in the Humber district the last week in September and early in October; the earliest dates in my notebook are September 26, October 3, October 5 (twice), October 10. Mr. Haigh has known them appear as early as August 26, in 1893, in excessively hot weather. During the day they haunt the stubbles and clover fields on the wolds and open districts, rising about the same hour in the evening and wend their way, in the long extended order, to the islands and sand banks in the Humber, to return as punctually to their feeding grounds at the break of day. They are the wildest and most unapproachable of all the geese.

Within the recollection of certainly three generations and probably since the inclosure of the wolds, if not before, flocks of wild geese, coming up from the coast, have been in the habit of passing over the town of Louth in the early morning on their way to their feeding grounds on the high wolds. The large barley walks are the places which are most frequented, not so much, as I have found by an examination of the stomach, for scattered grain as young white clover and trefoil plants, of which they are immoderately fond. Considering the persistency with which geese day by day resort to the same locality it is surprising so few are shot. The fields on the wolds are very extensive, and geese keep near the center; on coming in from the coast they fly high, and it is only in stormy weather that their flight is low enough for a shot from a heavy gun to do execution, fired from the vantage ground of a solitary barn, shed, or stack on a hilltop, where at the same time the shooter remains concealed till the skein of geese are well above him.

Mr. Howard Saunders says: "The voice of the pink-footed goose differs from that of the bean goose in being sharper in tone, and the note is also repeated more rapidly." It is extremely difficult to express the note or the difference between the calls of birds on paper. I can, however, testify from experience that there is a very distinct difference between the call note of these two species.

DISTRIBUTION

Breeding range.—Breeds in Spitsbergen, probably in Franz-Josef Land and possibly in Iceland.

Winter range.—Northern Europe, Scandinavia, Holland, Belgium, Great Britain, Iceland, France, Germany, and northern Russia.

Casual records.—Accidental in eastern Greenland and Massachusetts.

Egg dates.—Spitsbergen: Six records, June 16 to 27.

BRANTA CANADENSIS CANADENSIS (Linnaeus)

CANADA GOOSE

HABITS

The common wild goose is the most widely distributed and the most generally well known of any of our wild fowl. From the Atlantic to the Pacific and from the Gulf of Mexico nearly to the Arctic coast it may be seen at some season of the year, and when once seen its grandeur creates an impression on the mind which even the casual observer never forgets. As the clarion notes float downward on the still night air, who can resist the temptation to rush out of doors and peer into the darkness for a possible glimpse at the passing flock, as the shadowy forms glide over our roofs on their long journey? Or, even in daylight, what man so busy that he will not pause and look upward at the serried ranks of our grandest wild fowl, as their well-known honking notes announce their coming and their going, he knows not whence or whither? It is an impressive sight well worthy of his gaze; perhaps he will even stop to count the birds in the two long converging lines; he is sure to tell his friends about it, and perhaps it will even be published in the local paper, as a harbinger of spring or a foreboding of winter. Certainly the Canada goose commands respect.

Spring.—The Canada goose is one of the earliest of the water birds to migrate in the spring. Those which have wintered farthest south are the first to feel the migratory impulse, and they start about a month earlier than those which have wintered at or above the frost line, moving slowly at first but with a gradually increasing rate of speed. Prof. Wells W. Cooke (1906) has shown, from his mass of accumulated records, that beginning with an average rate of 9 miles a day, between the lowest degrees of latitude, the speed is gradually increased through successive stages to an average rate of 30 miles a day during the last part of the journey. Following, as it does, close upon the heels of retreating ice and snow, the migration of these geese may well be regarded as a harbinger of spring; for the same reason it is quite variable from year to year and quite dependent on weather conditions.

The first signs of approaching spring come early in the far south, with the lengthening of the days and the increasing warmth of the sun; the wild geese are the first to appreciate these signs and the first to feel the restless impulse to be gone; they congregate in flocks and show their uneasiness by their constant gabbling and honking, as if talking over plans for their journey, with much preening and oiling of feathers in the way of preparation; at length a flock or two may be seen mounting into the air and starting off northward, headed by the older and stronger birds, the veterans of many a similar trip; flock after flock joins the procession, until the last have gone, leaving their winter homes deserted and still. The old ganders know the way and lead their trustful flocks by the straightest and safest route; high in the air, with the earth spread out below them like a map, they follow no coast line, no mountain chain, and no river valley; but directly onward over hill and valley, river and lake, forest and plain, city, town, and country, their course points straight to their summer homes. Flying by night or by day, as circumstances require, they stop only when necessary to rest or feed, and then only in such places as their experienced leaders know to be safe. A thick fog may bewilder them and lead them to disaster or a heavy snowstorm may make them turn back, but soon they are on their way again, and ultimately they reach their breeding grounds in safety.

Courtship.—The older geese are paired for life, and many of the younger birds, which are mating for the first time, conduct their courtship and perhaps select their mates before they start on their spring migration. Audubon (1840) gives a graphic account of the courtship of the Canada goose, as follows:

It is extremely amusing to witness the courtship of the Canada goose in all its stages; and let me assure you, reader, that although a gander does not strut before his beloved with the pomposity of a turkey, or the grace of a dove, his ways are quite as agreeable to the female of his choice. I can imagine before me one who has just accomplished the defeat of another male after a struggle of half an hour or more. He advances gallantly toward the object of contention, his head scarcely raised an inch from the ground, his bill open to its full stretch, his fleshy tongue elevated, his eyes darting fiery glances, and as he moves he hisses loudly, while the emotion which he experiences causes his quills to shake and his feathers to rustle. Now he is close to her who in his eyes is all loveliness; his neck bending gracefully in all directions, passes all round her, and occasionally touches her body; and as she congratulates him on his victory, and acknowledges his affection, they move their necks in a hundred curious ways. At this moment fierce jealousy urges the defeated gander to renew his efforts to obtain his love; he advances apace, his eye glowing with the fire of rage; he shakes his broad wings, ruffles up his whole plumage, and as he rushes on the foe hisses with the intensity of anger. The whole flock seems to stand amazed, and opening up a space, the birds gather round to view the combat. The bold bird who has been caressing his mate, scarcely deigns to take notice of his foe, but seems to send a scornful glance toward him. He of the mortified feelings, however, raises his body, half opens his sinewy wings, and with a powerful blow, sends forth his defiance. The affront can not be borne in the presence of so large a company, nor indeed is there much disposition to bear it in any circumstances; the blow is returned with vigor, the aggressor reels for a moment, but he soon recovers, and now the combat rages. Were the weapons more deadly, feats of chivalry would now be performed; as it is, thrust and blow succeed each other like the strokes of hammers driven by sturdy forgers. But now, the mated gander has caught hold of his antagonist's head with his bill; no bull-dog can cling faster to his victim; he squeezes him with all the energy of rage, lashes him with his powerful wings, and at length drives him away, spreads out his pinions, runs with joy to his mate, and fills the air with cries of exultation.

Nesting.—Reaching their breeding grounds early in the season and being in most cases already paired, these geese are naturally among the earliest breeders; their eggs are usually hatched and the nests deserted before many of the other wild fowl have even laid their eggs, the dates varying of course with the latitude. When I visited North Dakota in 1901 there were still quite a number of Canada geese breeding there; probably many of them have since been driven farther west or north, as they love solitude and retirement during the nesting season. We found them nesting on the islands in the lakes and in the marshy portions of the sloughs, building quite different nests in the two locations. On May 31 we found a nest on an island in Stump Lake, which had evidently been deserted for some time; the island was also occupied by nesting colonies of double-crested cormorants and ring-billed gulls and by a few breeding ducks; the goose nest was merely a depression in the bare ground among some scattered large stones lined with a few sticks and straws and a quantity of down. In a large slough in Nelson County we found, on June 2, a deserted nest containing 3 addled eggs, the broken shells of those that had hatched being scattered about the nest. It was in a shallow portion of the slough where the dead flags had been beaten down flat for a space 50 feet square. The nest was a bulky mass of dead flags, 3 feet in diameter and but slightly hollowed in the center. Within a few yards of this, and of a similar nest found on June 10, was an occupied redhead's nest; the proximity of these two ducks' nests to those of the geese may have been merely accidental, but the possibility is suggested that they may have been so placed to gain the protection of the larger birds. This suggestion was strengthened when I saw a skunk foraging in the vicinity; undoubtedly these animals find an abundant food supply in the numerous nests of ducks and coots in these sloughs.

Somewhat similar nests were found by our party in Saskatchewan, including two beautiful nests on an island in Crane Lake, found on June 2, 1905. The largest of these was in an open grassy place on the island, about 25 yards from the open shore; it consisted of a great mass of soft down, "drab gray" in color, measuring 16 inches in outside diameter, 7 inches inside, and 4 inches in depth; it was very conspicuous and contained 6 eggs. As I approached it and when about a hundred yards from it, the goose walked deliberately from the nest to the shore and began honking; her mate, away off on the lake answered her and she flew out to join him. Both of these nests had been robbed earlier in the season and the birds had laid second sets.

According to Milton S. Ray (1912) the Canada goose nests quite commonly at Lake Tahoe in California; he found a number of nests there in 1910 and 1911. The nesting habits in this region are not very different from what we noted in northwest Canada. Referring to the nests found in 1910, Mr. Ray writes:

Anxious to learn something of their nesting habits, and hoping I might be in time to find a nest or so, May 23 found me rowing up the fresh-water sloughs of the marsh, unmindful of the numerous terns, blackbirds, and other swamp denizens, in my quest for a prospective home of the goose. Nor was I long without reward, for when about 100 feet from a little island that boasted of a few lodge-pole pine saplings and one willow, a goose rose from her nest, took a short run, and rising with heavy flight and loud cries, flew out to open water, where she was joined by her mate. The cries of the pair echoed so loudly over the marsh that it seemed the whole region must be awakened. Landing on the island I found on the ground, at the edge of the willow, a large built-up nest with 7 almost fresh eggs. The nest was composed wholly of dry marsh grasses and down, and measured 22 inches over all, while the cavity was 11 inches across and 3 inches deep.

After a row of several miles I noticed a gander in the offing, whose swimming in circles and loud honking gave assurance that the nesting precincts of another pair had been invaded. A heavily timbered island, now close at hand, seemed the most probably nesting place. This isle was so swampy that most of the growth had been killed, and fallen trees, other impedimenta, and the icy water, made progress difficult. I had advanced but a short distance, however, when a goose flushed from her nest at the foot of a dead tree. This nest was very similar to the first one found, and, like it, also held 7 eggs, but these were considerably further along in incubation. On the homeward journey, while returning through the marsh by a different channel, I beheld the snake-like head of a goose above the tall grass (for the spring had been unusually early) on a level tract some distance away. Approaching nearer, the bird took flight, and on reaching the spot I found my third nest. As it contained 5 eggs all on the point of hatching, I lost no time in allowing the parent to return.

Of his experiences the following year, he says:

I found the goose colony to consist of but a single nest, placed on the bare rock at the foot of a giant Jeffrey pine near the water's edge. It was made entirely of pine needles, with the usual down lining, and held an addled egg, while numerous shells lay strewn about. The parents were noticed about half a mile down the bay. Two days later at Rowlands Marsh I located another goose nest with the small compliment of 2 eggs, 1 infertile and 1 from which the chick was just emerging. The nest was placed against a fallen log, and besides the lining of down was composed entirely of chips of pine bark, a quantity of which lay near. From the variety of material used in the composition of the nests found, it seems evident that the birds have little or no preference for any particular substance, but use that most easily available.

A long day's work at the marsh on June 9 revealed three more nests. The first of these, one with 6 eggs, well incubated, was the most perfectly built nest of the goose that I have ever seen, being constructed with all the care that most of the smaller birds exercise. It was made principally of dry marsh grasses. The second nest held a set of 5 eggs, and was placed by a small willow on a little mound of earth rising in a tule patch in a secluded portion of the swamp. Dry tules entered largely into its composition. In this instance the bird did not rise until we were within 25 feet, although they usually flushed at a distance varying from 40 to 100 feet.

In the Rocky Mountain regions of Colorado and Montana the Canada goose has been known to build its nest, sometimes for successive seasons, on rocky ledges or cliffs at some distance from any water or even at a considerable height. In the northwestern portions of the country it frequently nests in trees, using the old nests of ospreys, hawks, or other large birds; it apparently does not build any such nest for itself, but sometimes repairs the nest by bringing in twigs and lining it with down. John Fannin (1894) says that in the Okanogan district of British Columbia, "Canada geese are particularly noted for nesting in trees, and as these valleys are subject to sudden inundation during early spring, this fact may have something to do with it." He also relates the following interesting incident:

Mr. Charles deB. Green, who spends a good deal of his spare time in making collections for the Museum, writes me from Kettle River, Okanogan district, British Columbia, to the effect that while climbing to an osprey's nest he was surprised to find his actions resented by not only the ospreys but also by a pair of Canada geese (*Branta canadensis*), the latter birds making quite a fuss all the time Mr. Green was in the tree. On reaching the nest he was still further surprised to find 2 osprey eggs and 3 of the Canada goose. He took the 2 osprey's eggs and 2 of the geese eggs. This was on the 1st of May. On the 12th of May he returned and found the osprey setting on the goose egg; the geese were nowhere in sight. Mr. Green took the remaining egg and sent the lot to the Museum.

A. D. Henderson has sent me the following notes on the nesting habits of the Canada goose, in the Peace River region of northern Alberta, as follows:

The geese breed on the small gravelly islands in the Battle River and its two tributaries, known at that time as the Second and Third Battle Rivers. Another favorite breeding place is in old beaver dams, where they nest on the old sunken beaver houses, which in course of time have flattened down into small grass-covered islets. Even inhabited beaver houses are used as nesting sites, as my half-breed hunting partner, on one of our trips, took 5 eggs from a nest on a large beaver house in an old river bed of the Third Battle, which we repeatedly saw entered and left by a family of beaver, showing that the geese and beaver live together in unity.

On May 18 I found a nest containing 7 eggs on a low grassy islet, probably a very old beaver house, in the same flooded beaver meadow. The nest was made of grass lined with finer grasses and feathers. The sitting bird permitted a near approach, with her head and neck stretched out straight in front of her and lying flat along the ground, watching my approach. This appears to be the usual behavior when the nest is approached during incubation. We saw two other nests on this day, one on a small grassy islet in the same beaver meadow, containing 3 eggs, and another on an island in the Third Battle with 6 eggs.

Eggs.—The Canada goose lays from 4 to 10 eggs, usually 5 or 6. They vary in shape from ovate to elliptical ovate, with a tendency in some specimens toward fusiform. The shell is smooth or only slightly rough, but with no gloss. The color is creamy white or dull, dirty white at first, becoming much nest stained and sometimes variegated or nearly covered with "cream buff." The measurements of 84 eggs, in various collections average 85.7 by 58.2 millimeters; the eggs showing the four extremes measure **99.5** by 56, 87.6 by **63.6**, **79** by 56.5, and 86.5 by **53.5** millimeters.

Young.—The period of incubation varies from 28 to 30 days; probably the former is the usual time under favorable circumstances. The gander never sits on the nest, but while the goose is incubating he is constantly in attendance, except when obliged to leave in search of food. He is a staunch defender of the home and is no mean antagonist. Audubon (1840) relates the following:

It is during the breeding season that the gander displays his courage and strength to the greatest advantage. I knew one that appeared larger than usual, and of which all the lower parts were of a rich cream color. It returned three years in succession to a large pond a few miles from the mouth of Green River, in Kentucky, and whenever I visited the nest it seemed to look upon me with utter contempt. It would stand in a stately attitude until I reached within a few yards of the nest, when suddenly lowering its head and shaking it as if it were dislocated from the neck, it would open its wings and launch into the air, flying directly at me. So daring was this fine fellow that in two instances he struck me a blow with one of his wings on the right arm, which for an instant I thought was broken. I observed that immediately after such an effort to defend his nest and mate he would run swiftly toward them, pass his head and neck several times over and around the female, and again assume his attitude of defiance.

The same gifted author writes regarding the care of the young as follows:

The lisping sounds of their offspring are heard through the shell; their little bills have formed a breach in the inclosing walls; full of life and bedecked with beauty they come forth, with tottering steps and downy covering. Toward the water they now follow their careful parent; they reach the border of the stream; their mother already floats on the loved element; one after another launches forth and now the flock glides gently along. What a beautiful sight. Close by the grassy margin the mother slowly leads her innocent younglings; to one she shows the seed of the floating grass, to another points out the crawling slug. Her careful eye watches the cruel turtle, the garfish, and the pike that are lurking for their prey, and, with head inclined, she glances upward to the eagle or the gull that are hovering over the water in search of food. A ferocious bird dashes at her young ones; she instantly plunges beneath the surface, and in the twinkling of an eye her brood disappear after her; now they are among the thick rushes, with nothing above water but their little bills. The mother is marching toward the land, having lisped to her brood in accents so gentle that none but they and her mate can understand their import, and all are safely lodged under cover until the disappointed eagle or gull bears away.

More than six weeks have now elapsed. The down of the goslings, which at first was soft and tufty, has become coarse and hairlike. Their wings are edged with quills and their bodies bristled with feathers. They have increased in size and, living in the midst of abundance, they have become fat, so that on shore they make their way with difficulty, and as they are yet unable to fly, the greatest care is required to save them from their numerous enemies. They grow apace, and now the burning days of August are over. They are able to fly with ease from one shore to another, and as each successive night the hoarfrosts cover the country and the streams are closed over by the ice, the family joins that in their neighborhood, which is also joined by others. At length they spy the advance of a snowstorm, when the ganders with one accord sound the order for their departure.

Samuel N. Rhoads (1895) published the following interesting note, based on the observations of H. B. Young in Tennessee:

At Reelfoot Lake the goose nearly always builds in the top of a blasted tree over the water, sometimes nesting as high as 50 feet or even higher. When the young are hatched the gander soon gets notice of it and swims around the foot of the tree uttering loud cries. On a signal from mother goose he redoubles his outcries and, describing a large circle immediately beneath the nest, beats the water with his wings, dives, paddles, and slashes about with the greatest fury, making such a terrible noise and commotion that he can be heard for several miles. This effectually drives away from that spot every catfish, spoonbill, loggerhead, hellbender, moccasin, water snake, eagle, mink, and otter that might take a fancy to young goslings, and into the midst of the commotion mother goose, by a few deft thrusts of her bill, spills the whole nestful. But a few seconds elapse ere the reunited family are noiselessly paddling for the shores of some secluded cove with nothing to mark the scene of their exploits but a few feathers and upturned water plants and above them the huge white cypress with its deserted nest.

While the family party is moving about on the water the gander usually leads the procession, the goslings following, and the goose acting as rear guard. The old birds sometimes lead their young for long distances over large bodies of water. While cruising on Lake Winnipegosis on June 18, 1913, we came upon a family party fully 5 miles from shore and evidently swimming across the lake. The two old birds when hard pressed finally took wing and flew away, leaving the three half-grown young to their fate. The young were still completely covered with down, and their wings were not at all developed, although their bodies were as large as mallards. They could swim quite fast on the surface, could dive well, and could swim for a long distance under water. They were surprisingly active in eluding capture, and when hard pressed they swam partly submerged, with their necks below the surface and their heads barely above it, in a sort of hiding pose.

P. A. Taverner (1922) describes an interesting pose assumed by a family party on Cypress Lake, Saskatchewan. When pursued by a motor boat—

they put on more speed and arranged themselves in a long single file, one parent leading, the other bringing up the rear, swimming low, and both with their long necks outstretched and laid down flat on the water, making themselves as inconspicuous as possible. The young, coaxed from ahead and urged from behind, paddled along vigorously between, one close behind the other. From our low and distant point of view the effect was interesting. They looked like a floating stick. Certainly they would not impress the casual eye as a family of Canada geese, and if we had not first seen them in a more characteristic pose they would undoubtedly have been passed without recognition.

Dr. Alexander Wetmore tells me that, in the Bear River marshes in Utah where these geese breed, both old and young birds resort during the summer to the seclusion of the lower marshes. Here he found numerous places where the thick growth of bullrushes had been beaten down to form roosting places for family parties, well littered with cast-off feathers and other signs of regular occupancy. Here they live in peace and safety while the young are attaining their growth and their parents are molting. Before the shooting season begins they gather into larger flocks, now strong of wing and ready for their fall wanderings.

James P. Howley (1884) gives the following account of the behavior of these geese in Newfoundland:

During the breeding season they molt the primary wing and tail feathers, and are consequently unable to fly in the months of June, July, and the early part of August. They keep very close during this molting season and are rarely seen by day; yet I have frequently come across them at such times in the far interior and on many occasions have caught them alive. When surprised on some lonely lake or river side they betake themselves at once to the land and run very swiftly into the bush or tall grass to hide. But they appear somewhat stupid, and if they can succeed in getting their heads out of sight under a stone or stump imagine they are quite safe from observation. When overtaken in the water and hard pressed they will dive readily, remaining a considerable time beneath, swimming or running on the bottom very fast. About the 15th of August the old birds and most of the young ones are capable of flight, and from thence to the 1st of September they rapidly gain strength of wing. Soon after this they betake themselves to the seaside, congregating in large flocks in the shallow estuaries or deep fiords, to feed during the nighttime, but are off again to the barrens at earliest dawn, where they are generally to be found in daytime. Here they feed on the wild berries, of which the common blueberry, partridge berry, marsh berry, and a small blackberry (*Empetrum nigrum*) afford them an abundant supply. They are exceedingly wary at this season, and there is no approaching them at all on the barrens.

Mr. Henderson has given the following observation regarding the young:

On June 4, while walking up the river bank looking for bear, we met a pair of geese and four goslings on shore and got within 20 yards before they moved. The old birds made a great fuss and flew down to the foot of a rapid and waited on the still water about 60 yards below. The goslings took to the water, which was tumbling and boiling over the stones; swimming and diving, they went down the rapid, under water most of the time, and joined their fond parents below.

On the 28th, while walking up the gravel banks of the Third Battle, hunting bear, I came on a pair of geese with six goslings, also three other geese about 100 yards upstream from them. The three geese flew on my approach, and the female took her brood across the stream to about 30 yards distant. Her mate went upstream, flopping along the water pretending to be crippled. He would allow me to approach to about 40 yards and then flap along the water again for a few yards and wait for me again. He repeated this performance several times, until he thought he had enticed me far enough around the next bend, when he had a marvelous recovery, flying away and giving me the merry honk! honk! for being so easy. I am sure he enjoyed the ease with which he fooled me, and I enjoyed watching him and letting him think so.

Plumages.—The downy young when recently hatched is brightly colored and very pretty. The entire back, rump, wings, and flanks are "yellowish olive," with a bright, greenish-yellow sheen; a large central crown patch is lustrous "olive"; the remainder of the head and neck is bright yellowish, deepening to "olive ocher" on the cheeks and sides of the neck and paling to "primrose yellow" on the throat; the under parts shade from "deep colonial buff" on the breast to "primrose yellow" on the belly; the bill is entirely black. Older birds are paler and duller colored, "drab" above and grayish white below.

When about 4 weeks old the plumage begins to appear, the body plumage first and the wings last; they are fully grown when about 6 weeks old, and they closely resemble their parents in their first plumage. There is, however, during the first summer and fall at least a decided difference. The plumage of young birds looks softer and the colors are duller and more blended. The head and neck are duller, browner black; the cheeks are more brownish white, and the edges of the black areas are not so clearly cut; the light edgings above are not so distinct; and the sides of the chest and flanks are indistinctly mottled, rather than clearly barred. During the fall and winter these differences disappear by means of wear and molt, so that by spring the young birds are practically indistinguishable from adults.

Food.—Canada geese live on a variety of different foods in various parts of their habitat and at different seasons, but they seem to show a decided preference for vegetable foods where these can be obtained. They usually feed in flocks in certain favored localities where suitable food can be found in abundance, feeding during the daytime if not too much disturbed, or at night, if necessary, in localities where it would be unsafe to feed in daylight. The feeding flocks are guarded by one or more sentinels, which are ever on the alert until they are relieved by some of their companions and allowed to take their turns at feeding. Their eyesight is very keen and their sense of hearing very acute. They are very wary at such times and among the most difficult of birds to approach; at a warning note from the watchful sentry every head is raised and with eyes fixed on the approaching enemy they await the proper time for taking their departure. Geese are very regular in their feeding habits, resorting day after day to the same feeding grounds if they are not too much disturbed; they prefer to feed for a few hours in the early morning, flying in to their feeding grounds before sunrise and again for an hour or two before sunset, spending the middle of the day resting on some sandbar or on some large body of water.

While on their spring migration overland wild geese often do considerable damage to sprouting grain, such as wheat, corn, barley, and oats; nipping off the tender shoots does no great harm, but they are not always content with such careful pruning and frequently pull up the kernel as well. They also nibble at the fresh shoots of growing grasses and other tender herbage, nipping them off sideways, cleanly and quickly.

Aububon (1840) says that "after rainy weather, they are frequently seen rapidly patting the earth with both feet, as if to force the earthworms from their burrows." Farther north, where they meet winter just retreating, they find the last year's crop of berries uncovered by the melting snow in a fair state of preservation and various buds are swelling fresh and green. Later on some animal food becomes available, insects and their larvae, crustaceans, small clams and snails, and probably some small fishes. In the marshes they feed on wild rice, arrowhead, sedges, marsh grasses, and various aquatic plants, eating the roots as well as the leaves and shoots. On the fall migration they again frequent the grain fields to pick up the fallen grain, pull up the stubble, and nibble at what green herbage they can find. They resort to the shallow ponds and borders of lakes to feed after the manner of the surface-feeding ducks, reaching down to the bottom with their long necks and even tipping up with their feet in the air, in their attempts to reach the succulent roots and the tender water plants. On the coast in winter they prefer to feed in fresh or brackish water on the leaves, blades, and fruits of marine plants, such as *Zostera marina*, the sea lettuce (*Ulva lactuca*) and various *Algae*. Probably some small mollusks, crustaceans, and other small marine animals are taken at the same time.

Behavior.—In flight, Canada geese impress one as heavy, yet powerful birds, as indeed they are. In rising from the water or from the land they run along for a few steps before rising, but Audubon (1840) says that "when suddenly surprised and in full plumage, a single spring on their broad webbed feet is sufficient to enable them to get on the wing." When flying about their feeding grounds or elsewhere on short flights, they fly in compact or irregular bunches. Their flight then seems heavy and labored, but it is really much stronger and swifter than it seems, and for such heavy birds they are really quite agile. It is only when traveling long distances that they fly high in the air in the well-known V-shaped flocks, which experience has taught them is the easiest and most convenient for rapid and protracted flight. In this formation the leader, cleaving the air in advance, has the hardest work to perform; the lead is taken by the strongest adult birds, probably the ganders, which change places occasionally for relief; the others follow along in the diverging lines at regular intervals, so spaced that each has room enough to work his wings freely, to see clearly ahead, and to save resistance in the wake of the bird ahead of him. As the wing beats are not always in perfect unison, the line seems to have an undulatory motion, especially noticeable when near at hand; but often the flock seems to move along in perfect step. Flight is not always maintained in the stereotyped wedge formation; sometimes a single, long, sloping line is formed or more rarely they progress in Indian file. The speed at which geese fly is faster than it seems, but it has often been overestimated; the following statement by J. W. Preston (1892) is of interest in this connection:

> The Canada goose presses onward, borne up by strong and steady pinions. For forceful, solid business he has few rivals. I remember once, while traveling by rail at a rate of 30 miles an hour, our way lay for a time along the course of a swollen creek. A flock of geese, among them one little teal, came alongside the train and kept almost within gunshot for fully 10 miles, seemingly at an ordinary rate; and the teal was at no loss to keep his place among his larger companions.

There are exceptions to the orderly method of procedure outlined above. Audubon (1840) says that:

> When they are slowly advancing from south to north at an early period of the season, they fly much lower, alight more frequently, and are more likely to be bewildered by suddenly formed banks of fog, or by passing over cities or arms of the sea, where much shipping may be in sight. On such occasions great consternation prevails among them, they crowd together in a confused manner, wheel irregularly, and utter a constant cackling resembling the sounds from a disconcerted mob. Sometimes the flock separates, some individuals leave the rest, proceed in a direction contrary to that in which they came, and after awhile, as if quite confused, sail toward the ground, once alighted on which they appear to become almost stupefied, so as to suffer themselves to be shot with ease, or even knocked down with sticks. Heavy snowstorms

also cause them great distress, and in the midst of them some have been known to fly against beacons and lighthouses, dashing their heads against the walls in the middle of the day. In the night they are attracted by the lights of these buildings, and now and then a whole flock is caught on such occasions.

When preparing to alight the whole flock set their wings and drift gradually down a long incline until close to the surface, then scaling or flying along they drop into the water with a splash. They swim gracefully on the water after the manner of swans and can make rapid progress if necessary. That they can dive and swim under water, if need be, is well illustrated by the following incident, related by Audubon (1840):

I was much surprised one day, while on the coast of Labrador, to see how cunningly one of these birds, which, in consequence of the molt, was quite unable to fly, managed for awhile to elude our pursuit. It was first perceived at some distance from the shore, when the boat was swiftly rowed toward it, and it swam before us with great speed, making directly toward the land; but when we came within a few yards of it, it dived, and nothing could be seen of it for a long time. Every one of the party stood on tiptoe to mark the spot at which it should rise, but all in vain, when the man at the rudder accidentally looked down over the stern and there saw the goose, its body immersed, the point of its bill alone above water, and its feet busily engaged in propelling it so as to keep pace with the movements of the boat. The sailor attempted to catch it while within a foot or two of him, but with the swiftness of thought it shifted from side to side, fore and aft, until delighted at having witnessed so much sagacity in a goose, I begged the party to suffer the poor bird to escape.

Mr. Henderson describes in his notes an interesting habit of pose assumed by this species, as follows:

I rode down the river a short distance to where I had noticed a pair of geese alight and soon saw one standing on a gravelly island. Making a short detour and riding closer, I saw both birds lying flat on the gravel, head and neck outstretched along the ground, precisely as they do on the nest. They were hiding right in the open without the slightest cover. Though I have what is called the hunter's eye pretty well developed, it is doubtful if I would have noticed them if I had not previously known they were there. They remained perfectly motionless and resembled pieces of water-worn driftwood so perfectly that I now understand how it was that in descending rivers in a canoe I had so often failed to observe them until they took wing. It was the most beautiful example of protective coloring I have ever seen. As I rode up to the river bank in plain sight and making a good deal of noise, one bird remained perfectly still and the other moved its head slightly to watch me. I then rode out into the river to within 35 yards before they broke the pose and took to flight.

M. P. Skinner has noticed similar habits in Yellowstone Park. He says in his notes:

Geese have a curious habit of "playing possum." Instead of flying away, they squat flat with head and neck stretched out straight before them in a most ungooselike attitude. After one has passed by three or four hundred yards they raise their heads slowly an inch or two at a time and finally get to their feet again. They do this on the ice, on stony banks of streams, on bowlders, on sandbars, in the grass, and I have even seen a sitting bird do it on her nest. On the ice it makes them inconspicuous, on stony shores or bowlders the deception is perfect, for the rounded gray back looks just like a stone; as sand beaches may have stones, the method is good hiding there; but on the grass "playing possum" fails because of the contrast. In the water's edge the deception is good, as the inert, idly rocking body looks very little like a live bird. And this method is carried even further, for I have seen geese swim the Yellowstone River with heads and necks at the surface and have had them sneak off through the grass in the same way. This subterfuge is used more in spring than in summer, but is practiced sometimes in September and October.

Geese are social and like to be together, although the flocks are usually small unless there is strong reason for their gathering temporarily. Pages can be written of the sagacity and wisdom of these birds. Wary as they are, they are one of the first to realize the protection given them and are quick to lose their suspicions of man and his ways. But it is interesting to observe that although they pay no attention to autos passing along a road near them, they are at once on the alert and suspicious if a car stops near. Often we find the geese tamer than the pintail and mallard they are associating with. And their sagacity extends to wild animals as well; they know just how near it is safe to let a coyote approach, and one September day I watched a flock on a meadow seemingly indifferent to a black bear near by, although they never let him get within 20 feet, first walking away, then flying, if he came too near.

The well-known resonant honking notes of a flock of geese flying overhead on the migration are familiar sounds to every observant person; they are characteristic and distinctive of such migrating flocks and are sometimes almost constant. The Canada goose is also a noisy bird at other times, indulging freely in softer, lower-toned, conversational honking or gabbling notes while feeding or in other activities. Ora W. Knight (1908) gives a very good description of the notes of this species, as follows:

The cry uttered when on the wing is a clear trumpetlike "honk," seemingly uttered by various individuals in the flock. When the weather is foggy their "honk" seems uttered more frequently and in a querulous tone. When a flock has alighted and is sporting in the water without apprehension of trouble they swim gracefully about, plunging their heads and necks under the water to feed. Now and then some lusty or exuberant individual (probably a gander) will stretch itself up in the water, flap its wings over its back, and utter a series of resonant honks, the first loudest, longest drawn out, and highest pitched, and gradually lessening in loudness and length and decreasing in pitch, about as follows: "*h——o——n——k, h——o——n——k, h—o—n—k, h-o-n-k, honk, onk, uf*," the last note being a mere expelling of the breath. This proceeding I have only observed with one flock, never having been able to observe others while they were unconscious of my whereabouts and feeding, but judge that it is a characteristic habit.

The attitude of the Canada goose toward other species seems to be one of haughty disdain; although it often frequents the same breeding grounds and the same feeding resorts with various other species of geese, ducks, and other waterfowl, it never seems to mingle with them socially or to allow them to join its flocks. Toward man and other animals it shows remarkable sagacity in discriminating between harmless friends and dangerous enemies, and the latter must be very crafty to deceive it. On this point Audubon (1840) writes:

At the sight of cattle, horses, or animals of the deer kind, they are seldom alarmed, but a bear or a cougar is instantly announced, and if on such occasions the flock is on the ground near water, the birds immediately betake themselves in silence to the latter, swim to the middle of the pond or river, and there remain until danger is over. Should their enemies pursue them in the water, the males utter loud cries, and the birds arrange themselves in close ranks, rise simultaneously in a few seconds, and fly off in a compact body, seldom at such times forming lines or angles, it being in fact only when the distance they have to travel is great that they dispose themselves in those forms. So acute is their sense of hearing that they are able to distinguish the different sounds or footsteps of their foes with astonishing accuracy. Thus the breaking of a dry stick by a deer is at once distinguished from the same accident occasioned by a man. If a dozen of large turtles drop into the water, making a great noise in their fall, or if the same effect is produced by an alligator, the wild goose pays no regard to it; but however faint and distant may be the sound of an Indian's paddle, that may by accident have struck the side of his canoe, it is at once marked, every individual raises its head and looks intently toward the place from which the noise has proceeded, and in silence all watch the movements of their enemy.

These birds are extremely cunning also, and should they conceive themselves unseen, they silently move into the tall grasses by the margin of the water, lower their heads, and lie perfectly quiet until the boat has passed by. I have seen them walk off from a large frozen pond into the woods, to elude the sight of the hunter, and return as soon as he had crossed the pond. But should there be snow on the ice or in the woods, they prefer watching the intruder, and take to wing long before he is within shooting distance, as if aware of the ease with which they could be followed by their tracks over the treacherous surface.

Fall.—The beginning of the fall migration in Ungava is described by Lucien M. Turner, in his unpublished notes, as follows:

The birds first seen in the fall in the vicinity of Fort Chimo are those asserted to have been reared in the Georges River district and repair to this locality in search of fresh feeding grounds. They appear about August 12 to 20, but are in very lean condition. By the first of September the earlier birds hatched north of the strait begin to appear and become quite numerous by the latter week of September. By this time they are in tolerable condition and rapidly become fat by the first of October, feeding on vegetable matter growing in the ponds, in the swamps and flats along the river banks. They remain until the latter part (24th) of October and follow up the rivers which flow from the south. In the year 1882 immense numbers of these geese flew southward on the 19th of October. Hundreds of flocks of various sizes, from 15 to 80 birds, passed over. A cold snap immediately succeeded, although a flock of 6 settled in the river a few yards from the houses on October 24.

From the foregoing it will be seen that the fall movement from the breeding grounds begins early in the season, the flocks gradually gathering on the coasts or on the larger bodies of water in large numbers, moving about slowly and deliberately and reveling in the milder temperatures and abundant food to be found in such places. But the shortening days and the sharpening frosts of autumn accelerate their movements and they prepare for their long journey; at length the leaders summon their hosts to meet on high; and forming in two long converging lines, pointing toward the already feeble rays of the noonday sun, they start. High in the air they travel on, cheered by the clarion call of the leader, answered at frequent intervals by his followers, far above all dangers and straight along the well-known path. When bewildered by fogs or storms or when overtaken by darkness the flight is lower and full of dangers. But usually toward night a resting place is sought; perhaps some well-known lake is sighted and the weary birds are glad to answer the call of some fancied friend below them; so setting their wings the flock glides down in a long incline, circling about the lake for a place to alight, and greeting their friends with loud calls of welcome. Too often their friends prove to be domesticated traitors, trained to lure them to the gunner's blind, and it is a wary goose indeed that can detect the sham. But, if all goes well, they rest during the hours of darkness and are off again at daybreak, for now they must push along fast until they reach their winter haven. Dr. John C. Phillips (1910 and 1911*a*) has published two very interesting papers on the migrations of Canada geese in Massachusetts which are well worth reading; but as they are principally of local interest and are too lengthy to quote in full, I would refer the reader to them rather than attempt to quote from them.

Game.—Many and varied are the methods employed by gunners to bring to bag the wily wild goose. On account of its large size and generally good table qualities it has always been much in demand as a game bird; it is so wary, so sagacious, and so difficult to outwit that its pursuit has always fascinated the keen sportsman and taxed his skill and his ingenuity more than any other game bird. According to Henry Reeks (1870) the settlers of Newfoundland were formerly adepts in tolling geese with the help of a dog; he describes the method, as follows:

The sportsman secretes himself in the bushes or long grass by the sides of any water on which geese are seen, and keeps throwing a glove or stick in the direction of the geese, each time making his dog retrieve the object thrown; this has to be repeated until the curiosity of the geese is aroused, and they commence swimming toward the moving object. If the geese are a considerable distance from the land, the dog is sent into the water, but as the birds approach nearer and nearer the dog is allowed to show himself less and less; in this manner they are easily tolled within gunshot. When the sportsman has no dog with him he has to act the part of one by crawling in and out of the long grass on his hands and knees, and sometimes this has to be repeated continuously for nearly an hour, making it rather a laborious undertaking, but I have frequently known this device to succeed when others have failed. The stuffed skin of a yellow fox (*Vulpes fulvus*) is sometimes used for tolling geese, and answers the purpose remarkably well, especially when the geese are near the shore, by tying it to a long stick and imitating the motions of a dog retrieving the glove or stick.

On the coast of New England in winter geese have been successfully pursued by sculling upon them among the drift ice in a duck float. The float sits low in the water, with pieces of ice on her bow and along her sides; the gunner, clad in white clothing, crouches out of sight, and if properly handled the whole outfit can scarcely be distinguished from a floating ice cake. But a much more successful and more destructive, though less sportsmanlike, method is used on the inland lakes and larger ponds of eastern Massachusetts. This is the duck-stand method, which I have so fully described under the black duck that it is necessary only to refer to it here. Perhaps it should have been described under this species, for, although more ducks than geese are usually killed in such stands, the goose-shooting part of it is the more highly developed. Large numbers of live decoy geese are raised and trained for annual use in these stands and the most efficient teamwork is employed. The old mated pairs and their young are separated and made to call to each other in such a way that the wild birds are attracted. An old gander may be tethered out on the beach, while its young are kept in a "flying pen" back of the stand; when wild geese appear the goslings are released by pulling a cord; they fly out to meet the incoming flock; their parents call to them and they return to the beach, bringing the wild birds with them. When the geese are near enough and properly bunched a raking volley from a battery of guns is poured into them and other shots are fired as the survivors rise, with the result that very few are left to fly away. Even some of these may return and be shot at again if the leaders or the parents of the young birds have been killed. Such slaughter can hardly be regarded as sport.

Farther south on the Atlantic coast, in Virginia and North Carolina, geese are shot from open blinds in a much more sportsmanlike manner. A box, large enough for a man to lie down in or deep enough for a man to sit in and barely look over the top, is sunken into the ground on some sand spit or bar where the geese are wont to come for gravel or to rest, or perhaps it is placed on some marshy point on their feeding grounds where it can be concealed in the tall grass or covered with grass to match its surroundings. The decoys, either live birds or wooden imitations, are strung out in front of the blind, and the hunter crouching in the box eagerly awaits the inspiring sight of a flock of oncoming birds. At last a long line of dark, heavy birds in a wedge-shaped phalanx is seen approaching, with apparently labored flight. The well-known challenge note of the leader, repeated along the line of his followers, arouses the decoys to answering notes of invitation to alight. The flock wheels and swings in to the decoys, anxiously scanning the surroundings for any suspicious object. Seeing nothing to alarm them, they all set their wings and scale down to join their fellows. This is the sportman's opportunity for a flight shot; the pothunter would prefer to wait until they had all alighted and gathered in a dense bunch near the decoys. But in either case the birds have a better chance than in front of a concealed battery of heavy guns.

Goose shooting on the western grain fields is perhaps the most sportsmanlike method, as it is practically all wing shooting. The birds frequent the grain fields in large numbers to feed on the tender shoots of growing grain in the spring or on the stubble and fallen grain in the autumn. They are very regular in their feeding habits, flying in to the fields from their roosting grounds on the lakes and sloughs about daylight and feeding for a few hours after sunrise; they rest during the middle of the day and come in again to feed for a few hours before sunset. Gunners take advantage of these regular habits to shoot them on their lines of flight. A hole is dug in the ground deep enough to conceal the gunner entirely, and the decoys, usually wooden ones, are set out around it. Or a convenient and effective blind is made by hollowing out the center of a corn shock, with which the geese are already familiar. Concealed in such a blind before daylight, the hunter is well prepared for some excellent shooting when the flight begins, especially if he is an expert in calling the birds by imitating their notes. It must be exciting sport to shoot these large birds flying over and often within easy range.

I suppose that the Canada goose has been more persistently hunted, over a wider range of country and for a longer period of years, than any other American game bird, for in the earlier days, when all game was so abundant, only the largest species were considered worth the trouble. In spite of this fact it has shown its ability to hold its own and is even increasing in numbers in many places to-day. Messrs. Kumlien and Hollister (1903) report it in Wisconsin as—

abundant, increasing rather than diminishing in numbers during the fall, winter, and spring. To such an extent has this species changed its habits that it is no longer looked upon as a sure harbinger of spring, as in most sections of southern and even south-central Wisconsin it remains all winter, flying back and forth from its favorite cornfields to some lake or large marsh for the night. When snow is plenty it even remains in the fields for days at a time. Twenty-five to fifty years ago the flocks which first made their appearance were noted by everyone, and spring was not far distant. Now, the flocks which return from the north in October are continually added to until they are often several hundred strong, and remain thus until the beginning of spring.

On a recent (1916) visit to the great shooting resorts on the coasts of Virginia and North Carolina, I was told by the members of some of the gun clubs that geese were more abundant than ever before and are increasing every year. I certainly saw more geese in the north end of Currituck Sound on one of the rest days than I had ever seen in my life before; great rafts of them were gathering to feed in the shallow water on the fox-tail grass and wild celery which abounds in that region; the water was black with them as far as I could see; flock after flock was constantly coming in from the sea; and sometimes it seemed as if they came in flocks of flocks. They winter here in large numbers; probably this vicinity is the greatest winter resort on the Atlantic coast, for here they find abundant food in the fresh-water bays and sounds and ample security from pursuit on the broad waters of Chesapeake Bay or even on the open sea in calm weather. They feed largely at night, as they are often driven out of the bays during the days when shooting is allowed.

Winter.—Canada geese spend the winter quite far north in the interior, where they can find suitable food and large bodies of open water.

M. P. Skinner says in his notes:

In winter the reduced number remain at the outlet of Yellowstone Lake and on numerous waters kept open by hot springs and geysers. A number of our meadows are underlain by springs sufficiently warm to melt the snow and even furnish a little green grass all winter. These are frequented by geese as well as mallard and green-winged teal.

On the coast they winter abundantly as far north as Massachusetts; probably the greatest winter resort on the New England coast is on Marthas Vineyard, where the large fresh-water ponds are not always frozen and where there are open salt-water ponds which never or very rarely freeze.

That the Canada goose winters abundantly in northern Florida is well illustrated by the following notes sent to me by Charles J. Pennock:

The numerous shallow bays, bayous, and broad river mouths of the counties of Wakulla, Jefferson, and Taylor, lying south and southeast from Tallahassee, offer attractive feeding for winter visiting *Branta canadensis canadensis*, while not infrequently a short distance inland, just back of the bordering salt marshes, numerous sand flats and burnt-over semimarsh areas afford irresistible attractions to a hungry goose. Fresh shoots of grass with plenty of gravel and a clear, clean sand bed on which to take a siesta seems to be a combination most alluring, and in February and early March, with weather conditions favorable, numerous bands of these sturdy birds may be found constantly on the move, flying in as the tide rises and stops their feeding along shore or, if undisturbed after a hearty feeding on the freshly grown grass, they betake them to a long stretch of bare sand, where they evidently feel secure from surprise by virtue of sentinels most alert with keenest senses of sight and hearing, some hunters even claiming them to have a like keen sense of smell; at any rate they are most difficult to approach at such times and usually beat off up wind just before an approaching hunter gets within range.

DISTRIBUTION

Breeding range.—Northern North America, south of the barren grounds. East to Labrador (Okak, Nain, Hopedale, etc.), and Newfoundland (Grand Lake). South to the Gulf of St. Lawrence (Anticosti Island), James Bay, South Dakota, northern Colorado (Boulder County), northern Utah (Bear River), northern Nevada (Halleck), and California (Lake Tahoe). Formerly, and perhaps occasionally now, as far south as western Tennessee (Reelfoot Lake) and northeastern Arkansas (Walker Lake). West to northeastern California (Eagle Lake and Lower Klamath Lake), central Oregon (Lake County), central Washington (Douglas County), central British Columbia (Cariboo District), probably to the coast in southern Alaska, and to the upper Yukon (Fort Yukon). North to the northern limit of trees in Mackenzie (Providence and Fort Anderson) and northern Quebec (Whale River).

Winter range.—Nearly all of the United States. East to the Atlantic coast. South to Florida (Wakulla and Marion Counties), the Gulf coasts of Louisiana and Texas, Mexico (San Fernando, Matamoras, etc.), and southern California (San Diego). West nearly or quite to the Pacific coast. North to southern British Columbia (Chilliwack, Shuswap Lake, and the Okanogan Valley), northwestern Wyoming (Yellowstone Park), South Dakota, southern Wisconsin (Sauk County), southern Ohio, southern New England (Long Island and Marthas Vineyard), and northeastward to Nova Scotia (Barrington Bay, Port Joli, etc.).

Spring migration.—Early dates of arrival: Rhode Island, Block Island, February 21; central Massachusetts, February 25; southern New Hampshire, March 11; southern Maine, March 5; Quebec City, March 1; Prince Edward Island, March 9; Labrador, Sandwich Bay, April 30; southern Iowa, February 4; Minnesota, Heron Lake, February 23; North Dakota, Argusville, March 8; Manitoba, Aweme, March 9; Saskatchewan, Reindeer Lake, April 17; Mackenzie, Fort Simpson, April 22, and Fort Anderson, May 15. Average dates of arrival: Central Pennsylvania, March 17; central New York, March 13; central Massachusetts, March 17; southern Maine, March 24; Quebec City, March 27; southern Iowa, March 1; southern Wisconsin, March 13; southern Ontario, March 16; Manitoba, Aweme, March 29; Mackenzie, Fort Simpson, April 28. Late dates of departure: Florida, Marion County, May 22; Texas, Grapevine, April 15; southern Mississippi, April 20; Kentucky, Bowling Green, May 7; central Maryland, April 22; central New Jersey, May 9; Massachusetts, Cape Cod, May 26; California, Gridley, April 11.

Fall migration.—Early dates of arrival: Central Massachusetts, September 4; Long Island, Montauk Point, September 30; Virginia, Alexandria, October 5; South Carolina, Anderson, October 10; Florida, Wakulla County, October 9; northern Nebraska, September 7; central Iowa, September 16; central Missouri, September 23; Kentucky, Bowling Green, September 22; California, Gridley, November 5. Average dates of arrival: Central Massachusetts, October 11; Long Island, Montauk Point, October 20; central New Jersey, October 18; Virginia, Alexandria, October 20; northern Nebraska, October 7; central Wisconsin, October 12; central Indiana, October 19; southern Mississippi, November 12. Late dates of departure: Quebec, Hatley, November 25; Prince Edward Island, December 22; southern Ontario, November 10; southern Michigan, November 25; central Minnesota, December 1; Manitoba, Aweme, December 2; Montana, Columbia Falls, November 24.

Casual records.—Accidental in Bermuda (fall, 1874, and January and February, 1875) and the West Indies (Jamaica).

Egg dates.—Northern Canada: Eighteen records, May 18 to July 14; nine records, June 19 to July 9. Utah and Nevada: Sixteen records, March 29 to May 19; eight records, April 18 to 27. North Dakota and Saskatchewan: Thirteen records, April 29 to July 19; seven records, May 9 to June 3. Labrador and Newfoundland: Eleven records, May 24 to July 7; six records, June 4 to 13.

BRANTA CANADENSIS HUTCHINSI (Richardson)

HUTCHINS GOOSE

HABITS

After writing such a full life history of the Canada goose, it seems unnecessary to go over the same ground again in writing about this small northern subspecies, which, though it differs somewhat in habits from its larger relative, has many characteristics in common with it. There seems to be little doubt that *hutchinsi* is a true subspecies of the *canadensis*, for it seems to be exactly like it except in size, and perhaps in the number of tail feathers, which is variable in both forms. The other two, so called, subspecies can not be so satisfactorily placed.

Spring.—The Hutchins goose is a later migrant than the Canada goose, probably because it goes so much farther north to breed. It is said to pass through the Hudson Bay region at about the same time that the snow geese are migrating; probably both of these hardy northerners know enough not to migrate until their summer homes become habitable. The migration is about due northward on both sides of the Rocky Mountains, through the interior valleys to the Arctic coast, and along the Pacific slope to northern Alaska.

Nesting.—MacFarlane (1891) found this goose breeding abundantly on the Arctic coast, of which he says:

A large number (50) of nests of the smaller Canada goose was found on the Lower Anderson, as well as on the shores and islands of the Arctic Sea. All but one were placed on the earth, and, like that of the preceding species, it was composed of hay, feathers, and down, while 6 was the usual number of eggs. The exceptional case was a female parent shot while sitting on 4 eggs in a deserted crow's or hawk's nest built on the fork of a pine tree at a height of about 9 feet. At the time the ground in the vicinity thereof was covered with snow and water, and this may have had something to do with her nesting in so unusual a place.

In a letter to the Smithsonian Institution he writes:

I have no doubt about Hutchins goose being a good species; its mode of nesting alone would go far to prove it distinct from the Canada goose, which it so greatly resembles. The former, so far as I have been able to ascertain, *invariably* nests on the small islands which occur on the small lakes of the islands situated on the shores of the Arctic Sea, while the latter generally builds in the neighborhood of the lakes and rivers of the wooded country. The former also scoops a hole in the sand or turf, lining its sides with down, while the nest of the latter is composed of a large quantity of feathers and down placed on or supported by some dry twigs or willow branches.

I have had several sets of eggs of the Hutchins goose sent to me from Point Barrow, which were evidently taken from nests on the tundra, for the nesting down, which came from them, was mixed with tundra mosses, bits of grass, leaves, and other rubbish. Nests of the Canada goose generally contain pure, clean down.

Eggs.—The Hutchins goose lays from 4 to 6 eggs, usually 5. These are in no way distinguishable from the eggs of the Canada goose except that they are smaller, as they should be. The measurements of 83 eggs, in various collections, average 79.2 by 53.1 millimeters; the eggs showing the four extremes measure **85.4** by 57, 78.5 by **58**, **72.1** by 53.1, and 76 by **50** millimeters.

Food.—The food and feeding habits of this goose are similar to those of its well-known relative. Nuttall (1834), however, calls attention to the fact that their habits "are dissimilar, the Canada geese frequenting the fresh-water lakes and rivers of the interior, and feeding chiefly on herbage; while the present species are always found on the seacoast, feeding on marine plants, and the mollusca which adhere to them, whence their flesh acquires a strong fishy taste." Dr. J. G. Cooper (1860) says: "They feed principally on the mud flats at low tide, eating vegetable and animal food which they find there," during their annual visits to the coasts of Oregon and Washington.

Dr. Joseph Grinnell (1909) was informed by the natives of Alaska "that some years these geese stop in large numbers for a short time to feed upon the herring spawn which is to be seen all along the beach at low tide, where it sticks to the rocks."

While sojourning in California these geese associate with the white-fronted and snow geese and feed largely in the grain fields and grassy plains. In spring they do considerable damage by pulling up freshly sprouting grain; formerly, when they were much more abundant, it was customary for farmers to hire men and furnish them with guns and ammunition to keep the geese away from the grain; but the geese have decreased in numbers so decidedly in recent years that this is no longer necessary.

Game.—The importance of this bird as a game bird in California is well illustrated by the following statement made by Grinnell, Bryant, and Storer (1918):

The Hutchins goose, although not quite so desirable a bird for the table as are some other species, is the goose which has afforded the greatest amount of sport for the hunter because of its abundance. It has usually been a common goose on the market, where it is known as the "brant." In 1909–10 one transfer company in San Francisco sold the following numbers of brant: October, 1,442; November, 2,196; December, 1,592; January, 1,479; February, 1,226; March, 251. Cackling as well as Hutchins geese are probably included in these numbers. This makes a total of over 8,000 geese of only two varieties sold by the one transfer company. That season the same company sold more than 20,000 geese of all kinds. In 1906–7, it sold only 7,431. In 1895–96 there were sold on the markets of San Francisco and Los Angeles 48,400 geese, of which 16,319 were brant. There is little wonder that geese have decreased in numbers more than most other game birds. The markets of San Francisco during 1910–11 paid from $2.50 to $8 a dozen for geese other than the snow geese. On the Los Angeles markets during 1912–13 the same geese sold at from 65 cents to $1 a pair.

Winter.—A very good account of the winter home and habits of this goose is given by Coues (1874), as follows:

We must, however, visit the regions west of the Rocky Mountains to find the Hutchins goose plentiful in its favorite winter residences, and observe it under the most favorable circumstances. On river, lake, and marsh, and particularly along the seacoast, it is found in vast numbers, being probably the most abundant representative of its family. It enters the United States early in October, or sometimes a little earlier, according to the weather, and in the course of that month becomes dispersed over all its winter feeding grounds. It is generally in poor condition on its arrival, after the severe journey, perhaps extending from the uttermost Arctic land; but it finds abundance of food and is soon in high flesh again. During the rainy season in California the plains and valleys, before brown and dry, become clothed in rich verdure, and the nourishing grasses afford sustenance to incredible numbers of these and other geese. Three kinds, the snow, the white-fronted, and the present species, have almost precisely the same habits and the same food during their stay with us, and associate so intimately together that many, if not most, of the flocks contain representatives of all three. At least, after considerable study of the geese in Arizona and southern California, I have been unable to recognize any notable differences in choice of feeding grounds.

The following extract on Hutchins goose, from Doctor Heermann's report, will be found interesting: "While hunting during a space of two months in the Suisun Valley I observed them, with other species of geese, at dawn, high in the air, winging their way toward the prairies and hilly slopes, where the tender young wild oats and grapes offer a tempting pasturage. Their early flight lasted about two hours, and as far as the eye could reach the sky was spotted with flock after flock, closely following in each other's wake, till it seemed as though all the geese of California had given rendezvous at this particular point. Between 10 and 11 o'clock they would leave the prairies, first in small squads, then in large masses, settling in the marshes and collecting around the ponds and sloughs, thickly edged with heavy reeds. Here, swimming in the water, bathing and pluming themselves, they keep up a continual but not unmusical clatter. This proves the most propitious time of the day for the hunter, who, under cover of the tall reeds and guided by their continual cackling, approaches closely enough to deal havoc among them. Discharging one load as they sit on the water and another as they rise, I have seen 23 geese gathered from two shots, while many more, wounded and maimed, fluttered away and were lost. About 1 o'clock they leave the marshes and return to feed on the prairies, flying low, and affording the sportsman again an opportunity to stop their career. In the afternoon, about 5 o'clock, they finally leave the prairies, and, rising high up in the air, wend their way to the roosting places whence they came in the morning. These were often at a great distance, as I have followed them in their evening flight until they were lost to view. Many, however, roost in the marshes. Our boat, sailing one night down the sloughs leading to Suisun Bay, having come among them, the noise they made as they rose in advance of us, emitting their cry of alarm (their disordered masses being so serried that we could hear their pinions strike each other as they flew), impressed us with the idea that we must have disturbed thousands. Such are the habits of the geese during the winter. Toward spring they separate into small flocks and gradually disappear from the country, some few only remaining, probably crippled and unable to follow the more vigorous in their northern migration."

DISTRIBUTION

Breeding range.—Barren grounds of North America. East to southern Baffin Land. South to Southampton Island, west coast of Hudson Bay (Cape Fullerton and Churchill), northern Mackenzie (Fort Anderson and Fort Good Hope) and northern Alaska (northern coast and south to Kowak River). Said to breed on the Bering Sea coast of Alaska and on the Aleutian Islands, but such reports need confirmation; it may, however, breed on the extreme western Aleutians (Agattu Island) as it is reported as breeding on the Commander and Kurile Islands. North to Victoria Land (Cambridge Bay) and Boothia Peninsula (Felix Harbor). Intergrades with *minima* in Alaska and with *canadensis* in northern Canada.

Winter range.—Mainly western United States. East regularly to the Mississippi Valley, rare in the Eastern States, and only casual on the Atlantic coast. South regularly to the Gulf coasts of Louisiana and Texas and probably Mexico and Lower California (San Rafael Valley). West nearly or quite to the Pacific coast. North to southern British Columbia (mouth of Fraser River), northern Colorado (Barr Lake), Nebraska, southern Illinois, and rarely to southern Wisconsin. On the Asiatic coast south to Japan.

Spring migration.—Early dates of arrival: Manitoba, Aweme, April 2 (average April 12); Saskatchewan, Indian Head, April 29; British Columbia, Sumas, April 10; Alaska, Admiralty Island, April 18, Kuskokwim River, April 30, and Kowak River, May 14. Late dates of departure: Texas, Houston, April 18; California, Gridley, April 26; British Columbia, May 20.

Fall migration.—Early dates of arrival: Manitoba, Aweme, September 13; Montana, Terry, September 22; British Columbia, Sumas, October 4; Wisconsin, Delavan, October 12; Utah, Bear River, October 9; California, Gridley, October 9. Late dates of departure: Hudson Straits, Wales Sound, September 6; Great Bear Lake, September 25; Mackenzie, Fort Wrigley, October 12; Manitoba, Aweme, November 20; Alaska, Kowak River, September 14; British Columbia, November 25.

Casual records.—Has wandered east to Greenland (Disco and Godhaven), Maine (Cape Elizabeth, November 13, 1894), and Virginia (Cobb Island, winter of 1888–89), and south to Florida (Wakulla County, March 12, 1918), and Mexico (Vera Cruz and probably Lake Chapala, Jalisco).

Egg dates.—Arctic Canada: Eighteen records, May 17 to July 14; nine records, June 14 to July 5. Alaska: Thirteen records, May 25 to June 28; seven records June 1 to 11.

BRANTA CANADENSIS OCCIDENTALIS (Baird)

WHITE-CHEEKED GOOSE

HABITS

This large, dark-breasted form of the Canada goose seems to be a well-marked race of decidedly local distribution, occupying the northwest coast region from Prince William Sound, Alaska, to British Columbia. It is practically nonmigratory and does not wander far inland at any season. It is another one of the many saturated forms confined to this humid coast strip. Its specific status has been much discussed and is by no means definitely settled; this will be referred to under the next subspecies. It was formerly recorded by several observers as breeding in the lakes of the interior as far south as northern California, but these records have been shown to refer to the eastern Canada goose. The breeding range of the white-cheeked goose is now known to lie wholly north of the United States boundary.

Nesting.—Although the white-cheeked goose is quite common throughout its restricted range, and even numerous in certain parts of it, its nest has not often been found and very little has been published about its habits. The published reports of the Alexander expeditions to southern Alaska contain the most important contributions to its life history, and even these are meager enough.

Joseph Dixon (1908), a member of this expedition, writes:

The country about Canoe Passage on Hawkins Island was low and rolling, with large open parks bordered by wooded creeks. There were a number of lagoons almost shut off from the bay by long grassy gravel bars. One mountain in the interior of the island was 1,900 feet above the sea, according to the aneroid. Hutchins geese were nesting about these lagoons, and about the 20th of June goslings were everywhere. It was strange how they all hatched out so near the same time. I was wandering home one evening about 10 o'clock. It was just after sundown, but the deeper woods were beginning to darken slowly. It was high tide, so that I had to make a cut clear around the head of a slough. Just as I came out of the thick huckleberry underbrush in the strip of timber I stumbled over a log and almost fell on top of an old goose that was sitting on a nestful of eggs. She made a terrible racket as she went flopping and squawking off the nest, and I do not know which of us was the worst scared for a minute. The nest was placed in the open close to the trunk of a large tree just at the edge of the wood. It was lined with moss and down and held 6 eggs, which I afterwards regretted were almost ready to hatch.

Although he called them Hutchins geese at that time, the geese of that region all proved to be of the present form.

Eggs.—A set of five eggs of this subspecies in the United States National Museum was collected by Dr. Wilfred H. Osgood on Prince of Wales Island, Alaska, on May 22, 1903. These eggs are practically indistinguishable from average eggs of the eastern Canada goose. They measure 86 by 59, 87.4 by 59.4, 86 by 55.8, 87.2 by 57.8, and 87 by 58.2 millimeters.

Young.—Dr. Joseph Grinnell (1910) refers to two broods of young as follows:

On June 21, also on Hawkins Island, Miss Kellogg flushed an old goose from a nest in the tall grass near the beach. There were five newly hatched young. One of these, taken as a specimen (No. 1131), is identical in coloration with a downy young one from the Sitkan district. On June 22, Dixon records as follows: "In crossing some marshy flats we came upon six geese, five of which flew noisily away; but the sixth came gabbling toward us. We soon saw that her unusual tameness was due to her anxiety in regard to six or eight newly hatched goslings that scrambled from under our feet and disappeared with a splash into a near-by pond. I walked up to within 25 feet of the mother as she came with her head down in the usual manner of an irate goose. She followed us for some distance when we left."

Plumages.—Doctor Grinnell has kindly loaned me two specimens of the downy young of this subspecies. These and a specimen of downy young cackling goose in the United States National Museum resemble each other very closely, but are quite unlike the downy young of the Canada goose. The young of the eastern birds when first hatched are bright greenish olive above and bright yellow below, with no dark markings on the sides of the head. On the other hand, the young of the two western forms, at the same age, are much browner above and much duller and more buffy below, with more or less distinct head markings. The central crown patch and the upper parts of the body are lustrous "brownish olive," darkest on the head and rump; the lores are washed or striped with the same dark color, which surrounds the eye and extends in a postocular stripe down the neck; the under parts, including the forehead and the sides of the head and the neck, are dull yellowish or "colonial buff," washed on the sides of the head and neck with "honey yellow" or "yellow ocher," paling on the belly and flanks to "ivory yellow" and deepening on the breast to "deep colonial buff."

Judging from what little material there is available for study, I should say that the molts and plumages are similar to those of the Canada goose.

Behavior.—Very little has been published about the habits of this goose and practically nothing about its food. Doctor Grinnell (1909) gives the following general account of it, taken from Mr. Littlejohn's notes:

When Mole Harbor, Admiralty Island, was reached, on April 16, large flocks were seen about the creek mouth at the head of the bay. On the 18th many were found at Windfall Harbor, and by the 27th nearly all had paired and could be seen passing back and forth to the inland waters every day, remaining a good share of the time in the open water, where their loud notes could be heard at all times, but when night came on I think most, if not all, came to land to roost. They seemed to feed about the shores, especially where small streams and springs were flowing across the gravel. One large creek near our camp was a favorite place to assemble, and each evening they could be seen coming in from all directions to pass the night. At low tide they would remain on the gravel flats at the creek mouth, but when the tide came in they would retreat to the acres of ice inland, which had been formed during the winter; here they remained until morning if not disturbed, and then would break up in pairs, as a rule, and go off again for the day. Several pairs had chosen the lakes back of Mole Harbor for a nesting ground and were seen together when we first went there; but a few days later some old gander was apt to be seen in a secluded cove, or as happened several times, flushed from the thick timber at some distance from the water. At such times he would fly about, scolding away at a great rate, as if he were alarmed at our presence so near his mate, who was undoubtedly near by, but in the almost impenetrable forest and underbrush was not to be found.

Harry S. Swarth (1922) found these geese abundant at the mouth of the Stickine River in southeastern Alaska; he writes:

In our descent of the river, the first white-cheeked geese were seen at the boundary, August 16. From there on down an occasional small flock was noted, but not until the mouth of the river was reached were they seen in any numbers. At Sergief Island they were abundant. Flocks of large size frequented the marshes at that point, changing their feeding ground as the tides advanced and receded. These local movements covered but a few miles at most, and, of course, were gone through with daily as regularly as the tides. Aside from this hourly shifting, which kept some flocks on the wing practically throughout the day, there was no appearance of migration. Flocks of white-cheeked geese were never seen to depart in a manner suggestive of the beginning of along flight, nor were any seen arriving as though from a distance.

During the last two weeks in August the geese were still molting extensively. In some the breast and belly were almost entirely devoid of feathers, only the down remaining, and nearly all were renewing the tail feathers. Flight feathers were fully grown, or at any rate sufficiently so for flying. Presumably the birds would not gather upon these open and exposed marshes until they could fly; nesting and the beginning of the molt, including loss of the remiges, probably takes place in more sheltered localities.

Winter.—Of their winter habits he (1913) says:

Since so many of the water birds of the coast of southern Alaska and British Columbia are resident the year through in that general region, it is very probable that the white-cheeked goose belongs in the same category. In a letter recently received from Mr. Allen E. Hasselborg, a resident of Juneau, Alaska, and familiar with the native birds and mammals, he confirms this view, saying that the geese are about as abundant in the Sitkan district in winter as in summer. During the winter they frequent the more sheltered south and west facing bays and inlets, avoiding localities exposed to the cold land winds, while in summer they are of more general distribution. That this subspecies does not perform as extensive migrations as other members of the group is evident from its nonoccurrence in California. If it occurs in this State at all it should be found along the extreme northern coast.

DISTRIBUTION

Breeding range.—Pacific coast region of southeastern Alaska, from the vicinity of Prince William Sound southward, to British Columbia (Queen Charlotte Islands). Intergrades on the north with *minima* and on the east and south with *canadensis.*

Winter range.—Apparently the same as the breeding range. This seems to be a local form of very restricted habitat and non-migratory.

Egg dates.—Southern Alaska: Three records, May 22 to June 18.

BRANTA CANADENSIS MINIMA Ridgway

CACKLING GOOSE

HABITS

The characters which warrant the separation of the Canada goose group into four subspecies have been so generally misunderstood and so poorly designated in most of the manuals, and the nomenclature of the group has been so variable and puzzling, that much confusion has existed as to the true relationships of the various forms and their distribution. Various theories have been advanced, none of which seem to fit the known facts exactly. Rather than attempt to discuss this complicated systematic question, which would require more space than is warranted in a work on life histories, I would refer the reader to the interesting papers on the subject published by Harry S. Swarth (1913) and J. D. Figgins (1920 and 1922). It seems best, under the circumstances, to follow the classification and nomenclature of the American Ornithologists' Union Check List until some better arrangement is suggested and proven to be correct by the collection of large enough series of breeding birds to demonstrate the relationships and to outline the breeding ranges of the four forms. It is, however, difficult to explain, under this generally accepted theory, why the breeding ranges of *hutchinsi* and *minima* should overlap so extensively in Alaska and the Aleutian Islands, as they are said to do, which is contrary to the rule with subspecies. It is also of interest to note that the downy young of *occidentalis* and *minima* resemble each other very closely and are quite different from the downy young of *canadensis;* this suggests the possibility of a distinct, dark-breasted, western species. But perhaps both of these matters will be cleared up when more material is collected.

Spring.—Dr. E. W. Nelson (1887) gives the following interesting account of the arrival of these geese at St. Michael, Alaska:

As May advances and one by one the ponds open, and the earth looks out here and there from under its winter covering, the loud notes of the various wild fowl are heard, becoming daily more numerous. Their harsh and varied cries make sweet music to the ears of all who have just passed the winter's silence and dull monotony, and in spite of the lowering skies and occasional snow squalls everyone makes ready and is off to the marshes.

The flocks come cleaving their way from afar, and as they draw near their summer homes raise a chorus of loud notes in a high-pitched tone like the syllable "luk," rapidly repeated, and a reply rises upon all sides, until the whole marsh reechoes with the din, and the newcomers circle slowly up to the edge of a pond amid a perfect chorus raised by the geese all about, as in congratulation.

Even upon first arrival many of the birds appear to be mated, as I have frequently shot one from a flock and seen a single bird leave its companions at once and come circling about, uttering loud call notes. If the fallen bird is only wounded, its mate will almost invariably join it, and frequently allow itself to be approached and shot without attempting to escape. In some instances I have known a bird thus bereaved of its partner to remain in the vicinity for two to three days, calling and circling about. Although many are mated, others are not, and the less fortunate males fight hard and long for possession of females. I frequently amused myself while at the Yukon mouth by watching flocks of geese on the muddy banks of the river, which was a favorite resort. The females kept to one side and dozed or dabbed their bills in the mud; the males were scattered about and kept moving uneasily from side to side, making a great outcry. This would last but a few minutes, when two of the warriors would cross each other's path, and then began the battle. They would seize one another by the bill, and then turn and twist each other about, their wings hanging loosely by their sides meanwhile. Suddenly they would close up and each would belabor his rival with the bend of the wing, until the sound could be heard two or three hundred yards. The wing strokes were always warded off by the other bird's wing, so but little damage was done; but it usually ended in the weaker bird breaking loose and running away. Just before the males seize each other they usually utter a series of peculiar low growling or grunting notes.

Nesting.—Of their breeding habits in that vicinity he writes:

The last week of May finds many of these birds already depositing their eggs. Upon the grassy borders of ponds, in the midst of a bunch of grass, or on a small knoll these birds find a spot where they make a slight depression and perhaps line it with a scanty layer of grasses, after which the eggs are laid, numbering from 5 to 8. These eggs, like the birds, average smaller than those of the other geese. As the eggs are deposited the female gradually lines the nest with feathers plucked from her breast until they rest in a bed of down. When first laid the eggs are white, but by the time incubation begins all are soiled and dingy. The female usually crouches low on her nest until an intruder comes within a hundred yards or so, when she skulks off through the grass or flies silently away, close to the ground, and only raises a note of alarm when well away from the nest. When the eggs are about hatching, or the young are out, both parents frequently become perfectly reckless in the face of danger.

Both the cackling goose and the Hutchins goose are said to breed on the Aleutian Islands, but it seems hardly likely that these two subspecies should occupy the same breeding range. It seems more likely that some of the records are based on erroneous identifications or on misunderstandings as to the characters, both puzzling and variable, which separate these two forms. Lucien M. Turner (1886) reported both forms as breeding abundantly on the western islands of this chain, mainly on Agattu and Semichi, but I can not find any specimens of *hutchinsi* to substantiate his claim.

Mr. Austin H. Clark (1910) reported that the Hutchins goose—

is the most abundant bird on Agattu, where it breeds by thousands. When we approached the shore we saw a number of geese flying about the cliffs and bluffs and soaring in circles high in air. On landing I walked up the beach to the left and soon came to a small stream which enters the sea through a gap in the high bluffs, when I saw 50 or more of these birds along the bank preening their feathers. From this point I walked inland over the rough pasture-like country toward a lake where this stream rises. Geese were seen on all sides in great abundance, walking about the grassy hillsides in companies of six or eight to a dozen, or flying about from one place to another. When on the ground they were comparatively shy; at about 100 yards distant they would stop feeding and watch my movements; at about 50 yards they generally took wing, but instead of flying away they would circle about and fly toward me, often not more than 10 feet over my head, as if to see what sort of a strange beast it was which thus intruded on their domains.

He shot nine of the birds, but unfortunately was unable to preserve any of them; he did, however, write down descriptions and take the measurements of four of them. Although the measurements are rather large for *minima*, the descriptions seem to fit this form rather than *hutchinsi*, and perhaps if we had the specimens before us we might decide to refer them to the former subspecies.

On our expedition to the Aleutian Islands in 1911 we saw geese of this group on Kiska, Adak, Atka, and Attu Islands, but the only specimen taken was a female *minima* shot by R. H. Beck as she flew from her nest on Attu Island. The nest was located on the slope of a grassy hillside; it consisted of a mass of down in a hollow in the ground and contained 4 eggs on June 23.

Mr. Turner (1886) says that "the clutch of eggs varies from 7 to 13 and are laid in a carelessly arranged nest composed of dead grasses and few feathers."

Eggs.—There are 13 sets of eggs of the cackling goose in the United States National Museum, in which the numbers vary from 4 to 7; there are also four nests collected by C. H. Townsend on Agattu Island on June 5, 1894. The nests are large masses of "light drab" or "drab-gray" down, mixed with bits of white or whitish down, numerous breast feathers and bits of straw. The eggs are similar to those of the Canada goose, but smaller. The measurements of 110 eggs, in various collections, average 72.7 by 48 millimeters; the eggs showing the four extremes measure **85** by 55, 78.5 by **55.5**, and **60.3** by **37** millimeters.

Young.—Mr. Turner (1886) says:

The young remain with the parents until the latter molt, by the 20th of August, by which time the young are able to fly. This date witnesses a few of the older young and adult males coming from the breeding grounds on the Semichi Islands to the island of Attu. The geese have exhausted, by that time, the food supply of that place and repair to Attu to feast on the berries of the *Vaccineum* that are rapidly ripening. Attu Island has a great many blue foxes (*V. lagopus*) on it, hence is resorted to only by adult birds. The birds arrive poor and lean, but by the 10th of September they abound in thousands and are very fat at this time. The birds usually alight on the hillsides, and quickly strip the lower areas of the berries that have ripened earlier. Toward the evening the geese resort to the shallow pools (destitute of vegetation, with gravelly bottoms) on the sides of the mountains.

Plumages.—The downy young is exactly like that of *occidentalis.* The molts and plumages are probably similar to those of the other subspecies of this group. Doctor Nelson (1887) says:

> The first plumage of this bird is a dull grayish umber-brown; the head and neck almost uniform with the rest of the body and without any trace of the white cheek patches. As is common to the young of many waterfowl, the feathers of head, neck, and much of the rest of body are bordered with a lighter shade than the main part of the feathers. The old birds molt their quill feathers from the 20th of July until late in August, and flocks begin forming as soon as the birds are on the wing again. From that time until the last of September and first of October, when they migrate, they are found scattered over the country, feeding on various berries, which are ripe on the hillsides.

Behavior.—Very little has been published about the habits of this subspecies, but they are probably similar in the main to those of its close relatives. Mr. Turner (1886) writes:

> As an illustration of the parental solicitude exhibited by these birds, I will relate that several years ago a heavy fall of snow occurred in the latter part of June at the islands of Agattu and Semichi and covered the ground with more than 3 feet of snow. At that date the geese were incubating. The geese did not quit their nests and were suffocated. The natives found scores of the birds sitting dead on their nests after the snow had melted.
>
> During the summer the geese are not molested. The natives take many of the young and domesticate them. I have seen as many as 50 young ones at a time at Attu Island owned by the natives, to whom the goslings become much attached, especially those who attend them. The goslings remain at large during the winter, but have to be fed during severe spells of weather. The housetops being covered with sod, the excessive heat within causes the grass roots to continually send out new blades of grass. The geese are constantly searching every housetop to find the tender blades. One man had a pair of adult geese which he assured me had been reared from goslings, and that they were then entering the sixth year of their captivity. These two geese did not breed the second year of their life, but that every year thereafter they had reared a brood of young and brought them home as soon as hatched. The wings and half of the tail feathers had to be clipped every season to prevent them migrating.

Messrs. Grinnell, Bryant, and Storer (1918) say: "The high pitch of its call note, which resembles the syllables *luk-luk*, is about the best character to use in the field after recognizing the bird to be of the Canada type." This note is "oft repeated" and has caused the bird to be called the cackling goose. It is easily distinguished from the notes of the Canada goose and the Hutchins goose.

The same writers say further:

> In habits the cackling goose so nearly resembles the Hutchins goose that no one has been able to point out differences. As with the latter species, the cackling goose feeds largely on grass and grain during its stay in California.

Along with other geese, this species used to do much damage to young wheat in Colusa, Butte, Sutter, and Uba Counties. But the ranks of the birds are so thinned at the present time that the injury they inflict now is negligible.

On the market this species is usually classified along with the Hutchins goose as "brant." Very large numbers of cackling geese are to be found at times in the markets of our larger cities. The cackling goose, once just as numerous, if not more so, than the Hutchins goose, is, like the Hutchins, rapidly decreasing in numbers from year to year. Old residents in some parts of the Sacramento Valley say that now there is "not more than one of these geese present where formerly there were hundreds." To the work of the market hunter can be attributed much of this decrease, for this goose is one which is easily procured and which finds a ready sale on the market. While still rated as common in restricted portions of the State, this goose is in a fair way to disappear completely unless enough of the birds are left each winter to guarantee the return of an adequate stock in the spring to the breeding grounds in the north.

Game.—The primitive hunters of the Aleutian Islands formerly killed large numbers of these geese by catching them in long nets set on the edges of ponds where they fed. Some of the natives were also quite expert at throwing at the passing flocks a bolas made of three stones attached to leather thongs, which became entangled with the necks and wings of the birds, bringing sometimes two and even three birds down to the ground. The birds were salted away for future use during the winter and must have served as an important addition to the food supply of the natives.

Mr. Turner (1886) describes the more modern method of shooting geese on Attu Island as follows:

The manner of shooting geese at Attu Island is different from that pursued in other localities. In the evening the geese repair to the shallow pools to preen their feathers and be secure from the attacks of foxes. These resorts leave unmistakable signs of the presence of geese of preceding nights. The native wanders over the hills until he finds a lake where "signs" are abundant. A hut is generally to be found near the favorite night haunts of the geese. To this one journeys in a canoe, and on arriving the *chynik* (teakettle) is hung on the soon-kindled fire to boil, as the *chypeet* (tea drinking) is a certain concomitant of all Alaskan jaunts, either of pleasure or of profit. The chypeet over, the approach of dusk is awaited. The hunters then seek the chosen ponds and secrete themselves in a gully or on the hillside near the place to watch the geese as they come in for the evening, for during the day the geese have been feeding on the smooth, sloping hillsides.

The hunter is careful to approach these lakes lest he leave a footprint or other sign of his presence, as the goose is ever on the alert for such traces and forsakes any lake that is suspected. They will in such cases hover round and round endeavoring to discover the danger, and when satisfied that the lake has been visited by man or that he is present their loud cries give warning to all the geese within hearing as they quickly stream off and away to the head of the ravine from which they came. After such an occurrence the hunter would just as well go home or seek some other locality, for no more geese will visit that lake until the next night.

A night on which the sky is partly clouded and a light wind is blowing is the best. If the air is calm and the night bright, the still water reflects too strongly the outlines of the surrounding hills, making the water inky black, and renders it impossible to distinguish a goose sitting on the water.

At the time the geese are expected each person has selected his place and remains quiet. On the approach of the first flock for the night a low whistle from the hunter to his companions gives signal. A low *hunk hunk* of the geese and a swirl of wings announce their approach. A straight dash or a few circles round the pond and they settle. Shoot just as they alight and again as they rise. Sometimes they become so confused as to enable the holder of a breechloader to get four shots at a single flock. The dead geese serve as decoys and soon many are added to those already killed. The gentle wind slowly blows them ashore while you are waiting for others. In a short time a sufficient number is obtained. At an appointed time another native comes from the hut to help bear home the geese.

Winter.—Cackling geese are abundant winter residents in the interior valleys of California, where they frequent the grain fields in company with Hutchins, snow, Ross, and white-fronted geese. As their habits are evidently similar, it is unnecessary to repeat what has already been written about the others.

DISTRIBUTION

Breeding range.—The Bering Sea coast of Alaska and the Aleutian Islands. South to the north side of the Alaskan Peninsula (Bristol Bay and Nushagak River). West to the western Aleutians (Agattu and Attu Islands). North to Norton Sound (St. Michael), probably Kotzebue Sound (Kowak River) and possibly as far as Wainwright, where it has been taken in July. Intergrades with *hutchinsi* in northern and with *occidentalis* in southern Alaska.

Winter range.—Western North America, west of the Rocky Mountains from southern British Columbia to southern California (San Diego County).

Spring migration.—Arrives at St. Michael, Alaska, April 25 to 30. Latest date of departure from California, Stockton, April 25. Taken on St. Paul Island, Alaska, May 14.

Fall migration.—Earliest date of arrival in California, Gridley, October 1. Late dates of departure in Alaska, Yukon Delta, October 1, Aleutian Islands, November 15; British Columbia, Okanogan, November 20.

Casual records.—Said to wander on migrations as far east as Wisconsin and Colorado (Loveland, April 10, 1898), but identifications are doubtful.

Egg dates.—Alaska: Eighteen records, May 20 to June 30; nine records June 5 to 15.

BRANTA BERNICLA BERNICLA (Linnaeus)

BRANT

HABITS

For the past 25 years the American brant has been called *Branta bernicla glaucogastra* (Brehm), white-bellied brant, ever since Dr. Elliott Coues (1897) called our attention to the supposed subspecific distinction between the common European brant and the North American bird of the Atlantic coast and proposed the adoption of Brehm's name for our bird. The European bird was supposed to breed in Spitsbergen, Franz-Josef Land, Nova Zembla, and the Taimyr Peninsula, and the American bird, *glaucogastra*, was supposed to breed in western Greenland and westward as far as the Parry Islands. Recent investigations and explorations have shown that this theory is untenable, that the European and the American birds are not separable, and that *Branta bernicla glaucogastra* (Brehm) is a *nomen nudum*. Even if there were such a subspecies as the white-bellied brant, the name, *glaucogastra*, could not be used for it, as it was applied by Brehm to a dark-bellied bird shot on the coast of Germany.

Rev. F. C. R. Jourdain, who has made two recent visits to Spitsbergen, in 1921 and 1922, writes to me:

> The pale-breasted form is the one which breeds in Spitsbergen. We got over 20 birds in 1921 and they were all pale, as also were all those examined in 1922. Chapman got 30 or more, and they were all pale too. Koenig speaks of getting dark-breasted birds there, but he figures what he calls the dark-breasted bird, and it does not in the least resemble the uniformly slaty-breasted individuals which visit us in winter. There is some variation among the light-breasted birds, and it is quite evident that the "dark" birds which Koenig refers to are merely specimens in which the brownish-gray markings extend over the whole of the lower breast to the vent, instead of being confined to the breast only. The only really dark-breasted *breeding* bird I have seen came from Franz-Josef Land, but we get lots of them here (England) in winter. I am inclined to think that the pale-breasted bird extends from America to Spitsbergen in breeding time, and that perhaps farther east the darker-breasted form predominates.

The two so-called species of brant, *Branta bernicla* and *B. nigricans*, together have a complete circumpolar breeding range. The western limit of *nigricans* on the Siberian coast is not definitely known, and future investigation may show that it intergrades with the so-called dark-bellied form of *bernicla* somewhere in the palaearctic region. This would explain the occurrence of both light and dark birds in western Europe in winter, for it is well known that certain birds, such as the yellow-billed loon and the Steller eider, migrate from their breeding grounds in Arctic Siberia to the Scandinavian coasts to spend the winter, and the brant might well be expected to do the same.

In later communications Mr. Jourdain refers to a third form, darker than our so-called *glaucogastra* but lighter than *nigricans*, breeding in western Siberia, Kolguev, and Nova Zembla. Such birds, it seems to me, are intermediates, and show intergradation between *bernicla (glaucogastra)* and *nigricans*.

There is considerable evidence which points to intergradation between *bernicla* and *nigricans* in Arctic America. P. A. Taverner writes to me that they have specimens of both forms from Melville Island and that he obtained two specimens at Comox, British Columbia, which are halfway between the two forms and furnish "almost positive proof of complete intergradation between them." Dr. R. M. Anderson writes to me:

I have seen the natives shoot a good many brant (supposed to be black brant) around the Mackenzie delta and in Arctic Alaska, and they were quite variable as to paleness. The females and young are lighter in color than the males as a rule, and, if my memory serves me right, some were as light as so-called *glaucogastra* which I have since seen.

Alfred M. Bailey writes to me regarding birds collected in northern Alaska:

The majority of our specimens taken in the fall were dark bellied and would be referable to the so-called *nigricans*, although we got many light-colored birds. In the spring at Wales and Wainwright the birds collected averaged considerably lighter than those taken in the fall, with the dark breasts sharply defined in many cases. The white collar was practically continuous in both the spring and fall birds.

I notice that European writers are now recognizing *nigricans* as a subspecies of *bernicla*, and I believe that they are correct in doing so.

Spring.—The brant which have spent the winter farthest south begin to move northward in February, joining the birds which have wintered on the Virginia coast late in February, and moving on to Great South Bay, Long Island, in March. By the middle of April they have moved farther east and are congregated in large numbers off Monomoy, on the south side of Cape Cod. In the old days before spring shooting was abolished this was a famous resort for brant shooting in April. Monomoy Island, now joined to the mainland, extends for 9 or 10 miles southward from the elbow of Cape Cod; the eastern or outer side is a long, high, sandy beach washed by the surf of the broad Atlantic; on the western or inner side, beyond the sand dunes, are large grassy meadows or salt marshes, invaded by shallow bays; at low tide broad mud flats, more or less covered with eelgrass, extend for miles along the shore, furnishing ideal feeding grounds for brant and other sea fowl. Between here and the flats around Muskeget Island, a few miles south, immense numbers of brant congregate in the spring; they have been moving gradually northward from their winter resorts, gaining in numbers, by picking up those that have wintered farther north, as they went along. The spring flight is leisurely and largely dependent on the weather and the direction of the wind; cold, northerly, or easterly winds hold them back; but warm days and favorable southwest winds are sure to start them moving. Dr. Leonard C. Sanford (1903) has well described the departure of the brant from the Monomoy flats, which usually takes place in the latter part of April, as follows:

Some time in April comes a pleasant day, warm and sunny, with a southwest wind. The several thousand brant in Chatham Bay feed greedily until the rising tide removes their food from reach. Now they assemble in deep water in the center of the bay, study the weather, and discuss the advisability of journeying toward their summer home. Soon 15 or 20 birds take wing, fly back and forth over the others, honking loudly, and circling ever higher until they have reached a considerable altitude; then the long line swings straight, headed northeast. Out over the beach, over the ocean it goes, and the birds in it will not be seen again. Then another flock follows, taking exactly the same course; flock after flock succeeds and the movement is kept up until dark. You may sit in the blind next day or sail across the bay, you will see no brant save a few stragglers; branting is through for the year.

Up to this time the flight has been wholly coastwise, marshaling the hosts for the main flight. But here there is apparently an occasional, if not a regular, offshoot from the main flight, which migrates overland from Long Island Sound to the St. Lawrence River, for Dr. Louis B. Bishop (1921) writes:

Prof. A. E. Verrill informed me that on May 17, 1914, he saw, with Mr. G. E. Verrill, many flocks of brant flying north up the Housatonic Valley near the mouth of the Housatonic River; that most were high in the air, but some almost within gunshot; also that he saw others flying northwest while at Outer Island, Stony Creek, about May 22.

But the main flight, when it leaves Cape Cod, flies northeastward to the Bay of Fundy, across the neck of Nova Scotia to the Gulf of St. Lawrence, in the vicinity of Prince Edward Island. Here, according to E. H. Forbush (1912)—

about June 1 those in the district around Charlottetown (which probably comprise a great part of the Atlantic coast flight) begin to assemble in Hillsborough Bay, outside of Charlottetown Harbor, on the south side of the island. Here they gather, between St. Peters and Governors Island, in preparation for their northern journey. From June 10 to 15 they leave in large flocks. Sometimes four or five such flocks follow one another, about a mile apart.

Richard C. Harlow writes to me that—

during the last half of May the brant fairly swarm in the sheltered bays and channels along the coast of Northumberland County, New Brunswick. Here in certain favored places are channels from 1 to 5 miles wide, lying between the mainland and the sandy coastal islands. Where the water is shallow an abundant growth of eelgrass occurs, and here, protected from the open seas beyond the islands, the brant find an ideal haven. There is a noticeable increase of the birds since spring shooting was abolished, and during the last 10 days of May I have seen the water dotted with them in flocks ranging from three or four to hundreds. During a 10-mile run in this region flocks were rising every few minutes, and the din from their honking was terrific, especially in the early morning and evening. The main flocks leave in a body about June 1, though stragglers linger up until June 10. As the time draws near for their departure they gather into larger flocks and grow still more noisy, but no premature mating tendencies are observed.

Harrison F. Lewis has sent me the following notes on the movements of brant on the north shore of the Gulf of St. Lawrence:

Warden P. G. Rowe, of Prince Edward Island, was sent to the Bay of Seven Islands last spring (1922) to protect the brant while they passed there, as they do every spring. He arrived there on May 17 and found at that time some hundreds of brant in the bay. He was unable to get near them. Natives of Seven Islands told him that they had arrived there about the 1st of May; that they came into the bay from the southwest by a "pass" never used by the great brant flight, as though they had descended the St. Lawrence; and that the same thing occurred each spring. These brant all left the bay some days before the usual big flight of brant began to arrive.

This may be the flight, referred to above, which migrates overland from Long Island Sound and which naturally would arrive earlier. When Dr. Charles W. Townsend and I reached the Bay of Seven Islands, May 23, 1909, no brant were there, nor did we see any on our cruise farther eastward down the coast; but when we returned to Seven Islands on June 22 we were told by several observers that there had been a big flight there in the meantime and all had gone but one. They came the last of May and were practically all gone, in the direction of Hudson Bay, before June 20. Mr. Lewis's notes confirm these statements very well, as follows:

I may also say that the big flight of brant from Prince Edward Island passed the Bay of Seven Islands in the early days of June (5–15) this year. They all enter the bay by the two easternmost "passes," chiefly by the most eastern "pass" of all. The number entering the bay was checked carefully by Warden Rowe, who was able to count them very well, as they passed at a low elevation, chiefly in flocks of two or three hundred or less. He sets the total number at at least 60,000. I believe that this includes practically the entire species.

Their route overland to Hudson Bay and from there to their northern Arctic breeding places is not so easily traced. They have not been noted in spring on the west coast of Hudson Bay, where they are reported as common in the fall. They may migrate up the east coast or they may fly straight north overland to Ungava Bay. Mr. Lucien M. Turner's notes, which seem to point to the overland route, say:

At Fort Chimo they arrive from the 20th of May to the 20th of June. They fly past the station of Fort Chimo over the water in the Koksoak. At times they are as high as 100 yards, and oftener only a few feet above the water or running ice. They come at a time when it is almost impossible to get at them on account of ice, and if this is not present they fly too high. They follow the sinuosities of the river and only cross such points that they can see over. Thousands of them are seen every spring and never one of them in the fall. They are reported by the Eskimo to fly southward over Hudson Bay. The route taken in the spring is to the west of Anticosti Island, thence north to the "Height of land" where the Koksoak River descends, along which they fly northward. They appear fatigued when they reach Hudson Strait, but with rapid beat of wing they pursue their course to the unknown regions beyond.

From here they apparently pass to the westward of Baffin Land and spread out over their breeding grounds from Melville Island to northern Greenland. Dr. Donald B. MacMillan tells me that they appear in large numbers in spring on the southwest coast of Baffin Land and apparently cross the center of that land, over a chain of lakes, to Baffin Bay and fly straight north to Ellesmere Land and northern Greenland, arriving at Etah about June 1. These early arrivals must be the vanguard of the earliest flight. He says that Sverdrup has seen them in Eureka Sound and Peary has seen them at the north end of Axel Heiberg Land, Cape Thomas Hubbard, and on the north coast of Grant Land, the northernmost land known.

Nesting.—The best account I have of the nesting habits of this species comes in some notes sent to me by the Rev. F. C. R. Jourdain; he writes:

Although closely allied to the barnacle goose, the brant differs widely from it in its breeding habits. While the barnacle goose has attained security from marauders by placing its nest in inaccessible spots on mountain sides, the brant prefers as a rule to breed on the little holms and outlying islets which fringe the coast. Here for centuries they have bred undisturbed save for an occasional visit from a sealing sloop, but of late years, since most of these vessels have been fitted with auxiliary oil engines, the sloops have taken to working systematically the eider holms for eggs and down, and the geese in consequence have been driven from many of their old haunts by indiscriminate shooting during the breeding season. No doubt a good many pairs still breed on some of the less accessible islands, but apparently many birds now nest also on the mainland, especially on the grassy islands formed by the many channels into which the rivers tend to divide when the valleys open out. The diminution in the number of Arctic foxes by trapping has probably rendered this type of site less dangerous than formerly, but we had evidence of nests of this kind being destroyed by foxes.

On Moffen Island, which is practically a huge shingle bed, we found several pairs breeding among the shingle and in one case on the "slob-land" not far off. As in the case of the other geese, incubation is performed by the goose alone, but the gander mounts guard by her side and is generally to be found on duty. The normal clutch is apparently 3 to 5, but Koenig records one instance of 6 and Kolthoff 7. Several cases of 2 and even 1 incubated egg occurred, but are probably due to the fact that a previous clutch had been taken. The nests vary a good deal in appearance. On low islets they may be found among the eider ducks' nests, and are substantially built of mosses, lichens, and other vegetable matter, with a plentiful lining of light grayish down. When flushed from incubated eggs, both birds show great anxiety, flying round and round with anxious "gaggling." On shingle banks, where little nesting material is available, the nest may be little more than a hollow among the stones, lined with down and perhaps a few bits of seaweed or lichen. While hiding in order to watch the bird back to the nest, it was interesting to notice the head and upper neck of the goose silently watching us from behind a bank of shingle, and all the time keeping as much as possible out of sight, so that the contest resolved itself into one of endurance.

The nests found by Doctor MacMillan's (1918) party in northern Greenland were apparently similar in location and construction. W. Elmer Ekblaw writes to me:

The brant, whether it comes from Europe or from America, arrives in northwest Greenland about mid-June. It appears in rather large flocks, from about 15 to 50 in number, generally sweeping along the coast in low, long files. When they arrive the ice is well broken and open, so that they find no lack of feeding grounds. Either they are mated when they arrive, or they mate soon after, without any distinct mating season, because they proceed at once to the business of nesting and incubation. They gather along the shores of the rocky islets, and usually group their nests somewhat gregariously. Lyttleton Island, McGarys Rock, and Sutherland Island are favorite nesting places. The nests, like those of the eiders, are placed among the tussocks of grass or sedge growing on the low flat ledges of the islets where they nest. The nests are heavily lined with dark down, and the full clutch of eggs seems to vary considerably. Some nests held but 4 eggs, while a few had 11 or 12. The period of incubation is between three and four weeks.

Col. John E. Thayer (1905) has a nest and four eggs taken by J. S. Warmbath on Ellesmere Land, of which he says:

The nest was found June 17, 1900, on a ledge of rock, 20 feet from the ground, among elder ducks' and glaucous gulls' nests. Both birds were shot. The female was shot on a slight elevation above the nest and the male in the water near it.

Eggs.—The brant ordinarily lays from 3 to 5 eggs, though as many as 6 or 8 have been recorded. The 4 eggs in the Thayer (1905) collection, referred to above, are described as "dull creamy white and smaller than eggs of the black brant." Four eggs in the author's collection, taken on Eider Duck Island, northern Greenland, June 18, 1917, by W. Elmer Ekblaw, are creamy white in color, clean and smooth, but not glossy, and elliptical ovate in shape. The measurements of 54 eggs, in various collections, average 71.7 by 47.1 millimeters; the eggs showing the four extremes measure **81.1** by 49.4, 77 by **51.3**, **60.4** by 38.8, and 61.8 by **36.5** millimeters.

Young.—The period of incubation is not definitely known, but it is probably about four weeks. The female alone incubates, but the male stands guard near by and helps defend the nest. Mr. Jourdain says, in his notes:

We first met with the pretty ashy gray goslings on July 14, but Koenig mentions having seen them on July 4 and 10, so that probably the normal time for fresh eggs is during the first fortnight of June. The sailors on our boat succeeded in rearing one gosling, providing freshly cut pieces of turf daily and allowing the bird to feed itself.

On July 17 a flock of about 60 geese, which had shed their flight feathers and were temporarily unable to fly, was met with on a lagoon up the Sanen River. They kept close together in a compact body, the barnacles usually leading and the brent and pink-footed following. When the flock had been fired at several times and about 18 specimens secured, the pink-footed geese separated themselves from the rest of the party and led the way toward land, running with considerable speed and soon putting themselves out of danger, an example soon followed by the rest.

Mr. Ekblaw writes to me:

As soon as the young are hatched they hurry to sea. They dive and swim agilely almost as soon as they reach the water. Several broods of young are wont to congregate together, the mothers aiding one another in vigilant guidance and guard over the flock. They are exceedingly shy, and it is well-nigh impossible to approach them except by surprise. They grow and develop fast, so that by mid-September they are ready for their departure. As soon as the ice begins to form in the fjords the brant begin to leave. By October 1 the last long, low-flying files of migrating brant have passed the outer capes; almost nine months slip by before they appear again.

Plumages.—I have never seen the downy young of the brant, but it is described in Witherby's Handbook (1921) as follows:

Down of crown, center of nape, upper parts and sides of body pale mouse gray, some tipped grayish white; patch on lores and lines above eye sepia; chin, throat, and rest of neck white; remaining under parts ashy white, down with dusky-brown bases.

The juvenal plumage during the first fall is similar in a general way to that of the adult, except that the black areas are duller and more brownish, the white neck patches are lacking, the feathers of the scapulars, and median wing coverts are broadly edged with buffy white, and those of the back and other coverts are more narrowly edged with the same; the secondaries are also narrowly edged with white. Molting and wear of the contour feathers, especially about the head and neck, during the winter produce an advance toward maturity; the white neck patches appear in January, but the wings retain the juvenal characters until the following summer. A complete molt in summer produces a plumage indistinguishable from that of the adult.

Food.—According to Witherby's Handbook (1921) the food of the brant on its breeding grounds consists of "grass, algae, moss, and stalks and leaves of arctic plants (*Eriophorum*, *Ranunculus*, *Cerastium*, *Oxyria*, and *Saxifraga*)." The "young feed on *Gramineae* and *Oxyria*."

While on our coasts their chief food is eelgrass (*Zostera marina*), which grows so extensively in our shallow bays and estuaries. At certain stages of the tides, the last half of the ebb or the first half of the flood, when the beds of eelgrass are uncovered or covered with shallow water, the brant resort to them in large numbers to feed. They prefer the roots and the whitish lower stems, but they eat the green fronds also. As soon as the water is shallow enough for them to reach the grass by tipping up they begin to feed, and they keep at it until the tide again covers the flats too deeply. While most of the birds are feeding with heads and necks below the surface there are always a few sentinels on watch to warn them of approaching danger. They pull up much more eelgrass than they can eat at once; this floats off with the tide and often forms small floating islands, far off from shore, to which the brant resort at high tide to feed again. John Cordeaux (1898) says that the longer pieces of *Zostera* "are neatly rolled up, like ribbons, in their stomachs"; they also devour the fronds of some species of algae, crustaceans, mollusca, worms, and marine insects. Gätke says that at Heligoland, when the sea is calm, small companies will approach the cliffs and pick off the small mollusca and crustaceans.

I have at times been greatly entertained in watching a flock of brant feeding in shallow water, close inshore, the greater portion of the birds upside down, their rumps and tails showing the white coverts, only visible as they greedily tear at the blades and roots of the grass wrack, whilst others are seizing the floating fragments of the plant, broken off and dislodged by their mates; and on the outside there are always some with heads held high, ever on the watch, and ready to give alarm. All the time they keep a continuous, noisy gabbling and grunting, the rear birds constantly swimming forward to get in advance of their fellows, a procedure which I have known, more than once, bring them within range of an ordinary sporting gun.

Brant in captivity are especially fond of barley and will eat corn and other grains. George H. Mackay (1893) says:

Two wing-tipped birds I have in confinement eat with avidity the alga (*Ulva lactuca*). They also eat *Zostera marina*, preferring the white portion farthest from the extremity of the blade. They cut this up by chewing first on one side and then on the other of their mandibles, which cuts the grass as clean as if scissors had been used. The motion reminds one strongly of a dog eating, the bird turning its head much the same way. They are fond of whole corn and common grass. These confined birds drink after almost every mouthful, from a pan of fresh water. The wild birds living in this neighborhood have no opportunity of obtaining fresh water.

Doctor MacMillan tells me that he has seen brant, when they first arrive in Baffin Land in the spring, feeding on the black lichens which grow on the rocks on the uplands.

Behavior.—Brant do not ordinarily fly in V-shaped flocks, like Canada geese, but in long undulating lines, spread out laterally in straight company-front formation, or in a curving line, or in an irregular bunch, and without a definite leader. When migrating overland they fly high, but when traveling along the coast they usually fly within a few feet of the water. Their flight is apparently slow and heavy, but it is really swifter than it seems. A flock of on-coming brant is a thrilling sight to the expectant gunner; he can recognize afar the long wavy line of heavy black birds; as they draw near, the white hind parts show up in marked contrast to the black heads and necks; and soon he can hear their gabbling, grunting notes of greeting to his well-placed decoys. They are naturally shy birds, and we seldom got a shot at the passing flocks when anchored off the shore in small boats. But on their feeding grounds they are more fearless and will decoy well to live or even wooden decoys around a well-concealed blind. Brant can swim well, but do not dive unless hard-pressed. They prefer to skulk and hide by stretching the neck out on the water or in the grass. They are very fond of sand and like to rest on sandy points and sand bars.

Mr. Cordeaux (1898) describes the voice of the brant very well, as follows:

The common cry or call note of the brant is a loud metallic *chronk, chronk.* The confused gabbling and mixed cries of a flock can be heard at an immense distance at sea. They have another, and double, note, which has been likened to the word *torock*, constantly repeated on the wing; and the alarm cry is a single word, *wauk.*

Dr. D. G. Elliott (1898) says:

It has a peculiar guttural note, which is frequently uttered, resembling *car-r-r-rup*, or *r-r-r-rouk*, or *r-r-rup*, and with a rolling intonation, and, when a large number of these birds are gathered together, the noise they make is incessant and deafening. I have been in the vicinity of a bar on which were congregated many thousands of brant, and their voices made such a din that it was difficult to hear one's own in speaking, and when they rose at the report of a gun the sound of their myriad wings was as the roar of rushing waters.

Fall.—Winter comes early in the far North, and the brant are forced to start on their fall migration early in September or even late in August. The route differs only slightly from that taken in the spring. They now migrate down the west coast of Hudson Bay, cross eastern Canada to the Gulf of St. Lawrence, which they reach late in September, cross the neck of Nova Scotia to the Bay of Fundy, and then head straight for Cape Cod, where they usually begin to arrive about the middle of October. So far their flight has been more rapid than in the spring, but from here on their movements are more leisurely and they scatter along the coast, lingering at favorable spots until well into the winter.

Game.—From the standpoint of the epicure the brant is one of our finest game birds, in my opinion *the* finest, not even excepting the far-famed canvasback. I can not think of any more delicious bird than a fat, young brant, roasted just right and served hot, with a bottle of good Burgundy. Both the bird and the bottle are now hard to get; alas, the good old days have passed.

A few brant are shot, as they migrate along the coast, by gunners anchored offshore for coot shooting; the long undulating line of big black birds with white hind quarters is easily distinguished from the irregular flocks of the scoters; a thrill of anticipation runs along the line as word is passed from boat to boat; the brant will not decoy to the wooden blocks of coot shooters, but they may pass over or near one of the boats or swing and fly along the line near enough for a shot; but more often they give the boats a wide berth, and the disappointed gunner fires in vain at the coveted birds, which are probably farther away than they seem to be.

Real brant shooting may be had at only a few favored localities where the birds are wont to congregate, and then only with an elaborate equipment. It was a sorry day for the brant shooter when spring shooting was abolished, for the brant is the one bird above all others for which spring seems to be the natural shooting season. Brant seem to be more plentiful in the spring than in the fall and to linger longer on their favorite feeding grounds at that season. April brant shooting at Monomoy has long been famous in the annals of Massachusetts sportsmen, when formerly splendid sport could be enjoyed. E. H. Forbush (1912) says:

> Hapgood gives a record of 44 birds killed from one of these boxes *at one shot*, and states that 1,000 or 1,500 were killed in a season. This was many years ago, before the formation of the brant clubs. No such number has been killed in recent years. The average number killed by the members of the Monomoy Branting Club for 34 years, during the Hapgood régime, is a trifle over 266 birds per year.

Now that we can shoot only in the fall, no such sport is obtainable; there are still plenty of brant, as the spring flights are enormous, but on the Monomoy Flats, where my limited experience has been gained, the brant seem to be comparatively scarce in the fall, and they are so constantly disturbed by the busy fleet of scallop fishermen's power boats that they do not remain to feed.

On Monomoy our brant shooting is done from boxes located on favorable points or sand bars near the feeding grounds. The box is well made and water-tight, 6 feet long, 4 feet wide, and 4 feet deep; big enough for three men; it is sunken into the sand deep enough to be covered at high tide; numerous bags, sometimes 50 or 60, of sand are piled around it to hold it in place; and if it is in a grassy place, which helps to conceal it, the sloping sides of the pile may be thatched with marsh grass woven into the meshes of poultry netting, held in place by stakes and weighted with sand. Unless there is a natural sand bar near the box, one must be made, on which the live decoys are located and where the wild birds may alight. Live decoys are preferable, but brant will come to good wooden decoys if properly placed; a supply of both is desirable. The brant feed at low tide away off on the eelgrass beds; but as the rising tide covers the grass too deeply, they are driven to seek other feeding grounds or sanding places, and in flying about will often come to the decoys. The best shooting then is for a short time only at about half-flood tide and again at about half ebb, while the birds are moving. The morning tides are considered the best, so it is often quite dark when we tramp down through the marsh to our box, heavily laden with decoys, guns, and ammunition and encumbered with rubber boots and oilskins, for it is cold and wet work. We set out the decoys, bale out the box, and sit low on a wet seat, our eyes just above the rim of the box, and scan the flats for distant flocks of brant. Occasional shots at passing birds or small bunches on their way seem like fair sport. But when a large flock swims up to the decoys on the rising tide or flies up and settles on the bar among them, it is exciting enough, but it seems like wanton slaughter to fire a battery of guns at a given signal into a dense mass of birds. Perhaps a dozen or a score of birds are killed or wounded and we jump out of the box and go splashing off through the mud and water to retrieve the cripples. When the rising tide finally drives us out of our box, we may have a large bunch of birds to lug back to the club house, but have we given them a fair show for their lives?

Brant shooting in Great South Bay, Long Island, and in southern waters is usually done from batteries, such as are used for canvas-back-duck shooting. As this means wing shooting most of the time, it seems more sportsmanlike. The battery may be anchored near their feeding ground or near a floating mass of eelgrass, known as a "sea-weed bank," to which the brant resort to feed.

Winter.—The winter home of the brant is along the Atlantic coast from New Jersey southward. Probably most of the birds spend the winter on the coast of Virginia and North Carolina. T. Gilbert Pearson (1919) writes:

In Pamlico Sound the long extended lines of submerged sandbars and mud-flats, with their abundant supplies of eelgrass, make an ideal winter resort for the brant. They arrive from the north usually early in November, but the exact date depends much upon weather conditions. In flight they usually go in compact flocks without any apparent leader. They move slowly and often appear loath to leave a favorite feeding ground, even returning to it many times after being disturbed.

On clear winter days, as one sails along the reefs in the region about Ocracoke or Hatteras, flocks of brant, disturbed from their feeding areas, arise in almost constant succession for miles, their numbers running far into the tens of thousands. When heavy winds arise these large rafts are broken up.

and later when the birds are flying singly or in small companies they readily draw to decoys. It is then that the gunners get in their most telling work, bags of 75 or 100 birds being sometimes taken in a day. Near Cape Hatteras I once lay in a battery near a local gunner, who shot 50 brant between the hours of 10 a. m. and 2 p. m., and the size of his kill in four hours occasioned no particular comment in the neighborhood.

DISTRIBUTION

Breeding range.—Arctic regions north of eastern North America, Europe, and Western Asia. East to Spitsbergen, Franz Joseph Land, Nova Zembla, Kolguev, and the Taimyr Peninsula. On both coasts of Greenland, south to about 70° N., and north to at least 81° or 82° N. South to about 74° on islands around the Gulf of Boothia, Prince Regent Inlet, and Wellington Channel. West to about 100° or 110° W. Seen in summer and probably breeding as far north as explorers have been, on the Parry Islands, Axel Heiberg Land, Grant Land, and Greenland. Probably intergrades with *nigricans* at both the eastern and the western limits of its range.

Winter range.—Atlantic coast of United States. Regularly from New Jersey to North Carolina; more rarely eastward to Long Island and Massachusetts (Marthas Vineyard); and rarely south to Florida. Coasts of northwestern Europe, from the Baltic and North Seas, to the British Isles and France; occasionally to Morocco, the Mediterranean, and Egypt. Occurs more or less regularly on the Pacific coasts of Canada and United States.

Spring migration.—Fully described above. Early dates of arrival: New York, Long Island, February 15; Rhode Island, Block Island, March 8; Massachusetts, Vineyard Sound, March 10; Maine, Englishman Bay, April 22; Quebec, Bay of Seven Islands, May 1; latitude 79° N., May 30; northern Greenland, Etah, June 1; Grinnell Land, latitude 82°, N., June 7; Boothia Peninsula, June 8; Wellington Channel, June 2.

Late dates of departure: North Carolina, first week in April; Rhode Island, April 28; Massachusetts, Cape Cod, May 17; Prince Edward Island, June 12; Quebec, Bay of Seven Islands, June 15.

Fall migration.—Early dates of arrival: Massachusetts, Cape Cod, September 11; Rhode Island, September 17; New York, Long Island, September 8; New Jersey, Barnegat Bay, October 14. Main flight reaches the Gulf of St. Lawrence late in September and New England in October. Late dates of departure: Maine, December 8; Massachusetts, Cape Cod, December 14; New York, Long Island, December 20.

Casual records.—Has wandered east on the fall migration to Labrador (Nain, October, 1899) and Nova Scotia (Sable Island,

November 7, 1908). Stragglers, probably from a Hudson Bay migration route, have been recorded from Ontario (Toronto, December 2, 1895, and November 12, 1899), Manitoba (Shoal Lake and Lake Manitoba), Michigan (Monroe, November 8, 1888), Wisconsin (Hoy's record), and Nebraska (Omaha, November 9, 1895). Occurs often, perhaps regularly, on the Pacific coast: British Columbia (Comox, December 13, 1903), and California (Humboldt County, January 30, 1914). Brooks (1904) says that "about 8 per cent of the brant in Comox Bay are the eastern species." Accidental in the West Indies (Barbados, November 15, 1876), Louisiana, and Texas.

Egg dates.—Greenland: Four records, June 14 to July 13. Ellesmere Land: One record, June 17.

BRANTA BERNICLA NIGRICANS (Lawrence)

BLACK BRANT

HABITS

It seems strange that this bird (which is apparently only a subspecies of the eastern brant), so abundant on the Pacific coast and such a rare straggler on the Atlantic coast, should have been first recognized in New Jersey and described from a specimen taken at Egg Harbor. George N. Lawrence (1846), who described it, says:

When on a shooting excursion some years since, at Egg Harbor, I noticed a bird flying at some distance from us which our gunner said was a black brant. This was the first intimation I had of such a bird. Upon further inquiry of him, he informed me he had them occasionally, but they were not common. I have learned from Mr. Philip Brasher, who has passed much time at that place, that, speaking to the gunners about them, they said they were well known there by the name of black brant, and one of them mentioned that he once saw a flock of five or six together. Since then two others have been obtained at the same place, one of which I have in my possession.

Spring.—Chase Littlejohn, in some notes sent to Major Bendire, says of the migration of black brant across the Alaska Peninsula:

Thousands of these geese pass a mile or two offshore each spring on their way north; they follow the coast line from the eastward until they come to Morzhovia Bay, where they sheer off for the Bering Sea. There bay and sea almost meet, and as they have a great aversion to flying over the land they select a narrow portage from the bay to a long lake, which is separated from the Bering Sea by a very narrow sand bar, not over 100 yards wide, and instead of crossing the bar they fly to the opposite end of the lake, fully 2 miles, and follow the outlet of the lake into Bering Sea and back to where you would suppose they would have crossed in the first place, and then continnue on their way north. In years gone by I think there was a passage through, but by the action of the sea it has been closed. But the geese do not care to forsake their old route and consequently break through their aversion of flying over the land and cross the narrow portage. A few birds go farther on and fly through False Pass and probably passes farther on, but the great majority take the Morzhovia Bay route.

The spring migration of the black brant in northern Alaska and the circumstances surrounding it have been so attractively portrayed by Dr. E. W. Nelson (1881) that I can not refrain from quoting parts of what he says about it, as he observed it at St. Michael, Alaska; he writes:

The long reign of ice and snow begins to yield to the mild influence of the rapidly lengthening days; the middle of May is reached, and the midnight sky over the northern horizon blushes with delicate rose tints, changing to purple toward the zenith. Fleecy clouds passing slowly across the horizon seem to quiver and glow with lovely hues, only to fade to dull leaden again as they glide from the reach of fair Aurora. The land, so lately snow-bound, becomes dotted with pools of water, and the constantly narrowing borders of the snow soon make room for the waterfowl which, with eager accord, begin to arrive in abundance, some upon lagging wings, as if from far away, others making the air resound with joyous notes as they recognize some familiar pond where, for successive seasons, they have reared their young in safety, or, perhaps a favorite feeding ground. At this time the white-fronted and Hutchins geese take precedence in numbers, though, to be sure, they have been preceded for two weeks by the hardy pintail duck, the common swan, and, lastly, that ornithological harlequin, the sandhill crane, whose loud rolling note is heard here and there as it stalks gravely along, dining upon the last year's berries of *Empetrum nigrum*, when, meeting a rival, or perchance one of the fair sex, he proceeds to execute a burlesque minuet.

A few days later, upon the mirror-like bosoms of myriads of tiny lakelets, the graceful northern Phalaropes flit here and there or swim about in pretty companies. At length, about the 20th of May, the first barn swallow arrives, and then we begin to look for the black brant, the "*nimkee*," as it is called by the Russians, the "*luk-lug-u-nuk*" of the Norton Sound Eskimo. Ere long the *avant-courier* is seen in the form of a small flock of 10 or 15 individuals which skim along close to the ice, heading directly across Norton Sound to the vicinity of Cape Nome, whence their route leads along the low coast to Port Clarence where, I am told by the natives, some stop to breed; but the majority press on and seek the ice-bordered northern shore of Alaska and even beyond to unknown regions far to the north.

The 22d of May a native came in bringing a lot of geese and reporting plenty of black brant up the "canal." For the benefit of the unfortunate few who have not been at St. Michael I may explain that the "canal" is a narrow and shallow tidal channel which separates St. Michael Island from the mainland and is bordered on either side by a stretch of low, flat land abundantly dotted with brackish ponds and intersected by numerous small tide creeks. As would be surmised, this forms a favorite haunt for various kinds of waterfowl.

Preparing the tent and other paraphernalia, two of us, accompanied by a couple of natives, started out the next morning with a sled and team of five large dogs, driven tandem, just as the sun gilded the distant hilltops and gave a still deeper tint to the purple haze enveloping their bases. The sharp, frosty air and the pleasurable excitement of the prospective hunt, after months of inactivity, causes an unusual elation of spirits, and with merry jests we speed along until, in a short time, we approach a low, moundlike knoll rising in the midst of innumerable lakelets. A strange humming, for which we were at first unable to account, now becomes more distinct, and we perceive its origin in the united notes of scores of flocks of brant which are dispersed here and there over the half bare ground. Some sit along the edges of the snow banks or upon the ground, still sleeping, while others walk carelessly about or plume themselves in preparation for the work before them. Their low, harsh, gutteral *gr-r-r-r, gr-r-r-r* rises in a faint monotonous matinal whose tone a week later may waken the weird silence in unknown lands about the Pole.

Reaching the knoll before mentioned, we pitch our tent, and after tieing the dogs to keep them within bounds we separate to take positions for the morning flight. Each of the party is soon occupying as little space as possible behind some insignificant knoll or tuft of grass that now and then breaks the monotonous level. The sun rises slowly higher and higher until at length the long, narrow bands of fog hovering over the bare ground are routed. Now we have not long to wait, for, as usual at this season, the lakes, which are frozen over nightly, open under the rays of the sun between 7 and 9 in the morning and start the waterfowl upon their way. The notes, which until now have been uttered in a low conversational tone, are raised and heard more distinctly and have a harsher intonation. The chorus swells and dies away like the sound of an aeolian harp of one or two heavy bass strings, and as we lie close to the ground the wind whispers among the dead plants in a low undertone as an accompaniment; but while we lay dreaming the sun has done its work; the lakes have opened, and suddenly a harsh *gr-r-r-r, gr-r-r-r, gr-r-r-r* causes us to spring up, but too late, for, gliding away to the northward, the first flock goes unscathed. After a few energetic remarks upon geese in general and this flock in particular, we resume our position, but keep on the alert to do honor to the next party.

Soon, skimming along the horizon, flock after flock is seen as they rise and hurry by on either side. Fortune now favors us, and a large flock makes directly for the ambush, their complicated and graceful evolutions leading us to almost forget why we are lying here upon our face in the bog with our teeth rattling a devil's tattoo in the raw wind. On they come, only a few feet above the ground, until, when 20 or 30 yards away, we suddenly rise upon one knee and strike terror into the hearts of the unsuspecting victims. In place of the admirable order before observed, all is confusion and, seemingly in hope of mutual protection, the frightened birds crowd into a mass over the center of the flock, uttering, the while, their ordinary note raised in alarm to a higher key. This is the sportsman's time, and a double discharge as they are nearly overhead will often bring down from 4 to 10 birds. Scarcely have the reports died away when they once more glide along close to the ground; the alarm is forgotten; order is again restored, and the usual note is heard as they swiftly disappear in the distance. Thus they continue flying until 1 or 2 o'clock in the afternoon when, after a pause of three or four hours, they begin again and continue until after sundown.

Nesting.—At Point Barrow, Alaska, according to John Murdoch (1885)—

the black brant appear at the end of the main spring migrations of the waterfowl, but in no very considerable numbers, following the same track as the eiders. A few remain to breed and are to be seen flying about the tundra during June. The nest is placed in rather marshy ground and is a simple depression lined with down, with which the eggs are completely covered when the birds leave the nest. The birds sometimes begin to sit on 4 eggs and sometimes lay as many as 6.

MacFarlane (1891) sent to Washington 650 eggs of this brant from Fort Anderson, obtained by the Eskimos on the Arctic coast of Liverpool Bay, where it was exceedingly abundant, but he adds little to our scanty knowledge of its breeding habits. He wrote to Professor Baird, concerning a visit to an island in Liverpool Bay on July 4, 1864, as follows:

Bernicla brenta breeds on small islands on the small lakes occurring on this island. It scoops a hole in the sand or turf composing the island and lines it with down taken from the body of the female. They frequently nest in small parties, but a pair will sometimes select an island for themselves. Very few specimens were seen, though numerous nests containing broken eggs were met with; these had evidently been destroyed by white foxes, gulls, owls, and crows. But for the depredations committed by them we should have made a superb collection of eggs of *Somateria, Bernicla hutchinsii, Anser gambelli, Columbus articus,* gulls, and terns. A great many nests of each and all of these species were found without eggs, their contents having been destroyed. The entire damage inflicted on the poor birds in one season must be enormous.

Rev. A. R. Hoare sent me a fine nest and 5 eggs of the black brant, taken at Tigara, near Point Hope, Alaska, in June 1916; it consisted of a great mass of down on a small low island in an extensive marsh; judging from the two photographs, which came with it, it must have been quite conspicuous. He found other nests sunken in moss and grass near the lakes out on the tundra. There is a nest and 5 eggs in the collection of Herbert Massey, Esq., taken by W. E. Snyder, at Admiralty Bay, Alaska, on June 16, 1898; the nest which consists of a large quantity of down, was in a depression in the dry tundra. I received 11 sets of black brant's eggs, all consisting of 5 eggs, with the nests, from Point Barrow, Alaska, in 1916. The nests of this species are the most beautiful nests I have ever seen of any of the ducks and geese; they are great, soft, thick beds of pure, fluffy down, unmixed with the tundra rubbish so common in nests of other species; the down is a rich, handsome shade of "benzo brown" or "deep brownish drab," flecked with whitish; it must make a warm and luxurious blanket to cover the eggs.

Eggs.—The black brant lays from 4 to 8 eggs, but 5 seems to be by far the commonest number. The prevailing shapes are elliptical ovate or elliptical oval; some are elongate ovate and a few are nearly ovate, rounded at the small end. The colors are "cartridge buff," "ivory yellow," "pale olive buff," or "cream color." The original color is often much obscured by stains or mottlings of various buffy or brownish shades, such as "cinnamon buff" or "ochraceous buff"; some of the eggs seem to be wholly of these darker shades; in others the stains have apparently worn off or been scratched off, exposing the original color. The shell is smooth and sometimes quite glossy.

The measurements of 107 eggs in the United States National Museum and the author's collections average 71.1 by 47.4 millimeters; the eggs showing the four extremes measure **79** by 47.5, 68 by **51**, **66** by 45 and 75.5 by **43.5** millimeters.

Plumages.—The downy young black brant is thickly covered with soft down in dark colors; the upper half of the head, including the lores, to a point a little below the eyes is "fuscous" or "benzo brown"; the chin is white; the back varies from "benzo brown" to "hair brown," darkest on the rump; the flanks and chest shade from "hair brown" to "light drab," fading off nearly to white on the belly and throat.

I have seen no specimens showing the change from the downy to the juvenal plumage. But a bird in the juvenal plumage, taken September 5, has the whole head, neck and breast plainly colored, "fuscous" to "fuscous black," with only faint traces of the white markings on the neck; the feathers of the back are edged with grayish white or buffy white; the juvenal wing has broad, conspicuous, pure white edgings on the lesser and the greater coverts, the secondaries, and the tertials, broadest on the greater coverts; the belly and flanks are plain "fuscous," with no barring.

Molting begins first on the head and neck; the white neck patches are often, but not always, conspicuous in October; they are well developed by spring. Aside from a partial molt of body and tail feathers there is not much change during the winter and spring. Young birds in May and June are still decidedly juvenal, much worn and faded. At the following post-nuptial molt, when a little over a year old, the young bird undergoes a complete molt and becomes practically adult in plumage, with white neck patches, barred flanks, and plain dark wings without any light edgings.

Food.—The black brant, like its eastern relative, is a decidedly maritime bird, living on salt water and feeding on the grassy mud flats; it never comes in to the uplands to feed. It seems to feed almost entirely on the leaves and roots of various marine grasses, mainly *Zostera*. In connection with its vegetable food it picks up various small mollusks, crustaceans, and other forms of marine animals. Mr. W. Leon Dawson (1909) says that they "not only dip but dive as well."

Behavior.—Doctor Nelson (1881) gives such a good account of the flight of the black brant that I quote it in full, as follows:

The flight of this species is peculiar among North American geese and bears a close resemblance to that of the eider and other species of heavy-bodied short-winged sea ducks. It has a parallel in the flight of the emperor goose except that the latter is a far heavier bird and, in consequence, the wing strokes are less rapid. In *B. nigricans* the strokes are short, energetic, and repeated with great rapidity, carrying the bird with a velocity far greater than that attained by any other goose with which I am acquainted, though probably its eastern prototype equals it in this respect.

But this is not the point upon which the mind rests when the birds are in view, for then the eye is held in involuntary admiration of the varied and graceful evolutions of the flocks, which have a protean ability to change their form without ever breaking the array or causing confusion. They are very gregarious, and two flocks almost invariably coalesce when they draw near each other. This frequently occurs until, as I have seen, it results in a single flock numbering between 400 and 500 birds. The usual size is considerably less, generally comprising from 20 to 50 or more, and it is rare to see less than 10 or 15 in a party. At times 4 or 5 individuals become detached, and until they can unite with a stronger party they fly irregularly about as though bewildered, continually uttering their harsh notes, and hurry eagerly away to join the first flock that comes in view. The order of flight is invariably a single rank, the birds moving side by side in a line at right angles to their course so that the entire strength of a flock is to be seen at a glance along its front, which at times covers several hundred yards. There is barely room enough between the individuals to allow a free wing stroke. Thus ranged, the flock seems governed by a single impulse, which sends it gliding along parallel and close to the ground, then, apparently without reason, careering 30 or 40 yards overhead, only to descend to its former level as suddenly as it was left; now it sways to one side and then to the other, while at short intervals swift undulations seem to run from one end of the line to the other. These movements are repeatedly taking place; they are extremely interesting to observe but difficult, I fear, to convey an adequate idea of in words.

The entire flock, consisting of perhaps over a hundred birds arranged in single line, is hurrying on, straight as an arrow, toward its destination, when, without warning, it suddenly makes a wide curving detour of several hundred yards, then resumes its original course only to frequently repeat the maneuver, but always with such unison that the closest scrutiny fails to reveal the least break or irregularity in the line; nor does the front of the flock swerve, excepting an occasional slight obliquity which is corrected in a few seconds.

In addition to this horizontal movement is a still more interesting vertical one which often occurs at the same time as the other but generally by itself. A bird at either end of the flock rises or descends a few inches or several feet, as the case may be, and the movement is instantly followed in succession by every one of its companions till the extreme bird is reached and the entire flock is on the new level; or it may be that a bird near the middle of the line changes its position, when the motion extends in two directions at once. These latter changes are made so regularly and with such rapidity that the distance between the birds does not appear altered in the least, while a motion exactly like a graceful undulation runs the length of the flock, lifting or depressing it to the level of the originator of the movement. These changes present to one's eye as the flocks approach, keeping close to the ground, the appearance of a series of regular and swift waving motions such as pass along a pennant in a slight breeze.

The black brant never wings its way far up in the sky, as many other geese have the habit of doing, but keeps, as a rule, between 10 and 30 yards above the ground, with more flocks below these limits than above them.

Another idiosyncrasy of this bird is its marked distaste for passing over low ranges of hills which may cross its path. A striking case of this is shown here where a low spur runs out from the distant hills in the form of a grass-covered ridge projecting several miles into the flat marshy land. This ridge is from 50 to 200 feet above the surrounding country and bars the course of the black brant. So slight an obstacle as this is enough to cause at least 95 per cent of the flocks to turn abruptly from their path and pass along its base to round the end several miles beyond, and then continue their passage. In consequence of this habit it has been a regular practice for years for the hunters to occupy positions along the front of this ridge and deal destruction to the brant, which still hold as pertinaciously as ever to their right of way.

Doctor Nelson (1887) says:

While upon the ground or in flight they have a low guttural note something like the syllables *gr-r-r-r-r*, When alarmed this note, repeated often and with more emphasis, was the only cry heard.

W. Leon Dawson (1909) writes:

From the esthetic standpoint the most interesting phase of brant life is the mellow *cronk, cronk, cronk*, which the birds frequently emit whether in flight or at rest. From the back bay near Dungeness in April rises a babel like the spring offering of a giant frog-pond, a chorus of thousands of croaking voices, among which the thrilling basso of bullfrogs predominates.

Mr. Hoare writes me that, at Point Hope, the brant "suffer very much from the depredations of the gulls and hawks." He "found many eggs scattered and broken." He says:

This summer I witnessed a battle between a male brant and three gulls. The female brant was on the nest and did not move. The gulls kept returning to the attack and were very fierce. Usually they are cowardly. Eventually the brant drove them away, although badly mauled. I could not find it in my heart to disturb the female on the nest.

Fall.—According to Mr. Murdoch (1885) the fall migration at Point Barrow begins early; he writes:

After the middle of August they begin to fly across the isthmus at Pergniak, coming west along the shore of Elson Bay, crossing to the ocean, and turning southwest along the coast. Whenever during August the wind is favorable for a flight of eiders at Pergniak the brant appear also. They, however, frequently turn before reaching the beach at Pergniak, follow down the line of lagoons and cross to the sea lower down the coast. The adults return first. No young of the year were taken till the end of August. During the first half of September, a good many flocks cross the land at the inlets as well as at Pergniak, and are to be seen resting and feeding along the lagoons and pond holes. At this season they are very shy and hard to approach, and all are gone by the end of September.

Doctor Nelson (1887) says:

Some old whaling captains assured me that they have frequently seen these birds coming from over the ice to the north of Point Barrow in fall; and to the hardy navigators of these seas this is strong evidence in support of the theory that bodies of land lie beyond the impenetrable icy barrier which heads off their advance in that direction. Perhaps it was the droppings of this bird which we found on the dreary shores of Wrangel Island, when our party from the *Corwin* were the first human beings to break in upon its icy solitude. Mr. Dall writes that on his return to the coast of California in the latter part of October enormous flocks of these birds were seen about 100 miles offshore. They were flying south and frequently settled in the water near the ship.

Along much of the route followed by the brant in the spring in such enormous numbers, they are comparatively scarce in the fall, indicating that a different route is followed. In the spring, when open water shows first near the land, they would naturally follow the coast line; but in the fall, when Bering Sea is all open, they evidently prefer to migrate far from land, stopping to rest and perhaps to feed on the open sea.

Game.—Grinnell, Bryant, and Storer (1918) say of the game qualities of this bird:

The black brant evades the devices of the hunter better than any other duck or goose. In very early days on San Diego Bay it was never seen to alight on the shore or near it. By 1875 it was almost impossible to obtain a shot at the bird from a boat, and even with a box sunk in the mud and concealed by seaweed a good bag was secured with difficulty. In 1883 a floating battery with plenty of decoys alone would enable a hunter to obtain this much-prized bird. A few years later many of the birds failed to put in an appearance at all off San Diego, probably going farther south, along the Mexican coast. Because of its habit of occasionally cutting across low sand spits to avoid a long detour in its flight, most of the hunting has been done from blinds situated beneath such a line of flight. On Tomales Bay hunters have sailed down on flocks with "blind boats," when the birds were at rest during a fog, their whereabouts being disclosed by their "gabbling" noises. The black sea brant has not been sold on the markets to any extent for a good many years. About 20 years ago consignments were shipped to San Francisco from Humboldt Bay, and the birds sold for as little as 25 cents each. Even the high price that the bird would bring at the present time does not attract it to the market because of the difficulty now attached to obtaining it.

Winter.—The black brant winters abundantly on the Pacific coast from the Puget Sound region southward, living entirely on salt water, in the larger bays and channels. Mr. Dawson (1909) writes:

Black brants are the only geese one is quite sure of seeing from the deck of a steamboat on an average winter day on Puget Sound. While they have their favorite feeding grounds upon the mud flats and in shallow bays, they are widely distributed over the open water also, and their numbers during the spring migrations are such that not all other wild geese put together are to be mentioned in comparison. They sit the water in small companies; and although they are exceedingly wary in regard to rowboats, they often permit an approach on the part of steamers which is very gratifying to the student. An exaggerated use of their long wings as the birds get under way gives the beholder the impression of great weight—an impression which is not sustained in the hand, where the bird is seen to disappointingly light; all feathers, in fact, as compared with a chunky scoter, which does not equal it in extent of wing by a foot or more.

Mr. S. F. Rathbun has sent me the following notes in regard to the black brant, as seen on the coast of Washington:

Some time during November the arrival of the black brant may be expected in the Strait of Juan de Fuca and the lower Sound. During the winter months however, it will be found in all the waters of Puget Sound, but not so commonly as in the first mentioned localities, for it appears to prefer the open waters, in such at times assembling in large flocks. And this statement may be qualified to an extent, for there are certain spots to which the bird seems partial.

In and near the eastern end of the strait lies Smiths Island, this being an abrupt bit of land not many acres in extent. Running eastward from the island is a long ledge or spit almost a mile in length, an extension as it were, parts of which are exposed only at the times of low tides, and to the north is a wide expanse of water having numerous kelpbeds. This locality is a much favored one by the black brant, and here almost any time during the winter and early spring months it may be seen in large numbers. Another place where it may invariably be found is in the general vicinity of Dungeness on the strait, and here also kelpbeds exist, for this bird appears to favor the spots where this marine growth is to be found. These localities are favorite ones in which to hunt this bird, but even so this hunting involves at times considerable work and exposure, to say nothing of the time required. One never knows what character of weather will be experienced, for sometimes a number of days will elapse before one arrives on which sport can be had. As a rule the black brant is an easy bird to decoy, and when hunting it one needs but little concealment. Often the blind used is simply some log or uprooted tree that has drifted ashore, over which will be draped in a careless way seaweed or kelp, or both, the flotsam of the beach. On the windward side the decoys are placed in the water, to be changed from time to time according to the run of tide. If the brant are flying, ordinarily but little time will elapse before some passing flock sights the decoys and almost invariably comes to them. And frequently such birds that escape the fire will—as if governed by mere fatuousness—return to be shot at again, sometimes the result being that the entire flock is killed.

This species appears to be restricted to the salt water, and its food being in the nature of marine vegetation and the smaller forms of marine life, gives somewhat of a clue to the reason why it exhibits the partiality shown for the vicinity of the beds of kelp.

DISTRIBUTION

Breeding range.—Arctic coasts of western North America and eastern Asia. East on Arctic islands to about 110° or 100° W. (Melville Island, Banks Land, etc.) and on the mainland to Coronation Gulf. Westward along the coasts of Canada, Alaska, and Siberia to the Taimyr Peninsula and New Siberia Islands. North to Banks Land (Cape Kellett, etc.) and Melville Island (Winter Harbor). Probably intergrades with *bernicla* at the eastern and the western limits of its range.

Winter range.—Mainly on the Pacific coast of the United States. North to British Columbia (Comox, Vancouver Island). South to

Lower California (San Quintin Bay and Cerros Island). Inland in Oregon (Malheur and Klamath Lakes) and in Nevada (Pyramid and Washoe Lakes). On the Asiatic coast south to northern China (Tsingtau) and Japan.

Spring migration.—Main flight is northward along the coast, but some fly overland across Alaska to the Mackenzie Valley. Early dates of arrival in Alaska: St. Michael, May 5; Point Hope, May 15; Wainwright, May 24; Demarcation Point, May 20; Point Barrow, June 5. Late dates of departure: Lower California, Cerros Island, May 10; California, San Francisco, April 24; Alaska, Yukon delta, May 22.

Fall migration.—A reversal of the spring routes. Migrants pass through Bering Sea during the last half of September and first half of October and reach California in October and November. Latest date for Point Barrow is September 21, and Kolyma River, Siberia, October 5.

Casual records.—Has wandered east to New York (3 records), Massachusetts (Chatham, spring of 1883 and April 15, 1902), and New Jersey (Great Egg Harbor, January, 1846, and Long Beach, April 5, 1877); and from the Pacific coast as far inland as Utah (Bear River). Accidental on the Hawaiian Islands (Maui).

Egg dates.—Arctic Canada: Fifteen records, June 8 to July 7; nine records, June 20 to July 6. Alaska: Eleven records, June 15 to July 4; six records, June 22 to 26.

BRANTA LEUCOPSIS (Bechstein)

BARNACLE GOOSE

HABITS

This Old World species, which probably resorts regularly to Greenland for the purpose of breeding, has been taken a number of times at various places in eastern North America and is therefore well worthy of a place on our list, as a straggling migrant, mainly in the fall.

Spring.—It is a very common bird on the west coast of Scotland, whence it takes its departure about the end of April or beginning of May. John Cordeaux (1898) quotes Robert Gray, concerning its departure as follows:

> Previous to leaving, the barnacle geese assemble in immense flocks on the open sands, at low tide, in the sounds of Benbecula and South Uist; and as soon as one detachment is on the wing it is seen to be guided by a leader, who points the way with a strong flight northwards, maintaining a noisy bearing until he gets the flock into the right course. After an hour's interval, he is seen returning, with noisy gabble, alone, southwards to the main body, and taking off another detachment as before, until the whole are gone.

Nesting.—The breeding grounds of the barnacle goose have only recently been found; the following quotation from A. L. V. Manniche (1910) seems to indicate that the species breeds abundantly in northeastern Greenland. He writes:

June 8 and 9, 1908, I got my first opportunity to study the barnacles in their real nesting territory. Up to this time the geese had led a comfortable and by me unsuspected existence in a lonely marsh and moor territory far up country—10 to 15 kilometers from the nearest salt water—east of Saelsoen, imposing by its extent and grandness of scenery. This territory, the farthest extent of which is in a northerly direction, comprises an area of some 20 square kilometers; on the north it is bordered by a mountain range, the lower slopes of which are covered by a vegetation more luxuriant than I saw in any other place in northeast Greenland.

To the east and north the marshes lose themselves in barren stony plains sprinkled with sandy spots and a few deep lying fresh-water basins bare of all vegetation. To the south the steep and barren mountain of Trekroner rises to a height of 360 meters in small terraced projections.

In the marsh and moor itself the vegetation was extremely luxuriant; as well the alpine willow as other plants reached here a relatively gigantic size. All over the snow had melted, though it was early in the season, and the place offered an increased allurement to the swimmers and waders by the countless ponds of melting snow. The influence of the powerful sunlight on the dark turfy soil surely accounts for the unusually early melting of the snow in this place.

At my arrival the barnacles were standing in couples or in small flocks in the ponds or they were grazing near these; some were high up the mountain slopes. Almost all the geese used to leave the marsh every day at certain times and disappeared southwards toward the high middle part of Trekroner. I set out in this direction, thinking that a larger lake was lying near the mountain, and that the geese retired to this after their meal. I really found a pair of larger fresh-water basins and saw in these a few geese, which being frightened flew farther toward the mountains. Having come within a distance of one kilometer from Trekroner I solved the riddle. The barnacles were swarming to and fro along the gigantic mountain wall like bees at their hive, and I heard a continuous humming, sounding like a distant talk. I took a seat at the foot of the mountain and observed the behavior of the geese for some hours. Using my field glass I could without difficulty notice even the smallest details.

While some of the geese would constantly fly along the rocky wall and sometimes mounted so high in the air that they disappeared on the other side of the rocks, the majority of the birds were sitting in couples upon the shelves of the rocky wall, some of which seemed too narrow to give room for the two birds—much less for a nest. It was only on the steep and absolutely naked middle part of the mountain wall that the geese had their quarters and in no place lower than some 200 meters from the base of the cliff. As the wall was quite inaccessible, I had to content myself by firing some rifle balls against it in order to frighten the birds and thus form an idea of the size of the colony. The birds which were "at home" then numbered some 150 individuals. As far as I could judge, breeding had not yet commenced.

I feel sure that some of the geese resorted to the mountain without intending to breed. A pair of females, which I later on secured in the marsh, had but undeveloped eggs in their ovaries. In the mountain the geese were not very shy and paid no attention to my shots; the great height at which the birds always stayed made them confident. In the marsh, however, they were very wary and almost always had sentinels posted while they were seeking food.

The Oxford expeditions to Spitsbergen in 1921 and 1922 succeeded in locating a number of nests of this species and collecting a series of eggs. The Rev. F. C. R. Jourdain has kindly sent me the following interesting notes on the subject:

The barnacle goose in Spitsbergen seems to have its main breeding haunts in the valleys opening into Ice Fjord. We saw nothing of it on the north coast or at Liefde Bay, but met with it at Advent Bay, Sassen Bay, Klaas Bitten Bay, and Dicksons Bay during the breeding season, and have good reason to believe that it breeds in all three localities. Probably in order to escape the attentions of the Arctic foxes this species has acquired the habit of nesting on ledges and hollows in precipitous bluffs, and even on the tops of isolated pinnacles of rock. In many of the valleys in Spitzbergen one finds a continuous succession of steep outcrops of rock running along the side of the valley, and in some cases more than one tier running parallel with one another and separated by a steeply sloping range of hillside or talus. It is on these projecting masses of rock that the barnacle nests, and as the goose sits closely and is not readily disturbed from below, a nest halfway up the hillside on a projecting rock may escape notice altogether on the part of those traversing the bottom of the valley. The gander stands close by his mate on watch, and his white face may be detected with a good glass from a considerable distance. Even when approached from above both male and female are slow to leave the nest when the eggs are much incubated, and the gander in some cases will look up inquiringly at the intruder from 20 or 30 yards below before taking wing, while the goose will allow of even closer approach. In one case where a goose was shot from the nest, the report failed to flush another bird, which was also sitting on eggs, not more than 20 yards farther on. Evidently the same breeding places are resorted to year after year, as it is often possible to detect the sites of several old nests in the immediate neighborhood of that in use. The sites varied to some extent, some nests were on gently sloping declivities, sparsely covered with lichens and mosses, at the foot of a low cliff or on ledges on its face, and with a drop of 10 to 20 or 30 feet below; others were on projecting spurs of rocks reached by a narrow "knife-edge" from the main face of the cliff; one was in a small cave overhung from above and with a sheer drop below; and a bird was observed sitting on the flat top of a mushroom-shaped pinnacle of rock, at the very top of a high cliff. In practically every case where the female was incubating eggs, the male was standing close beside her, and the ground near the nest was covered with the accumulation of droppings from the parent birds, which evidently spend most of the incubation period here.

In 1922 we renewed our acquaintance with this species, but found to our disappointment that none of the sites occupied in 1921 were tenanted. This was probably due to shooting and egg collecting by the residents prior to our arrival; but anyhow the birds had disappeared and the high winds had removed all traces of the nests. A range of hills some 1,500 to 1,700 feet high, with bold cliffs and conspicuous stony ridges running down from the top, seemed likely to produce the desired result, so we proceeded to make our way up a steep gulley till at last we reached the top of the ridge. Working our way along the top we could look down on the steep slopes below and could see two or three pairs of barnacle geese, some 300 or 400 feet from the top, the geese quietly incubating while the ganders stood close at hand. The usual nesting place was a sort of saddle on a stony ridge where a space of a few feet of mossy ground was comparatively flat. We had brought with us ropes and with their help were enabled to reach a point about 150 feet above the sitting bird. By careful stalking, with camera and rope in hand, the leading climber descended the ridge till only 20 paces from the sitting bird. The gander had already taken wing, but, as the goose still sat steadily, he tried another stealthy approach and was within 15 paces before the goose, which had previously shown signs of restlessness, rose from the nest and joined her mate. Within the down-clad nest hollow was a clutch of 5 eggs, and below was a range of cliffs, steep screes and bluffs, reaching for quite 1,200 feet by aneriod to the innumerable streams at the foot of the valley. The other three nests we examined in 1922 were all in very similar positions.

Eggs.—According to Mr. Jourdain (1922), who has published an interesting paper on the nesting habits of this species and sent me some extensive notes on the subject, the barnacle goose lays from 3 to 6 eggs, usually 4 or 5. He describes the eggs and the down as follows:

When fresh laid they are pure white, and bear a great resemblance to eggs of the pink-footed goose, but are somewhat smaller. They are, however, larger than those of the brant on the average, besides being considerably heavier. The amount of down in the various nests varied considerably; probably in some cases a certain amount remained in the nest hollow from the previous year. We found no black feathers in the nest; all were either white or had only a faint grayish tinge.

The measurements of 49 eggs, furnished by Mr. Jourdain, average 76.35 by 50.32 millimeters; the eggs showing the four extremes measure **82.7** by 46.4, 78.4 by **53.6**, **68.7** by 49.5, and 82.7 by **46.4** millimeters.

Young.—Referring in his notes to the young he says:

Piecing together the information we obtained, it is clear that the goslings remain in the nest only long enough for their down to become thoroughly dry. They must then scramble or fall down the cliffs, probably being to some extent helped by the strong updraft of wind sweeping up the side of the valley, and then make their way down the screes till they reach the flat ground at the foot of the valley, when they take to the water and are carried down to the marshes at the head of the bay. As there is no vegetation except a few lichens and mosses near the nest, it is obvious that the young can not feed till they reach the foot of the valley, and from what we saw of the pink-footed goose in somewhat similar circumstances the goslings are quite capable of surviving a perpendicular drop of considerable height without injury. There seems however, to be no evidence that the parents give them any assistance, though it would seem improbable for a newly hatched bird to descend a perpendicular cliff nearly 100 feet in height.

Food.—Mr. Cordeaux (1898) says:

The food of the barnacle goose is both vegetable and animal; it is remarkably fond of the short sweet grasses which cover the holms and islets off the western coasts of Scotland, at low water also resorting to saltings, fitties, mud flats, and foreshores, left uncovered by the sea, and is as much a land feeder as its congener, the brent, is a sea goose. Mr. C. M. Adamson, of North Jesmond, had some tame barnacles which in the spring would eat worms, as exceptional diet.

The stomachs and crops of birds taken in Greenland by Mr. Manniche (1910) were "filled with twigs, leaves, and catkins of alpine willow, with seeds of different plants and also grasses."

Behavior.—Mr. Cordeaux (1898) quotes Macpherson and Duckworth, as follows:

It is interesting to wait upon the point of Burgh marsh, before daybreak, and listen to the cries of the barnacles, feeding upon the point of Rockcliffe marsh, just opposite. About an hour after daybreak they rise *en masse* from their feeding ground, and after wheeling up and down the Solway for a few moments, displaying their pretty barred gray, black, and white plumage against the mud flats, they fly seawards to the estuary of the Wampool, or, circling round, pitch in a long line upon the exposed mud half a mile to windward. Barnacle geese are constantly vociferous, especially when feeding, and Mr. A. Smith compares the volume of sound produced by a flock of several hundred feeding at night together, as heard at a distance, to a pack of harriers in full cry.

Witherby's Handbook (1921) says:

Favorite feeding localities are wide stretches of "machar" land of firm springy turf inside sandhills. A common trait in all geese, but more noticeably so in this species, is a continual series of friendly quarrels amongst a feeding flock. Less inclined to associate with other geese than any other species, and never nearly so unapproachable, it is the "fool" amongst geese, and only one which on occasion will fail to take alarm after having seen and distrusted a moving object within 100 yards. Occasionally goes to sea to rest on very calm days, but as a rule as much land loving as any "gray" goose. Note a series of rapidly repeated short barks—some higher than others. Combined chorus produced by big flock of any geese has been likened to "music" of pack of hounds running. Taking that comparison, barnacle's cry is represented by one end of scale, gray lag's by other; gray lag's is full-mouthed music of pack of foxhounds, barnacle's that of host of small terriers. Almost if not quite insensitive to scent of man, and rarely if ever takes alarm from it.

DISTRIBUTION

Breeding range.—Known to breed only in northeastern Greenland (Scoresby Land, Trekones) and on Spitsbergen. Said to breed also on Nova Zembla, the Lofoten Islands, and in northwestern Siberia.

Winter range.—Northwestern Europe. South regularly to the Baltic and North Seas and the British Isles, occasionally inland to Switzerland and Austria and exceptionally south to the Azores, Spain, Morocco, and Italy. Occurs on migrations in western Greenland and Iceland.

Casual records.—There are about 9 North American records: Labrador (Okak), James Bay (Rupert House), Massachusetts (Chatham, November 1, 1895), Long Island (Jamaica Bay, October 20, 1876, Great-South Bay, October 16, 1919, and Farmingdale, November 28, 1922), North Carolina (Currituck Sound, October 31, 1870, and November 22, 1892), and Vermont (Marshfield, 1878).

Egg dates.—Spitsbergen: Nine records, June 25 to 28.

PHILACTE CANAGICA (Sevastianoff)

EMPEROR GOOSE

HABITS

The handsomest and the least known of American geese is confined to such narrow limits, both in its breeding range and on its migrations, that it has been seen by fewer naturalists than any other goose on our list. On the almost inaccessible, low, marshy shores of Alaska, between the mouths of the Yukon and Kuskokwim Rivers, it formerly bred abundantly; but recent explorations in that region indicate that it has been materially reduced in numbers during the past 30 years. My assistant, Mr. Hersey, who spent the season of 1914 at the Yukon delta, saw less than a dozen birds, where Doctor Nelson found it so abundant in 1879. The decrease is partially, if not wholly, due to the fact that large numbers are killed every year and their eggs taken by the natives, even within the limits of what is supposed to be a reservation.

Spring.—For what we know about the life history of the emperor goose we are almost wholly indebted to that pioneer naturalist, Dr. Edward W. Nelson, who fortunately has given us a very good account of the habits of this species. I shall quote freely from his writings, mainly from his educational leaflet on this species, in which he (1913) writes:

At the border of the Yukon delta, Esquimos familiar with the country were employed to lead us to the desired nesting ground of the emperor goose. Nearly half a day's journey among the maze of ice-covered channels of the delta brought us to a low, flat island, where our guide assured me many *nachau-thluk* would soon arrive to rear their young. It was a bare, desolate spot, with only a few scattered alders on the upper side of the islands, and an unbroken view out over the frozen sea to the west. A tent was put up on a slight rise and, after a stock of driftwood had been gathered, the guides took the sledge and left me with my Esquimo companion to await the arrival of the birds. Later, when the ice went out, they returned for me with kyaks.

A few white-fronted and cackling geese gave noisy evidence of their presence, but it was not until May 22 that the Esquimo brought in the first emperor goose—a male in beautiful spring plumage. After this, small flocks came in rapidly until they were plentiful all about us. They arrived quietly, skimming along near the ground, quite unlike the other geese, which appeared high overhead with wild outbursts of clanging cries, which were answered by those already on the ground. The river channels and the sea were still covered with ice, and the tundra half covered with snow, at the time of the first arrivals.

Courtship.—Almost at once after their arrival on the islands, the emperor geese appeared to be mated, the males walking around the females, swinging their heads and uttering low love notes, and incoming flocks quickly disintegrated into pairs which moved about together, though often congregating with many others on flats and sand bars. The male was extremely jealous and pugnacious, however, and immediately resented the slightest approach of another toward his choice; and this spirit was shown equally when an individual of another species chanced to come near. When a pair was feeding, the male moved restlessly about, constantly on the alert, and at the first alarm the pair drew near one another, and just before taking wing uttered a deep, ringing *u-lugh, u-lugh;* these, like the flight notes, having a peculiar deep tone impossible to describe.

At low tide, as soon as the shore ice disappeared, the broad mud flats along shore were thronged with them in pairs and groups numbering up to 30 or 40 individuals. They were industriously dabbling in the mud for food until satisfied, and then congregated on bars, where they sat dozing in the sun or lazily arranging their feathers. By lying flat on the ground and creeping cautiously forward, I repeatedly approached within 30 or 40 yards of parties near shore without their showing any uneasiness.

Nesting.—The first of June they began depositing eggs in the flat marshy islands bordering the sea all along the middle and southern part of the delta. The nests were most numerous in the marshes, a short distance back from the muddy feeding grounds, but stray pairs were found nesting here and there farther inland on the same tundra with the other species of geese and numerous other waterfowl. Near the seashore, the eggs were frequently laid among the bleached and wave-torn scraps of driftwood lying along the highest tide marks. On June 5, a female was found on her eggs on a slight rise in the general level. A small gray-bleached fragment of driftwood lay close by. The goose must have lain with neck outstretched on the ground, as I afterward found was their custom when approached, for the Esquimo and I passed within a few feet on each side of her; but, in scanning the ground for nesting birds, the general similarity in tint of the bird and the obvious stick of driftwood had complete misled our sweeping glances. We had gone some 20 steps beyond when the sitting bird uttered a loud alarm note and flew swiftly away. The ground was so absolutely bare of any cover that the 3 eggs on which she had been sitting were plainly visible from where we stood. They were lying in a slight depression without a trace of lining. The same ruse misled us a number of times; but on each occasion the parent betrayed her presence by a startled outcry and hasty departure soon after we had passed her and our backs were presented. They usually flew to a considerable distance, and showed little anxiety over our visit to the nests. The nests I examined usually contained from 3 to 5 eggs, but the full complement ranged up to 8. When first laid, the eggs are pure white, but soon become soiled. They vary in shape from elongated oval to slightly pyriform, and are indistinguishable in size and shape from those of the white-fronted goose. As the complement approaches completion, the parent lines the depression in the ground with a soft, warm bed of fine grass, leaves, and feathers from her own breast. The males were rarely seen near the nest, but usually gathered about the feeding grounds with others of their kind, where they were joined now and then by their mates.

Eggs.—The emperor goose lays from 3 to 8 eggs; probably 5 or 6 is the usual number. The eggs that I have seen are elliptical ovate in shape, with variations toward ovate and toward elongate ovate. The shell is smooth or very finely granulated and not at all glossy. The color is creamy white or dull white at first, becoming nest stained or variegated or finely speckled with buff. The measurements of 96 eggs, in various collections, average 78.6 by 52.1 millimeters; the eggs showing the four extremes measure **86** by 49, 80.2 by **56.2**, **70.3** by 50.3, and 75.8 by **48.3** millimeters.

Young.—The period of incubation is 24 days, according to F. E. Blaauw (1916), who has succeeded in raising this species in captivity. Doctor Nelson (1913) says of the young:

The young are hatched the last of June or early July, and are led about the tundras by both parents until, the last of July and the first of August, the old birds molt their quill feathers and with the still unfledged young become extremely helpless. At this time, myriads of other geese are in the same condition, and the Esquimos made a practice of setting up long lines of strong fishnets on the tundras to form pound traps, or inclosures with wide wings leading to them, into which thousands were driven and killed for food. The slaughter in this way was very great, for the young were killed at the same time and thrown away in order to get them out of the way of the next drive. The Esquimos of this region also gather large numbers of eggs of the breeding waterfowl for food and, with the demand for them at the mining camps of the north, a serious menace to the existence of these and other waterfowl might ensue.

Plumages.—Mr. Blaauw (1916) says: "The chick in down is of a beautiful pearl-gray, darkest on the head and upper side and lighter below. The legs and bill are black." A larger downy young, about the size of a teal, in the United States National Museum, has probably faded some; the upper parts vary in color from "bister" to "buffy brown" and the under parts from "smoke gray" to "olive buff."

Mr. Blaauw (1916) says:

The chicks grew very fast, and in a few weeks were completely feathered. In the first feather dress the bird resembles the adults, but the gray is not so bluish. The black markings on the feathers are only indicated, and the coverts on the upper side are not so square, but more pointed. The black throat is wanting, and so is the white head and neck, these parts being gray like the rest of the body. The tail is white.

The bill is dusky bluish, flesh color at the base and black at the tip. The legs are yellowish black. As soon as the birds are full grown they begin to molt, shedding all the feathers except the large flight feathers. The tail feathers are also molted.

At the end of October the young birds are quite grown, and similar to the old birds. By this time the upper mandible has got the beautiful blue and flesh colors of the old birds, whilst the lower mandible has become black. The legs are now orange. When the bird is molting, the first white feathers of the head to appear are near the base of the bill.

The above gray-headed plumage must be the juvenal plumage, which I have never seen, and which is probably not worn for more than a few weeks. Evidently the change into the first winter plumage must be very rapid. All the young birds which I have seen, collected between September 9 and November 17 of their first year, are in the first winter plumage. In this the head and neck are largely white above and black below, much as in the adult, but the black area is browner and the white area is much obscured by dusky mottling, especially on the forehead, lores, and neck; the juvenal wing is similar in color pattern to the adult, but it has narrower, buffy white edgings instead of broad white edgings and dull, brownish-dusky, subterminal markings instead of pure clear black; the feathers of the back are similarly marked with dull patterns and narrow buffy edgings, which soon wear away; the under parts are dull and mottled; and the tail is largely white, as in the adult. This plumage is worn for a very variable length of time by different individuals. I have seen birds taken in November in which the adult plumage was well advanced on the back and scapulars; and I have seen others which were just begininng the molt in June. Perhaps both of these were exceptional; and probably a more or less continual molt of the body plumage takes place all through the winter and spring. I have a fine young male in my collection, taken June 30, which is just completing this molt and is practically adult. The wings are molted during the coming summer, July and August, after which young birds, during their second fall, become indistinguishable from adults.

Food.—Lucien M. Turner (1886) says that "the emperor goose visits the vicinity of Stewart and St. Michael Islands in great numbers to feed on the shellfish exposed by the low water." Grinnell, Bryant, and Storer (1918) say that "at times it resorts to heath berries, which are available on the tundras closely adjacent to the seashore." Other writers speak of it as feeding on mussels and other shellfish and, as it is known to feed on the beaches and mud flats, rather than on the grassy marshes or uplands, its food is probably mainly animal. Its flesh is said to be rank and strongly flavored, which is generally not the case with vegetable feeders.

Behavior.—Doctor Nelson (1913) says, of the flight and notes of this species:

When on the wing, they were easily distinguished from the other geese, even at considerable distances, by their proportionately shorter necks and heavier bodies, as well as by their short, rapid wing strokes, resembling those of the black brant. Like the latter, they usually flew near the ground, rarely more than 30 yards high, and commonly so close to the ground that their wing tips almost touched the surface on the down stroke. While flying from place to place, they give at short intervals a harsh, strident call of two syllables, like *kla-ha, kla-ha, kla-ha,* entirely different from the note of any other goose I have ever heard. A group of them on a sand bar or mud flat often utter lower, more cackling notes in a conversational tone, which may be raised to welcome new arrivals. They are much less noisy than either the white-fronted or cackling geese, which often make the tundra resound with their excited cries. Occasionally I could cause a passing flock to leave its course and swing in close to my place of concealment by imitating their flight notes.

Again (1887) he writes:

While a pair is feeding, the male keeps moving restlessly about, with eyes constantly on the alert, and at the first alarm they draw near together and just before they take wing both utter a deep, ringing *u-lugh, u-lugh.* As in the case of the call note, this has a peculiar, deep hoarseness, impossible to describe.

Game.—Mr. Turner (1886) says that these geese—

Form an important article of food in the Yukon district, alike to the white and native population. They are mostly obtained by means of the gun. The best localities near St. Michael are toward the western end of the canal, along the edge of the low grounds bordering the hills of the mainland, and near the village of Stephansky (Athwik, native name), on the western side of St. Michael Island. This area is low, intersected with innumerable swamps and connecting streams, forming a fine feeding ground for all kinds of waterfowl.

A regular camping outfit is taken by sledge and dogs to a chosen locality. In the early morning a site is selected where the geese fly around some ending of a hill range, for they fly low and prefer to sweep around the hills rather than mount over them. They are frequently so low in their flight that the hunter has to wait until the geese are well past before he can shoot them to an advantage. A nearly constant stream of geese fly around a certain point, just to the left of the Crooked Canal, on a slight eminence, formed from the deposit of soil torn up by some immense ice cake, which the high tides of some December in years long gone by had left as the water receded and the warm weather of spring had melted; now overgrown with patches of rank vegetation.

By 10 o'clock the geese were done flying for that morning. The low character of the ground did not favor approach to the geese feeding at the ponds. During the middle of the day a quiet sleep invigorated the hunter for the late evening shooting, the latter generally affording a less number of geese than the morning's shooting.

By the next morning a sufficient number of geese were obtained to heavily load a sledge; drawn by six lusty Eskimo dogs, assisted by two sturdy natives. This sport generally lasts from the arrival of the geese until the first week of June. At this time they repair to the breeding grounds. During the summer the geese are not hunted. The eggs are eagerly sought by the natives and whites and take the place of meat of the birds. In the latter part of August or the early part of September the fall shooting begins, as the geese have molted, the young are able to fly, and they are fattening on the ripening berries. The geese are now obtained by watching the ponds, or as they fly over in small flocks or singly. Should a flock not fly sufficiently near, a favorite method to attract their attention is for the hunter to lie on his back, swing his arms and hat, kick up his legs, and imitate the call of the geese. It rarely fails to bring them within distance, and may, if several be just shot from their ranks, be repeated, and even a third time. Later in the season, when cool and frosty nights are regular, great numbers of the geese are killed and disemboweled for freezing to keep throughout the winter. The feathers are left on the birds, for the flesh is said to keep in better condition. The body is washed out and the bird hung up by the neck in the icehouse to keep, even until the geese have arrived the next spring. The flesh, when thawed out slowly, has lost all the rank taste, and, in my opinion, is much improved by the freezing process.

I have eaten the flesh of all the various kinds of geese, frequenting those northern regions, and place them in value of flesh as follows: white-fronted goose, *A. albifrons gambelli;* white-cheeked goose, *B. canadensis hutchinsii* and *B. canadensis minima;* Canada goose, *B. canadensis;* black brant, *B. nigricans*, is always tough and lean, fit food only for a Russian; snow goose, *Chen hyperboreus*, is scarcely fit for food, except in cases of necessity. Its flesh is coarse, rank, and has a decidedly unpleasant odor; the emperor goose, *P. canagica*, is scarcely to be thought of as food. There is a disgusting odor about this bird that can only be removed in a degree, and then only by taking off the skin and freezing the body for a time. Even this does not rid the flesh entirely of strong taste.

Winter.—According to Grinnell, Bryant, and Storer (1918):

The principal winter home of the emperor goose is on the seacoast of southwestern Alaska, and only stragglers reach California. But it is probable that if all the emperor geese ever observed in California had been recorded, it would be found that almost every year one or two of the birds had made their way within our borders. At least 10 definite instances of the occurrence of the emperor goose in this state are known. In spite of the fact that this is a marine species, most of the records are from the interior valleys. Mr. Vernon Shepherd, a taxidermist of San Francisco, informs us that he has known of the capture of at least a dozen specimens of this goose since 1906.

The emperor goose flies in pairs or in small flocks of 4 or 5. A juvenile killed at Gridley, Butte County, was alone, being the second in a flock of white-fronted geese. One taken near Modesto, Stanislaus County, came to the blind alone. Another taken near Davis, Yolo County, had been noted alone in the same pond for three weeks previous to capture. This species is said to be shyer than any other goose except the black sea brant.

DISTRIBUTION

Breeding range.—On the northwest coast of Alaska, from the mouth of the Kuskokwim River northward to the north side of the Seward Peninsula (Cape Prince of Wales, Cape Espenberg, and Deering). On St. Lawrence Island and on the northeast coast of Siberia, from East Cape westward at least as far west as Koliutschin Bay.

Winter range.—Mainly in the Aleutian Islands. East along the Alaska Peninsula at least as far as Sanakh Island and Bristol Bay and probably to Cook Inlet, straggling farther south and casually to California. West to the Commander Islands and perhaps Kamchatka.

Spring migration.—Early dates of arrival in Alaska: Pribilof Islands, St. George, April 26; Nushagak River, May 5; Yukon Delta, May 22; Cape Prince of Wales, May 19.

Fall migration.—Early dates of arrival: Pribilof Islands, St. George Island, September 22; Aleutian Islands, August 31; Oregon, Willamette River, September 30; California, San Francisco market, October 8. Latest date for St. Michael is November 15.

Casual records.—Rare south of Alaska, but frequently wanders south to Washington (Snohomish County, January 1, 1922), Oregon (Netarts Bay, December 31, 1920), and California (10 or more records). One record for the Hawaiian Islands (Kalapan, December 12, 1902), and one doubtful record for the Great Slave Lake region.

Egg dates.—Alaska: Fifteen records, May 26 to July 4; eight records, June 2 to 20.

DENDROCYGNA AUTUMNALIS (Linnaeus)

BLACK-BELLIED TREE DUCK

HABITS

As I have never seen either of the tree ducks in life, I shall have to quote wholly from the writings of others; and very little has been published about this species, which is to be found in only a very limited area north of the Rio Grande. Writing of its habits near Fort Brown, Texas, Dr. J. C. Merrill (1878) says:

This large and handsome bird arrives from the south in April, and is soon found in abundance on the river banks and lagoons. Migrating at night, it continually utters a very peculiar chattering whistle, which at once indicates its presence. Called by the Mexicans *patos maizal*, or cornfield duck, from its habit of frequenting those localities. It is by no means shy, and large numbers are offered for sale in the Brownsville market. Easily domesticated, it becomes very tame, roosting at night in trees with chickens and turkeys. When the females begin to lay, the males leave them, and gather in large flocks on sandbars in the river.

Nesting.—Mr. George B. Sennett (1879) writes:

First noticed early in May, in pairs, at Lomita, looking for nesting places. Soon after it became quite common. During the mating season it is found about in trees of open woodland, and very tame. It nests in hollow trees without regard to nearness of water. I was shown the nest from which a set of 12 eggs was taken the season before. It was in an ebony tree in an open grove, near the houses of the ranch, and much frequented; was about 9 feet from the ground, in a hollow branch, with no lining but the chips from the rotten wood.

Four sets of eggs in the United States National Museum, collected near Brownsville, Texas, were taken from nests in holes in elms and willows, 10 or 12 feet from the ground. Another set is said to have been taken from a nest on the ground, among rushes, weeds, and grass, on the edge of a lake. The tree nests were mostly in "big woods," usually near a lake or creek, and no lining was found in the nests except the rotten wood in the hollows. Doctor Merrill (1878) says that " the eggs are deposited in hollow trees and branches, often at a considerable distance from water (2 miles), and from 8 to 30 feet or more from the ground."

Eggs.—Doctor Merrill (1878) says that " two broods are raised," but Mr. Sennett (1879) was " of the opinion that but one brood is reared in a season." By reference to the egg dates, given below, it will be seen that the nesting season is very much prolonged, which suggests the possibility that two broods might be raised.

The black-bellied tree duck does not lay such large sets of eggs as its relative, the fulvous tree duck; from 12 to 16 eggs usually constitute a full set, the smaller number being more often found. The eggs are ovate or short ovate in shape, the shell is sometimes smooth and not at all glossy, but in other specimens it is highly glossy and very finely pitted; the color is white or creamy white, with occasional nest stains. The measurements of 99 eggs, in various collections, average 52.3 by 38.3 millimeters; the eggs showing the four extremes measure **58.5** by 39.5, 54.2 by **42.5**, **41.5** by 28.7 and 43.7 by **28.6** millimeters.

Young.—According to Doctor Merrill (1878) " the parent carries the young to water in her bill." And Mr. George N. Lawrence (1874) quotes Col. A. J. Grayson as saying: " The young are lowered to the ground one at a time in the mouth of the mother; after all are safely landed she then cautiously leads her young brood to the nearest water."

Plumages.—Baird, Brewer, and Ridgway (1884) describe the downy young as follows:

Above, blackish brown, varied by large areas of sulphury buff, as follows: A supraloral streak extending over the eye; a wide stripe from the bill under the eye and extending across the occiput, the blackish below it extending forward only about as far as directly beneath the eye, and confluent posteriorly with the nuchal longitudinal stripe of the same color; a pair of sulphury buff patches on each side of the black, and another on each side of the rump; posterior half of the wing whitish buff, the end of the wing blackish; the black of the upper parts sends off two lateral projections on each side, the first on each side of the crop, the second over the flanks to the tibiae; the buff of the abdomen extending upward in front of this last stripe as far as the middle portion of the buff spot on the side of the back. Lower parts wholly whitish buff, paler and less yellowish along the middle.

A bird in my collection, taken September 11, is apparently in the juvenal or first winter plumage. The bright rufus of the upper parts is replaced by duller shades of pale browns; and the under parts are uniform pale grayish buff, with no traces of the rufous breast or black belly; the bill is dusky. I have seen a bird in similar plumage, collected February 7, in which the color pattern of the under parts of the adult is faintly indicated. Apparently this immature plumage is worn at least through the first fall and winter and perhaps until the first postnuptial molt, the next summer.

Food.—This species is locally known as the "cornfield duck," on account of its habits of frequenting the cornfields to feed on the corn, where it is said to do considerable damage. Mr. Sennett (1878) writes:

Late in August, the young not full grown are seen about the corncribs picking up the refuse corn, at which time Mr. Bourbois says they afford most excellent eating. This bird does not alight in the water as do other ducks but on the land, and wades about in shallow water for food. When corn is nearly ripe, it alights on the stalks, strips the ears of their husks, and pulls the grain from the cob, making this its chief food during the season. I never saw it skulk in the grass for cover, but always take wing and fly to the woods, or to some removed open point by the water. It is a pretty sight to see this bird on some dead stub, pluming itself, its color and shape being very handsome.

Behavior.—Colonel Grayson, in his notes, quoted by Mr. Lawrence (1874), says:

This duck perches with facility on the branches of trees, and when in the cornfields, upon the stalks, in order to reach the ears of corn. Large flocks of them spend the day on the bank of some secluded lagoon, densely bordered with woods or water flags, also sitting among the branches of trees, not often feeding or stirring about during the day. When upon the wing they constantly utter their peculiar whistle of *pe-che-che-né*, from which they have received their name from the natives. (The other species is called Durado.) I have noticed that this species seldom lights in deep water, always prefering the shallow water edges, or the ground; the cause of this may be from the fear of the numerous alligators that usually infest the lagoons.

When taken young, or the eggs hatched under the common barnyard hen, they become very domestic without being confined; they are very watchful during the night, and, like the goose, give the alarm by their shrill whistle when any strange animal or person comes about the house. A lady of my acquaintance possessed a pair which she said were as good as the best watchdog; I also had a pair which were equally as vigilant, and very docile.

Doctor Sanford (1903) writes:

In April, 1901, I found these birds abundant in the vicinity of Tampico, Mexico. They were most often seen in small flocks of from 4 to 10 on the banks at the edge of the lagoon. Their long legs gave them an odd look. At our approach they would run together, raising their long necks much like geese. The flight was peculiar and characteristic, low down and in a line, their large wings with white bands presenting a striking aspect, and giving the impression of a much larger bird. We saw them occasionally on the smaller ponds, and shot several, all of them males. In one or two instances the appearance of the breast indicated the bird had been sitting on eggs. While the males of this species are supposed to attend to their own affairs during the period of incubation, it would seem as if they occasionally assisted in nesting duties. Once or twice I saw them near small ponds in woods, apparently nesting, flying from tree to tree with perfect ease, exhibiting some concern at our presence.

Winter.—Prof. W. W. Cooke (1906) says:

It winters in Mexico at least as far north as central Vera Cruz (Vega del Casadero) and Mazatlan. North of this district it is strictly migratory, and throughout most, if not all, of its ranges in Central America there seems to be a shifting of location between the winter and the summer homes, but no data are available to determine the movements with accuracy.

Since writing the above life history I have visited the Brownsville region in the lower Rio Grande Valley and made a special effort to learn something about the two tree ducks, which were formerly so abundant there. I did not see a specimen of either species. Capt. R. D. Camp, who has spent some 13 years in studying and collecting birds in that region, told me that the black-bellied tree duck had entirely disappeared from the Brownsville region and that the fulvous tree duck had become very scarce. He took me to a resaca where he had seen a pair of the latter this spring, 1923, but we saw no trace of the birds.

DISTRIBUTION

Breeding range.—East to the Gulf coasts of Texas and Mexico. South to Panama (River Truando). West to the Pacific coast of Mexico (Mazatlan). North to southern Texas (lower Rio Grande Valley) and irregularly north to Corpus Christi and perhaps Kerrville. Known to breed in Porto Rico and Trinidad and probably breeds in some of the other West Indies.

Winter range.—Resident in most of its range. Winters at least as far north as Vera Cruz and Mazatlan.

Migrations.—Arrives in Texas in April and leaves in September, October, and November.

Casual records.—Has wandered to Arizona (Tucson, May 5, 1899) and southern California (Imperial Valley, fall, 1912).

Egg dates.—Texas: Sixteen records, May 3 to October 18; eight records, June 20 to July 14.

DENDROCYGNA BICOLOR (Vieillot)

FULVOUS TREE-DUCK

HABITS

Messrs. Grinnell, Bryant, and Storer (1918) introduce this species in a few well-chosen words as follows:

The term tree duck, as applied to the fulvous tree duck, seems to be an almost complete misnomer for the bird. As regards structure this species seems to be more closely related to the geese than to the ducks, and, at least in California, it seldom nests in trees but chooses the extensive tule marshes of our interior valleys. Birds apparently belonging to the same species of tree duck that occurs in this State are found in South America, in southern Uruguay and Argentina, and also in South Africa and in India—a very striking case of what is known as interrupted or discontinuous distribution. In North America the chief breeding ground of the species is in Mexico, but a considerable number of birds breed in the southwestern United States. The latter contingent is migratory, moving south for the winter season.

Spring.—Col. A. J. Grayson, in his notes quoted by Mr. George N. Lawrence (1874), says:

Although its geographical range is confined within the limits of the Tropics, yet this species has its seasons of periodical migrations from one part of the country to the other; during the month of April their well-known and peculiar whistle may be heard nightly as they are passing over Mazatlan in apparently large flocks, going northward. At first this phenomenon puzzled me not a little, as I well knew that they are not often found far north of the Tropics, except an occasional straggler. But I was at length enlightened as to their point of destination; by frequent inquiries of the natives, I was satisfied that they went no farther north than the Mayo and Yaqui Rivers, in Sonora, and the adjacent lakes and lagoons, where they breed. Some, however, remain and breed in the State of Sinaloa, and the adjoining localities.

Referring to the migration in South America, W. H. Hudson (1920) writes:

This duck, the well-known *Pato silvon* (whistling duck) of the eastern Argentine country, is found abundantly along the Plata and the great streams flowing into it, and northwards to Paraguay. Along this great waterway it is to some extent a migratory species, appearing in spring in Buenos Aires in very large numbers, to breed in the littoral marshes and also on the pampas. They migrate principally by night, and do not fly in long trains and phalanxes like other ducks, but in a cloud; and when they migrate in spring and autumn the shrill confused clangor of their many voices is heard from the darkness overhead by dwellers in the Argentine capital; for the ducks, following the eastern shore of the sealike river, pass over that city on their journey.

Nesting.—One of the best accounts of the nesting habits of the fulvous tree duck in California is given by A. M. Shields (1899), as follows:

Starting early next morning to search a different locality, the place selected was an extensive strip of high grass growing in the damp swampy ground and sometimes in several inches of water. The grass was from 2 to 3 feet high, of a variety commonly known as "sword" or "wire" grass, and covered an area of perhaps 100 acres of low land between the deep water and the higher ground a few hundred yards back. Just as we were alighting from the wagon on the edge of the swampy area I saw a fulvous tree duck flying from the swamp. After a few circles she dropped down among the dense grass not 300 yards distant, and I, not stopping to put on my wading boots but keeping my eye on the spot where she had settled, quickly approached and when within a few yards I was delightfully shocked by a flutter of wings and the sight of the old bird rising and winging a hasty retreat. I reached the nest and what a thrill at the sight, there in the midst of a little vacant square of 4 or 5 feet was a beautifully built nest, composed entirely of grass, about 6 inches in height and containing 19 beautiful white eggs. I immediately saw by comparison that my surmise as to the identity of the strange parasite eggs found the day before was correct.

The nest was situated in the center of a little open spot in the grass; the open area had evidently been created by the bird in her quest for building material, for she had proceeded to pull up or break off the grass immediately adjacent as her nest grew higher and larger, until the nest finally occupied a position in broad daylight as it were, although it is not improbable that when the spot was selected it was well hidden by overhanging and surrounding grass. I was not long in securing this nest and eggs, after which we began a systematic search through the high grass and in a short time I had found my second nest, constructed similarly to the first but a little better hidden, being under an overhanging bunch of grass which furnished a slight covering. This nest contained 30 eggs, deposited in a double layer; and if the first set of 19 was a surprise, what shall I say of this?

Dr. Harold C. Bryant (1914) describes other nests, similarly located, as follows:

One of our most interesting finds was a nest of the fulvous tree duck, discovered on May 12, 1914. The nest was situated on a hummock in the middle of a marsh between two ponds. The nest was a well-woven one of dry sedges placed about 6 inches above the ground in a tall clump of sedge and weeds. The cavity was about 5 inches deep and in it lay 12 ashy white eggs. A few days later the nest was raided by some predacious animal and all the eggs destroyed. On May 18 we discovered a second nest in the same swamp. This one was built about 6 inches above the water in a small clump of sedge and contained but 4 eggs. The sedges were arched over the cavity in such a way as to conceal it effectively. Two days later when we visited this nest we found it also raided. The only other nest of this species noted was a new one found on June 23. No attempt had been made at special construction of a nest, the two eggs simply lying in a crushed-down place among tall sedges.

The method of nesting described above seems to be the method regularly followed in California. I have two sets of fulvous tree duck's eggs in my collection from Merced County, California; one nest is described as made of grass and small tules, lined with fine grass and a little down, and placed on the ground among high grass in a swamp; the other nest was made of grass and tules and was placed in a clump of grass and tules in a ditch with 2½ feet of water under it; there were 29 eggs in the latter nest, 10 of which were in the upper layer.

That this duck probably does nest occasionally in trees in California, as it certainly does elsewhere, is suggested by the following observation by W. Otto Emerson, published by Mr. Shields (1899), in his paper referred to above:

On May 23, 1882, while collecting with William C. Flint at Lillie's ranch near Tulare Lake I noticed a fulvous tree duck sitting in the entrance hole of a large white oak near one of the ditches, but it was out of the question to reach it. Again on May 26 another was located sitting on the edges of a hole high up in a white oak.

Mr. D. B. Burrows found this species nesting in hollow trees, at heights varying from 4 to 30 feet, in the valley of the lower Rio Grande, near Roma, Texas. He says in his notes:

The fulvous tree duck is a common species in some localities along the Rio Grande River. In the breeding season the birds are frequently seen singly, perched in large trees in the heavily timbered bottom lands.

I secured two sets of their eggs, one of which contained 14 and the other 11 eggs. Both of these nests were found by Mexicans, and one of them I visited and examined. This nest was a natural cavity in the large trunk of a mesquite tree and about 4 feet from the ground; the cavity was about 2 feet in depth and the eggs, 11 in number, were placed on rotten chips at the bottom.

The eggs are deposited at a warm time of the year, and I am informed by the Mexicans that the birds are in the habit of leaving the nests for the greater part of the day, the heat of the sun continuing their incubation, but they return and remain upon the nest during the night and the cool part of the morning. How true this is I am not able to say, as I had no opportunity of watching them, but it seems quite reasonable. The nest was more than half a mile from the Rio Grande River, in a broad mesquite bottom.

Eggs.—Either the fulvous tree duck lays an extraordinary number of eggs or several females lay in the same nest. Mr. Shields (1899) says:

We subsequently found about a dozen nests, all similarly situated, and most of them containing from 17 to 28, 30, 31, and 32 eggs. The smallest set found was of 9 and another of 11 eggs, both evidently being incomplete, as the nests were not finished and incubation had not commenced.

Authentic sets of as many as 36 eggs are in existence and probably much larger numbers have been found according to F. S. Barnhart (1901), who writes:

From time to time since 1895 pothunters have told wonderful stories of finding large numbers of eggs piled up on bunches of dead grass and on small knolls that rose above the water in the swamps. The number of eggs in these nests ranged from 30 to 100 or more, according to report, and in not a few cases the finder has brought the eggs with him in order to prove that what he said was true.

Probably the large numbers referred to by Mr. Barnhart are surplus eggs laid by various individuals and never incubated; and perhaps some of the large sets in collections are the product of more than one female. Evidently the fulvous tree duck is careless in its laying habits, for Mr. Shields (1899) speaks of finding eggs of this species in the nests of the redhead and the ruddy duck. All of the slough-nesting ducks seem to be careless about laying in each other's nests and to have the habit of using "dumping nests" in which large numbers of eggs are laid and forgotten.

In shape the eggs are bluntly ovate, short ovate, or oval. The shell is usually smooth and without gloss, but in many specimens, probably those that have been incubated, the shell is quite glossy and minutely pitted. The color varies from white to buffy white, but the eggs are often much stained with deep shades of buff. The measurements of 212 eggs, in various collections, average 53.4 by 40.7 millimeters; the eggs showing the four extremes measure **59.9** by 40.4, 52.5 by **44.03**, **49.1** by 39.7 and 51.7 by **37.6** millimeters.

Plumages.—The downy young of the fulvous tree duck is described by Grinnell, Bryant, and Storer (1918) as follows:

Top of head clove brown; chin, throat, and sides of head dull white, a streak of the same color extending around back of head on each side and meeting its fellow on hind head; a short, dull white streak on each side of head from side of bill to above eye; bill (dried) dusky brown with prominent yellowish nail; hind neck clove brown, a streak of same color invading side of head below streak of white which encircles head; rest of upper surface of body uniform bister brown; whole under surface of body dull white; feet (dried) grayish yellow.

A series of young birds in my collection, about two-thirds grown, are strikingly like adults, except that the colors are all duller, the brown edgings on the back are narrower, there is less chestnut in the wing coverts, and the upper tail coverts are tipped with brown. I have no data as to subsequent molts and plumages, but suppose that the adult plumage is assumed at the first postnuptial molt when the young bird is a little over a year old.

Food.—Mr. Shields (1899) says that these ducks "are equally at home in an alfalfa patch (about dusk) or in a lake of water, and are entirely at home in an oak forest not far from the breeding swamp, where they are said to assemble for the purpose of feeding on acorns."

Referring to the food of this species, Grinnell, Bryant, and Storer (1918) write:

The fulvous tree duck feeds largely on the seeds of grasses and weeds. In Mexico and Texas it is said to visit the cornfields at night where it finds palatable provender. When feeding in muddy or marshy situations the birds thrust their bills deep in the soft mud on both sides and in front of them as they walk along. The stomach of an individual obtained at Los Banos, Merced County, in May, 1914, and examined by us, contained finely cut up grass and other vegetable matter.

Behavior.—The same writers say on this subject:

The fulvous tree duck is more easily approached than many other waterfowl, but nevertheless is often difficult to find as it congregates among the dense tules or far out on the marshy ponds. On occasion a flock has been easily approached and a number killed at one shot. Sometimes, when tree ducks are surprised on grassy ground, they simply stand rigidly with their heads and long necks straight up in the air, and at a distance look more like stakes than birds. When wounded they are said to escape not only by diving but also by running at great speed and hiding in the grass, and thus often baffle entirely the hunter's efforts to recover them.

Game.—Regarding its status as a game bird, they say:

The flesh of the fulvous tree duck is light colored and juicy, and also free from the rank flavor possessed by sea-faring ducks and geese. On their arrival in California the birds are fat and eminently fit for the table; but since they are here in greatest numbers during the close season, they largely escape the slaughter levied on other wild fowl. The numbers of this species are, at best, small in comparison with many other ducks and geese. They could ill afford a heavy toll by the hunter during the period of their stay here. Any levy upon them during the actual breeding season would be contrary to all recognized principles of game conservation and humanity. As it is, but a few tree ducks are to be shot each year at the opening of the season, October 15. Those who are anxious to hunt the fulvous tree duck in numbers must go to Mexico, where the birds are to be found regularly in winter and where a certain toll may be levied with safety.

Winter.—On the winter habits of the fulvous tree duck in Mexico Mr. Lawrence (1874) quotes Colonel Grayson as follows:

At the conclusion of the rainy season, or the month of October, they make their appearance in the vicinity of Mazatlan, San Blas, and southward, in large flocks; inhabiting the fresh-water ponds and lakes in the coast region, or *tierra caliente*, during the entire winter, or dry months, subsisting principally upon the seeds of grass and weeds, and often at night visiting the cornfields for grain. During these months I have found them in the shallow grass-grown ponds in very large numbers, affording excellent sport to the hunter, and a delicious game for the table; their flesh is white, juicy, and, feeding upon grain and seed, is free from the strong or rank flavor of most other ducks; they are rather heavy or bulky and usually fat. They are more easy to approach than our northern species; I have shot as many as 15 with the two discharges of my double-barrel. When only winged they are almost sure to make their escape, which their long and stout legs enable them to do, running and springing with extraordinary agility, and ultimately eluding pursuit by dodging into the grass or nearest thicket; if the water is deep they dive, and as they rise to breathe, having only the head above water, and that concealed among the water plants, they are soon abandoned by the hunter.

Mr. H. B. Conover writes to me, of the haunts of this species in Venezuela, as follows:

The only place we saw this duck was at Lagunillas. Here on a large savannah or swamp we saw thousands of tree ducks, about 5 per cent of which were fulvous and the rest gray breasted. The fulvous tree duck was very much wilder than the other, and when polling through the marsh, would be the first to leave, generally rising 100 to 150 yards away. On the water these birds were very easy to distinguish from the gray breasted because of their whitish rump, which showed up plainly. They flocked by themselves, as I never noticed them mixed in with flocks of gray breasted or vice versa. This marsh was a shallow place, I believe not over 3 feet deep in any spot, and consisted of large pieces of open water with patches and islands of high grass on the sides. The open water was covered for the most part with floating aquatic plants somewhat similar to our lily. It was in these open places, among the floating aquatic plants, that the tree ducks could be seen. This bird was well known to the natives there and went by the name of Llaguasa Colorado as against the plain Llaguasa for the gray-breasted tree duck.

DISTRIBUTION

Breeding range.—Southwestern North America, parts of South America, and southern Africa and India. In North America east to Louisiana (Lake Catharine and the Rigolets). South to central Mexico (Jalisco and Valley of Mexico). North to central California

(Merced County), central Nevada (Washoe Lake), southern Arizona (Fort Whipple), and eastern Texas (Nueces County). In South America from central Argentina (near Buenos Aires and in Ajo district), northwards to northern Argentina (Tucuman Fort, in Donovan), central Paraguay (Asuncion), and southern Brazil. Occurs in northern South America (Colombia, Venezuela, and British Guiana); it probably breeds in these localities and casually on the island of Trinidad.

Winter range.—Includes the breeding range and extends southwards in Mexico to Guerrero and Chiapas and northwards in California to the Sacramento Valley (Marysville) and Marin County (Inverness).

Migrations.—Migratory movements are not well marked, but occur mainly in April and October.

Casual records.—Has wandered east to North Carolina (Currituck Sound, July, 1886) and Missouri (New Albany, fall, 1890) and north to British Columbia (Alberni, Vancouver Island, September 29, 1905.

Egg dates.—California: Twenty-three records, April 28 to July 13; twelve records, June 7 to 25. Texas: Nine records, May 16 to September 10; four records, June 16 to July 12.

CYGNUS CYGNUS (Linnaeus)

WHOOPING SWAN

HABITS

The status of this fine swan as an American bird has rested mainly on its former occurrence as a breeding bird of southern Greenland, of which Andreas T. Hagerup (1891) says: "Formerly nested in South Greenland, but is now only a rare visitor." It is said to have been exterminated in Greenland by the natives, who pursued and killed the young birds and the adults, when molting and unable to fly. Probably what few stragglers now occur there are wanderers from Iceland, where it is known to breed regularly.

The whooping swan is a Palearctic bird of wide distribution across the northern portions of Europe and Asia, breeding mainly north of the Arctic Circle.

Nesting.—John Cordeaux (1898) writes:

"It is the earliest of the Arctic breeding birds to move toward its nesting quarters, and its loud trumpet calls are the first notice to the dwellers in high latitudes that the long dreary winter is nearing its end. Swans arrive at their nesting quarters as early as the end of March. The nest is a round mass of water plants and moss, fragments of turf and peat, of considerable elevation and often visible at a long distance. It is placed in some vast wilderness of bog or marsh, and sometimes on a small island in a lake. The eggs, from 3 to

5, and 7, are creamy white, and small for the great size of the parent. They are buried in down from the bird's breast, with which the nest is also lined.

The following account of the breeding of this species in the West Highlands of Scotland is published by Mrs. Audrey Gordon (1922), as follows:

On May 21, 1921, my husband and I went to a certain loch in the West Highlands hoping to photograph a black-throated diver whose nest we had located on May 16. However, to our great disappointment the diver's eggs had been washed out of the nest during the flood in the night of May 19–20. Near a neighboring island, where many herons nested on low birch trees, we saw a pair of whooper swans (*Cygnus cygnus*) swimming. Suspecting a nest, a search was made, and soon revealed the swan's nest with 4 eggs. It was situated on a patch of green grass amid a mass of hummocks of blueberry and heather and about 4 yards from the edge of the loch. The nest was composed of dead grasses and weeds and was raised some 15 inches from the ground level. A "hide" was constructed among the hummocks and from it a watch was kept on the 22d and 23d.

The swan always landed at exactly the same place, on a tiny sandy beach, and approached the nest slowly, drying her breast feathers by rubbing them with her head. While sitting she spent a good deal of time building up the nest by pulling the grasses up and around her from the base of the nest. Several times she stood up and laboriously turned the great eggs completely over. Once she left the nest to feed and before doing so carefully covered the eggs with the nesting material. On returning, however, she did not remove the covering but wriggled it off the eggs with her body. Often she went to sleep on the nest, her long neck lying along her back in tortuous curves. The photograph shows clearly the distinguishing features of the whooper swan—the large size, long straight neck held erect and not curved as in the mute species, and the absence of knob or berry on the bill. Further proof of the identity of the species was given by the hearing of the repeated calling of both birds, a musical call, rather resembling that of the wigeon.

Eggs.—This swan is said to lay from 3 to 7 eggs, but usually from 4 to 6. They are creamy or yellowish white at first but soon become nest stained. The measurements of 75 eggs, as given in Witherby's Handbook (1921) average 112.8 by 72.6 millimeters; the eggs showing the four extremes measure **126.3** by 71.3, 114 by **77.4, 105.2** by 72, and 117 by **68.1** millimeters. The period of incubation is given as from 31 to 42 days, or as about five weeks.

Food.—Mr. Cordeaux (1898) says on this subject:

Swans feed on vegetable substances, as grass, and shoots of shrubs and trees, and the roots and leaves of water plants, which their long necks enable them to tear up from the bottom of the rivers and shallows of the lakes they frequent. They will also eat grain when it can be got.

Witherby's Handbook (1921) adds to the list of food taken in summer, "fresh-water mollusca, worms, and acquatic insects."

Behavior.—Rev. F. O. Morris (1903) writes:

They fly in a long line, at times divaricated in the form of a wedge, and go in flocks or teams of from 4 or 5 to 30, which unite together to the number of several hundreds, at the times of migration. Their flight is easy and well sustained and usually conducted at a great height. It is exercised without much noise, except on first rising or alighting, when the sound may be heard to a considerable distance. It is said that they can fly at the rate of above 100 miles an hour. They walk well and can also run with considerable rapidity. In swimming about, except when feeding, the neck is carried in an upright posture and seldom in the arched manner characteristic of the other species. In walking the neck is bent backward over the body and the head lowered as if to preserve a proper balance.

The note resembles the word "*hoop*," repeated ten or a dozen times; hence the name of the bird. It is both loud, clear, and sonorous, and sounds aloft like the clang of a trumpet. Other inflections of their voice are expressed by Meyer, by the syllables "*hang, hang*," "*grou, grou*," and "*killelee*." Montagu writes that having killed one of these species out of a flock of 10 or 12 its companions flew around several times making a most melancholy cry before they flew off.

Mr. Cordeaux (1898) refers to the notes of this swan as follows:

There is no sound in nature more likely to attract attention than the aerial music of a herd of migrating swans passing high overhead; some speak of it as exhilarating to the highest degree, but to me there is always a touch of sadness in the sound—the sadness of Highland music in those long drawn, melancholy, and plaintive notes, which seem suggestive of the illimitable wilds of the great lone lands where the birds have passed the long day of the short Arctic summer.

He further says:

Mr. St. John has seen them arrive on Loch Spynie as early as September 30. He says: "While they remain with us, they frequent and feed in shallow pieces of water, of so small a depth that in many places they can reach the bottom with their long necks and pluck off the water grasses on which they feed. While employed in tearing up these plants, the swans are generally surrounded by a number of smaller water fowl, such as wigeon and teal, who snatch at and carry off the pieces detached by their more powerful companions. The rapidity of the flight of a swan is wonderful; one moment they are far from you, the next they have passed you like an arrow. This speed, however, is only attained when at a considerable height above the ground." Swans are most powerful swimmers and will swim out from the seashore in the teeth of a considerable gale with the greatest ease.

DISTRIBUTION

Breeding range.—Palearctic region. East to northeastern Siberia (Anadyr and Kamchatka) and the Commander Islands. West to Great Britain. South to about 65° N. in northern Siberia and about 62° N. in northern Europe. North of the Arctic Circle in Finland and Scandinavia. West to Iceland, and formerly to Greenland.

Winter range.—South to southern Europe (rarely to northern Africa), central Asia, Persia, China, and Japan.

Casual records.—Now only casual in Greenland (Atangmik, Godthaab, Ingtuk, and Arsuk).

Egg dates.—Iceland: Ten records, May 1 to June 18. Great Britain: Five records, April 20 to June 21.

CYGNUS COLUMBIANUS (Ord)

WHISTLING SWAN

HABITS

I had lived to be nearly 50 years old before I saw my first wild swan, but it was a sight worth waiting for, to see a flock of these magnificent, great, snow-white birds, glistening in the sunlight against the clear blue sky, their long necks pointing northward toward their polar home, their big black feet trailing behind, and their broad translucent wings slowly beating the thin upper air, as they sped onward in their long spring flight. If the insatiable desire to kill, and especially to kill something big and something beautiful, had not so possessed past and present generations of sportsmen, I might have seen one earlier in my life and perhaps many another ornithologist, who has never seen a swan, might have enjoyed the thrill of such an inspiring sight. No opportunity has been neglected to kill these magnificent birds, by fair means or foul, since time immemorial; until the vast hordes which formerly migrated across our continent have been sadly reduced in numbers and are now confined to certain favored localities. Fortunately the breeding grounds of this species are so remote that they are not likely to be invaded by the demands of agriculture; and fortunately the birds are so wary that they are not likely to be exterminated on migrations or in their winter resorts.

Spring.—Dr. D. G. Elliot (1898) says of the start on the spring migration:

At the advent of spring the swan begin to show signs of uneasiness, and to make preparations for their long journey to the northward. They gather in large flocks and pass much of their time preening their feathers, keeping up a constant flow of loud notes, as though discussing the period of their departure and the method and direction of their course. At length all being in readiness, with loud screams and many *who-who's*, they mount into the air, and in long lines wing their way toward their breeding places amid the frozen north. It has been estimated that swan travel at the rate of 100 miles an hour with a moderate wind in their favor to help them along. The American swan is monogamous, and once mated the pair are presumed to be faithful for life. The young keep with their parents for the first year, and these little families are only parted during that period by the death of its members.

Being early migrants, swans are often overtaken by severe storms with disastrous results, as the following incident, related by George B. Sennett (1880), will illustrate:

An unusual flight of swans occurred in northwestern Pennsylvania on the 22d of last March (1879). On the day mentioned, as well as the previous day and night, a severe storm prevailed, the rain and snow freezing as they fell. The swans, on their migration north, were caught in the storm and, becoming overweighted with ice, soon grew so exhausted that they settled into the nearest ponds and streams, almost helpless. Generally a single one was seen in some mill pond or creek, and the fowling piece, loaded with large shot, and not infrequently the rifle, was used to bring to bag the noble game, though, considering the plight they were in, in all probability anyone might have paddled up to the birds and taken them alive. In fact, in a number of instances they were reported as thus taken alive. Large flocks were seen in some districts in the same pitiable condition. A flock of from 33 to 35 American or whistling swans surprised the inhabitants of Plumer on Saturday forenoon by alighting in the waters of Cherry Run. One of the swans was almost immediately shot at and killed, and, to the surprise of the now large crowd of men and boys, the remainder of the flock, on account of the ice accumulating on their wings, was unable to fly, and a general rush was then made for the poor birds, and 25 were captured alive by the eager fellows.

The late E. S. Cameron has sent me some very full notes on the whistling swan, which seems to be a regular spring and fall migrant through central Montana. He mentions a flock of 344 birds seen by W. R. Felton on Mallard Lake on April 4, 1912, and a still larger flock of about a thousand birds seen by J. H. Holtman on Marshy Lake on April 10, 1911. He says:

The swans come in small flocks, at short intervals, until they sometimes aggregate several hundred individuals. While the swans are usually the earliest birds to arrive, geese may be still earlier, and this year a small flock of six Canada geese (*Branta canadensis canadensis*) preceded the swans. In 1913 the first swans were observed on April 4 by Bob Morrow (one of Mr. Williams's men), who counted 26. On April 6 W. P. Sullivan, of the Square Butte Ranch, enumerated 25 in one flock and saw another smaller bunch of about half that number, which were too far away to count without glasses. At the present time (April 8), when we reached the lake side there were 125 swans, as we ascertained after frequent counts. These were grouped upon the southwest shore of the lake immediately below the ranch where the fine mountain stream called Alder Creek flows in. Some were standing upon one leg in 2 or 3 inches of water, others floated asleep behind these, with their heads under their wings, and farther away watchful birds, constituting a rear guard, were sailing about. With very few exceptions the swans held one leg along the side either when swimming or resting upon the water. They allowed us to examine them through binoculars for a few minutes and then all began swimming slowly for the center of the lake. Mr. Williams informed us that no matter how much the swans might be disturbed they would always return to this place, on account of the fresh water running in from the mountains. He also said that unless shot at (when they would probably leave altogether) the swans might possibly remain until May 1.

Doctor Nelson (1887) writes:

This fine bird arrives on the shore of Bering Sea in the vicinity of St. Michael early in May, and in some seasons by the 27th of April, as in 1878, when several were seen on that date about a spring hole in the ice. At this time the ground was clothed with over a foot of snow, and the sea covered, as far as could be seen, with unbroken ice. During the next few days a terrible storm of wind and snow swept over the country, but did these birds no harm, as was seen directly after the storm ceased by their presence at the water hole as usual. Mr. Dall records their arrival on the Yukon about May 1, and notes the fact of their descending that stream in place of going up the Yukon, as most of the geese do at this season.

Courtship.—Alfred M. Bailey has sent me the following account of this ceremony which he witnessed in Alaska:

At Wales I saw swans rarely, the first noted being on June 5, when I witnessed as pleasing a performance as it has been my privilege to see. The tundra was still clothed in its winter's coat of white, although pools of brilliant colors had formed here and there by the melting snow. It was in the height of the spring migration, with hundreds of snow geese, little brown cranes, and shore birds in sight continually. Then, far out on the tundra I heard a different call, a clamoring, quavering call, first full and loud and gradually dying down. With the aid of the glasses I made out three swans, possibly two males performing for the benefit of the female. They walked about with arched necks proudly lifted, taking high steps, with wings outstretched, two birds occasionally bowing to each other, and as they performed, they continually kept calling. After a few minutes in a given place, they took to wing and drifted across the tundra a hundred yards, where the ceremony was then repeated.

Nesting.—Doctor Nelson (1887) describes the nest, as follows:

The birds arrive singly or in small parties on the coast, and directly after scatter to their summer haunts. The nest is usually upon a small island in some secluded lakelet, or on a rounded bank close to the border of a pond. The eggs are deposited in a depression made in a heap of rubbish gathered by the birds from the immediate vicinity of the nest, and is composed of grass, moss, and dead leaves, forming a bulky affair in many cases. On June 14, 1880, a swan was seen flying from the side of a small pond on the marsh near St. Michael, and a close search finally revealed the nest. The eggs were completely hidden in loose moss, which covered the ground about the spot, and in which the bird had made a depression by plucking up the moss and arranging it for the purpose. The site was so artfully chosen and prepared that I passed the spot in my search, and one of my native hunters, coming close behind, called me back, and thrusting his stick into the moss exposed the eggs. I may note here that whenever the Eskimo of Norton Sound go egging on the marshes they invariably carry a stick 3 or 4 feet long, which they thrust into every suspicious tussock, bunch of grass, or spot in the moss, and if a nest is there it is certain to be revealed by the stick striking the eggs. They are very expert in detecting places likely to be chosen by the ducks and geese. I have seen my hunters examine the borders of a lake, after I had given it what I considered a thorough search, and unearth in one instance three geese nests and one duck's. This was after I had acquired considerable skill in finding eggs, so it may readily be seen that the birds are very cunning in placing their nests.

Swainson and Richardson (1831) say of this species:

This swan breeds on the seacoast within the Arctic Circle, and is seen in the interior of the fur countries on its passage only. It makes its appearance amongst the latest of the migratory birds in the spring, while the trumpeter swans are, with the exception of the eagles, the earliest. Captain Lyon describes its nest as built of moss peat, nearly 6 feet long and 4¾ wide, and 2 feet high exteriorly; the cavity a foot and a half in diameter. The eggs were brownish-white, slightly clouded with a darker tint.

According to Rev. C. W. G. Eifrig (1905) the Canadian Neptune Expedition to Hudson Bay found this swan common on Southampton Island;

also in the flat land north of Repulse Bay. They breed in lowlands with lakes, where their nests, constructed of seaweed, grass, and moss, are very conspicuous. They are very bulky affairs, about 3 feet in diameter at the base, tapering to 18 inches at the top, and 18 inches high. A set of 2 eggs was taken on Southampton, July 4, 1904.

Mr. Bailey has contributed the following notes on a nest he found in Alaska:

While collecting near Mint River, which empties into Lopp Lagoon about 20 miles north of Cape Prince of Wales, I found a nest of this species with three downy young. It was early in the morning that we discovered it, on July 12. Both adults were seen sitting close to the edge of a pond, and, as we approached, they flew majestically away, only to circle and sail back directly over our heads. The female was more stained than the male. There, near the water's edge, from where the parent birds had taken flight, were three beautiful little downy young, which had just left the nest, some 25 feet away, and were doubtless ready to undertake their first swim. They were as fluffy as balls of yarn, with dark brown eyes, and bills and feet of pink flesh-color. They showed no fear, and cuddled contentedly when we held them in our hands.

The nest was a conspicuous, built-up mound of moss on a ridge overlooking the little lagoon, and was unlined with down. From the size of the young, it was evident that the swans made their nest on the first bit of bare tundra. The swans are probably among the first birds to nest in the vicinity of Wales; the geese eggs were but half incubated at this time, while the loons' eggs were fresh.

The swans owe their present-day numbers to the fact that they nest over a wide stretch of barren country, uninhabited even by natives. They are continually persecuted on their breeding grounds, and were it not for their habit of nesting early, when the snow is deep and too soft for traveling, they would have been exterminated long ago.

Eggs.—The foregoing brief accounts are about all we have regarding the nesting habits of this well-known species. MacFarlane collected about 20 sets of eggs, but said very little about the nests. The usual number of eggs seems to be 4 or 5, though as few as 2 and as many as 7 have been reported. The eggs resemble goose eggs except that they are much larger. In shape they are elliptical ovate or elliptical oval, with a tendency toward fusiform in some specimens. The shell is fairly smooth or finely granulated and not glossy. The color is creamy white or dull white at first, becoming much nest stained. The measurements of 94 eggs, in various collections, average 106.9 by 68.2 millimeters; the eggs showing the four extremes measure **115.7** by 68.5, 115 by **73**, and **90** by **58.7** millimeters.

Young.—The period of incubation is said to be from 35 to 40 days. Doctor Nelson (1887) says of the young:

The last of June or first of July the young are hatched, and soon after the parents lead them to the vicinity of some large lake or stream, and there the old birds molt their quill feathers and are unable to fly. They are pursued by the natives at this season, and many are speared from canoes and kyaks. Although unable to fly, it is no easy task single handed to capture them alive. The young men among the Eskimo consider it a remarkable exhibition of fleetness and endurance for one of their number to capture a bird by running it down.

Plumages.—The downy young is described by Dr. D. G. Elliot (1898) as "pure white, bill, legs, and feet yellow"; but the young of European swans are all either pale grayish-white or grayish-brown.

Doctor Nelson (1887) describes a young bird taken in September, apparently in juvenile plumage, as follows:

The young birds of the year frequently retain the immature plumage until the last of September. A specimen in this plumage, taken on September 19, had its bill purplish flesh color, the nail and a border along the gape black; the iris hazel, and the feet and tarsi livid flesh color. The plumage of this bird, which is now before me, is sooty brownish with a plumbeous shade about the top and sides of the head; neck and throat all around dull plumbeous ashy of a light shade; back, tertials, and wing coverts dull plumbeous ashy with a silvery gray luster, especially upon the wings. Rump white, lightly washed with ashy, which increases to dull plumbeous ashy on the tail coverts and rectrices. Quills white, heavily mottled with ashy gray on their terminal third, but almost immaculate toward bases. Under surface white, washed with dingy gray.

Doctor Sharpless, quoted by Audubon (1840), says:

The swan requires five or six years to reach its perfect maturity of size and plumage, the yearling cygnet being about one-third the magnitude of the adult, and having feathers of a deep leaden color. The smallest swan I have ever examined, and it was killed in my presence, weighed but 8 pounds. Its plumage was very deeply tinted, and it had a bill of a very beautiful flesh color, and very soft. This cygnet, I presume, was a yearling, for I killed one myself the same day, whose feathers were less dark, but whose bill was of a dirty white; and the bird weighed 12 pounds.

Doctor Elliot (1898) also writes:

The young of this species is gray, sometimes lead color during its first year, and the bill is soft and reddish in hue. In the second year the plumage is lighter, and the bill white, becoming black in the third year, when the plumage, though white, is mottled with gray; the head and neck especially showing but little white. It is probable that it takes fully five years before the pure white dress is assumed and the bird becomes such an ornamental object.

Although Baird, Brewer, and Ridgway (1884) make a similar statement, I can not believe that it takes a swan any such length of time to acquire its full plumage. Witherby's Handbook (1921) seems to imply that the pure white plumages of the whooping swan and the Bewick swan are acquired before the second winter.

Hon. R. M. Barnes tells me that young swans, reared by him in confinement, acquired their full plumage during the second summer or fall, when 14 or 15 months old. I believe that this is usually the case with wild birds, though some traces of immaturity may not disappear until some time during the following winter or even spring.

Food.—The food of the whistling swan is largely vegetable, which it obtains by reaching down with its long neck in shallow water, occasionally tipping up with its tail in the air when making an extra long reach. While a flock of swans is feeding in this manner, one or more birds are always on guard watching for approaching dangers, as the feeding birds often keep their heads and necks submerged for long periods. It apparently never dives for its food except in cases of great extremity. In Back Bay, Virginia, and in Currituck Sound, North Carolina, the swans feast on the roots of the wild celery and fox-tail grass; they are now (1916) so numerous that they do considerable damage by treading great holes in the mud and by rooting and pulling up the celery and grass; they thus waste large quantities of these valuable duck foods, much more than they consume, and consequently spoil some of the best feeding grounds for ducks, much to the disgust of the sportsmen in the various clubs, who are not allowed to shoot the swans and have to submit to this interference with their duck shooting. The swans are really such a nuisance in this particular locality that a reasonable amount of shooting might well be allowed; these birds are so wary that there is little danger of any great number being killed.

Major Bendire (1875) found in the stomach of a whistling swan, shot in Oregon, "about 20 small shells, perhaps half an inch in length, quite a quantity of gravel, and a few small seeds." Mr. Cameron in his Montana notes, says:

> The swans were engaged in feeding upon the soft-shelled fresh-water snails which abound in this lake and explain its great attraction for them. During the several days that I watched the swans I never saw them eat anything else, but doubtless they pick up vegetation as well, being accustomed to walk about in the grass at the mouth of Alder Creek. Marshy Lake is so shallow (only 2 feet deep over most of it, and 4 feet in the deepest part) that the long-necked birds can generally reach the mollusca without much tilting of their bodies in characteristic swan fashion.

Dr. F. Henry Yorke (1891) says: "They feed upon corn, and upon tender roots of wheat, rye, and grass, and upon bulbous roots, pushing about for them in the mud at the bottom of lakes and rivers. They also catch and eat tadpoles, frogs, and even fish." Other writers have mentioned, among the food of this species, the roots of the *Equisetacae*, *Sagittaria*, various grasses, and other succulent water plants, also worms, insects, and shellfish.

Behavior.—Considering its size and weight, a swan rises from the water with remarkable ease and celerity; it runs along the surface for 15 or 20 feet, flapping its wings and beating the water with its feet alternately, until it has gained sufficient headway to launch into the air; like all heavy-bodied birds it must face the wind in rising. When well awing it flies with considerable speed and power, with the long neck stretched out in front and the great black feet extending beyond the tail; the wing beats are slow, but powerful and effective. It has been said to fly at a speed of 100 miles an hour; probably no such speed is attained, however, except when flying before a heavy wind; it undoubtedly flies faster than it appears to on account of its great size, and it certainly flies faster than any of the ducks and geese. When traveling long distances swans fly in V-shaped wedges, in the same manner as geese and for the same reason; the resistance of the air is less, as each bird flies in the widening wake of its predecessor; the leader, of course, has the hardest work to do, as he "breaks the trail," but he is relieved at intervals and drops back into the flock to rest. On shorter flights they fly in long curving lines or in irregular flocks. They usually fly rather high, and when traveling are often way up above the clouds. Audubon (1840) quotes Doctor Sharpless, as follows:

In flying, these birds make a strange appearance; their long necks protrude and present, at a distance, mere lines with black points, and occupy more than one-half their whole length, their heavy bodies and triangular wings seeming but mere appendages to the prolonged point in front.

When thus in motion, their wings pass through so few degrees of the circle that, unless seen horizontally, they appear almost quiescent, being widely different from the heavy semicircular sweep of the goose. The swan, when migrating, with a moderate wind in his favor, and mounted high in the air, certainly travels at the rate of 100 miles or more an hour. I have often *timed* the flight of the goose, and found one mile a minute a common rapidity, and when the two birds, in a change of feeding ground, have been flying near each other, which I have often seen, the swan invariably passed with nearly double the velocity.

Mr. Cameron, in his notes, refers to the powers of flight of swans, as follows:

Small parties of the swans on the water spread their long wings at regular intervals and took lengthy flights, presumably to keep themselves in practice for their forthcoming journey. The control which such large birds (weighing from 17 to 20 pounds) possess over their flight on a perfectly calm day is to me quite marvelous, and must be seen to be appreciated. A compact flock of from 4 to 6 swift-flying swans will circle the whole basin of the lake several times, and then, as if tied together, alight in the closest proximity to each other, yet never collide. They will pitch upon the water in the most graceful manner imaginable, without bringing their long legs forward, or making any splash. At exceptional times, however, the swans do make a loud splash when they alight.

The ease and grace with which a swan swims on the surface of the water is too well known and too far famed to need any further comment; there is no prettier picture, no grander picture, than a party of these beautiful birds floating undisturbed on the mirror surface of some northern mountain lake against the rugged background of one of nature's wildest spots. But few people realize the speed and power of the swan as a swimmer until they have tried to chase one in a boat and seen how easily he escapes, even against wind and waves, without recourse to flight.

The notes of the whistling swan are varied, loud and striking at times and again soft and musical trumpetings. To me they are suggestive of the Canada goose's call in form, but are more like soft musical laughter, suggested by the syllables "*wow-how-ou*," heavily accented on the second note. Mr. Cameron says, in his notes:

Mr. Skelton describes the sounds uttered by his tame swan as "long whoops, or clucking croaks, according to its mood." The wild swans upon taking wing, or when arriving on migration, produce sounds like a slow shake of two notes upon a clarinet. If the flock is large, as in the present instance, so many throats yield a great volume of musical sound. When the quiescent swans become suddenly alarmed, and contemplate flight, a subdued chorus runs through the flock like different modulations from an orchestra of reed instruments. Under no circumstances could the swan voices be compared to brass instruments (such as a trumpet or hunting horn) in my opinion, and herein concur Mr. Felton, Mr. Williams, and Mr. Skelton, who have had frequent opportunities for listening to them. We could distinctly hear the swan cries at the ranch a mile from the lake, and they might have been heard at a much greater distance.

The old saying that "a swan sings before it dies" has generally been regarded as a myth, but the following incident, related by so reliable an observer as Dr. D. G. Elliot (1898), is certainly worthy of credence:

I had killed many swan and never heard aught from them at any time, save the familiar notes that reach the ears of everyone in their vicinity. But once, when shooting in Currituck Sound over water belonging to a club of which I am a member, in company with a friend, Mr. F. W. Leggett, of New York, a number of swan passed over us at a considerable height. We fired at them, and one splendid bird was mortally hurt. On receiving his wound the wings became fixed and he commenced at once his song, which was continued until the water was reached, nearly half a mile away. I am perfectly familiar with every note a swan is accustomed to utter, but never before nor since have I heard any like those sung by this stricken bird. Most plaintive in character and musical in tone, it sounded at times like the soft running of the notes in an octave, and as the sound was borne to us, mellowed by the distance, we stood astonished, and could only exclaim, "We have heard the song of the dying swan."

Fall.—Referring to the beginning of the fall migration, Mr. Lucien M. Turner (1886) says:

The young are able to leave the nest by the first week in July, and fly by the middle of September. They migrate about the middle of October, and at this time the migration is invariably to the northward from St. Michael, and directed toward the head of Norton Sound. As many as 500 may form a single line, flying silently just over the shore line at a height of less than 600 feet. I always suspected that these birds flew to the northward as far as the Ulukuk Portage, in about 65° 30′ north latitude, so as to get to the Yukon River at Nulato, about 120 miles in the interior of the Territory, and continue their flight up the Yukon River, which would in its course let these birds more easily cross the Rocky Mountain ridge with least effort. This is supposed by the fact that I never saw swans, at any season of the year, migrating to the southward.

From this statement and similar observations by Dall and Nelson, one would infer that the swans which breed in northern Alaska cross the Rocky Mountains to join the main migration route of the species, which is southward through the interior of Canada; perhaps the birds which breed in southern Alaska and in Canada west of the Rockies migrate down the coast to their winter homes on the Pacific coast. Swans are very abundant in the interior of Canada and the northern States in the fall migration. Large numbers were formerly killed by the fur traders for their skins which were dealt in as regular articles of commerce. The Hudson's Bay Co. sold 17,671 swan skins between the years 1853 to 1877; the number steadily decreased, however, from 1,312 in 1854 to 122 in 1877, and during the next two or three years the traffic practically ceased.

From the vicinity of the Great Lakes the heaviest flight seems to take a southeastward direction to the Atlantic coast, but there is also a southward flight to the Gulf of Mexico and probably a limited southwestward flight to the Pacific coast.

A striking example of the disasters which may befall even one of our largest and strongest species of wild fowl is shown in the destruction of swans in the Niagara swan trap. In one instance over 100 of these great birds met their death; being caught in the rapids they were swept over the falls; many were killed by the fall, others were killed or maimed by the rough treatment they received in the whirlpools and rapids, where they were hurled against the rocks or crushed in the ice; a few probably escaped by flying back over the falls, but most of them were unable to fly at all on account of their injuries or were too exhausted to rise high enough to clear the falls. But eventually many of them would have escaped if they had not been attacked by a crowd of men and boys, who shot, beat, and clubbed the poor struggling birds until not a living bird remained. For full accounts of two such catastrophes I would refer the reader to Mr. J. H. Fleming's (1908 and 1912) interesting papers on the subject.

Game.—As game birds, swans have never held a prominent place. They are not abundant anywhere except in a few favored spots, as migrants or winter sojourners. They have always been so wary and shy that attempts to shoot them in any considerable numbers generally resulted in making them wilder than ever or in driving them away altogether. The flesh of the younger birds is comparatively tender and palatable, but the older birds are very tough. Swans always have been attractive marks for sportsmen on account of their large size and spectacular appearance, but comparatively few have ever enjoyed the privilege of shooting at them. Swans are now protected in their winter resorts on the Atlantic coast, but formerly they were shot in considerable numbers in the vicinity of Chesapeake Bay and Currituck Sound. They were shot mainly from the marshy points where blinds were built for duck shooting; the swans were wont to feed along the shores of the marshy coves and bays; and in passing from one cove to another they frequently flew close around or over these points, offering tempting shots. It was an exciting moment for the sportsman when he saw a flock of these great white birds approaching and few could resist the temptation to shoot at them. On windy, stormy days it was often possible to creep up to them through the marsh near enough to get a shot at them when they rose. Approaching swans on the open water of the bay was a different proposition, especially if they were surrounded, as they often were, by the watchful geese. But even this was successfully accomplished by sailing down the wind upon them, which made it necessary for them to rise toward the boat. In winter, boats covered with blocks of ice and manned by gunners dressed in white could sometimes be paddled or allowed to drift within gunshot of a feeding flock.

Winter.—Doctor Sharpless, in his interesting account of this species published by Audubon (1840), thus describes the arrival of the swans in their principal winter home on the coasts of Virginia and North Carolina:

The swans, in traveling from the northern parts of America to their winter residence, generally keep far inland, mounted above the highest peaks of the Allegheny, and rarely follow the watercourses like the geese, which usually stop on the route, particularly if they have taken the seaboard. The swans rarely pause on their migrating flight, unless overtaken by a storm, above the reach of which occurrence they generally soar. They have been seen following the coast in but very few instances. They arrive at their winter homes in October and November, and immediately take possession of their regular feeding grounds. They generally reach these places in the night, and the first signal of their arrival at their winter abode is a general burst of melody, making the shores ring for several hours with the vociferating congratulations whilst making amends for a long fast, and pluming their deranged feathers. From these localities they rarely depart unless driven farther south by intensely cold weather, until their vernal excursion.

The Chesapeake Bay is a great resort for swans during the winter, and whilst there they form collections of from 100 to 500 on the flats, near the western shores, and extend from the outlet of the Susquehanna River almost to the Rip Raps. The connecting streams also present fine feeding grounds.

They always select places where they can reach their food by the lengths of their necks, as they have never, so far as I can learn, been seen in this part of the world to dive under the water, either for food or safety.

Whistling swans are still abundant in winter on Bay Back, Virginia, and Currituck Sound, North Carolina, where, according to recent accounts, they are holding their own or even increasing in numbers. I have seen from 1,000 to 1,500 birds there in a day, as recently as 1916, standing in long white lines along the grassy shore of some marshy island, or feeding in large flocks, sometimes of two or three hundred birds, in the shallow waters of the bay, always conspicuous as striking features of these great wild-fowl resorts. Their chief companions here are the Canada geese, with whom they are intimately associated on their feeding grounds and on whom they depend largely, as sentinels to warn them of approaching danger, for the geese are even more watchful than the swans. But here, as well as on the lakes visited on their migration, they are also associated with the various ducks which resort to similar feeding grounds. They usually flock by themselves, however, when on the wing and do not mingle in the flocks of geese and ducks. They move about largely in family parties of 6 or 7 birds, but often gather together in large flocks to feed or to rest; large flocks are also often seen moving about, sometimes high in the air, calling to their fellows with loud mellow trumpetings in their search for quiet and safe feeding or resting places.

Nathan L. Davis (1895), writing of their winter habits on the coast of Texas, says:

I first saw them on Galveston Bay on January 1, and observed them every day until March 20, when there seemed to be but a very few left; all remaining on that date I think were crippled birds, being unable to stand the fatigue in their long journey to the north. It is a great sight to watch a flock of these birds assembled on the water, curling their long necks around each other, all making a strange honking noise, peculiar to themselves. This they continue for some time, then all turn with military precision and form in line; when they swim up and down the coast, proudly swaying their heads from side to side. In this manner they spend most of the bright days. They can be easily seen far out on the bay, their large white bodies glistening in the sun, as the restless waves toss their corklike forms above the level of the water. At first sight I could not distinguish whether the silvery spots rising on the waves were swans or the water breaking over some treacherous sandbar, which are common both in Galveston and San Jacinto Bays. Each day as the sun begins to go down they turn and slowly approach the shore, each keeping a sharp lookout ahead. If frightened any way they will either turn and swim quietly away or all take wing and survey the country for miles around before they will again settle on the water. Often small flocks may be seen in company with ducks, geese, pelicans, and gulls, but usually they will be found alone at some distance from all other birds, as well as human habitation. They are very hard to approach on a bright day, and hunting for them in clear weather is like fishing for trout in a thunderstorm. The dense fogs which prevail along the coast are no doubt the worst enemies these birds have, for then if the hunter is careful he can approach within easy range before they attempt to escape.

In stormy weather they are very restless and are continually flying from place to place as if hunting for a quiet spot, where they may rest in peace till the storm passes. In this continuous change of positions they often come too near the shore, and many are killed by the hunters who lay hidden, awaiting their approach. I once saw five of these large birds killed at a single discharge of a heavy double gun.

The tragic end of a belated cripple on a Montana lake is thus described in Mr. Cameron's notes:

In the fall of 1908, a member of a large flock of whistling swans, which settled upon Marshy Lake, was slightly wounded in the wing by a bullet (or, as is more probable, had a flight feather cut away by it) and could not leave with its frightened companions. Mr. Sullivan observed the swan about a dozen times when driving cattle to another Milner ranch near Shonkin, and when returning by the same route. He informed me that after the lake became frozen over, the swan, which was an adult in pure white plumage, by constantly swimming in a circle, kept open a small pond, about 25 feet wide. Until December 1 he regularly saw the swan upon this pond, which it was able to maintain open even when the ice was 3 inches thick upon the rest of the lake. The swan frequently dived, but was, of course, always obliged to come up in the same place on account of the ice; and Mr. Sullivan supposed that the poor bird eked out a scanty subsistence by means of the weeds or other food which it found at the bottom of the lake. The fate of this swan, though not absolutely known, can easily be surmised. Numerous coyotes, which crossed upon the ice, persistently menaced, and would have devoured the unfortunate bird but for its self-made asylum; hence, with the advent of colder weather, and consequent freezing up of the water, it would have undoubtedly become their prey. The above suggests a wintry scene which would be a fitting subject for an artists's brush; the famished prisoner swimming around the dark refuge pool, the scarcely less hungry jailers patrolling the ice edge and licking their expectant lips, the white world, and the onward creeping ice, grim with inexorable fate.

DISTRIBUTION

Breeding range.—Nearctic region, mainly north of the Arctic Circle. East to Baffin Land. South to Nottingham and Southampton Islands, the barren grounds of northern Canada, the Alaska Peninsula (Becharof Lake, Chulitna River and Morzhovia Bay), and St. Lawrence Island in Bering Sea. Northward in Alaska to the Arctic coast (Cape Prince of Wales and probably Point Barrow). North on the Arctic islands to about 74°, the North Georgia Islands, and Victoria Land (Cambridge Bay).

Winter range.—Mainly on the seacoasts of United States. On the Atlantic coast most abundantly from Maryland (Chesapeake Bay) to North Carolina (Currituck Sound); less commonly north to New Jersey; rarely north to Long Island (Shinnecock) and Massachusetts (Nantucket). Rarely south to Florida and the Gulf coasts of

Louisiana and Texas. On the Pacific coast from southern Alaska (Dall, Long, and Prince of Wales Islands) southward to southern California (San Diego). A few may winter irregularly in the interior as far north as large bodies of open water may be found.

Spring migration.—Toward the interior and generally northward. Early dates of arrival: Pennsylvania, Erie, March 11, and Williamsport, March 20; New York, Lockport, March 20; Ontario, Toronto, April 8; Michigan, Detroit, March 14; Wisconsin, Delavan, April 1; Minnesota, Heron Lake, March 31 and Elk River, April 8; Manitoba, Shoal Lake, April 30; northern Alberta, Athabasca Lake, May 17; Mackenzie, Fort Simpson, May 5, and Fort Anderson, May 18; Melville Island, May 31. Dates of arrival in Alaska: St. Michael, April 27; Kowak River, May 11; Point Hope, May 21. Late dates of departure: Maryland, Baltimore, May 4; Pennsylvania, Williamsport, May 30.

Fall migration.—Reversal of spring routes. Early dates of arrival: Mackenzie, Great Bear Lake, September 15, and Mackenzie River, October 6; Quebec, Cape St. Ignace, October 11; Maine, Crawford Lake, September 10; New Hampshire, Seabrook, October 18; Massachusetts, Nantucket, October 16; Rhode Island, Quonocontaug Pond, November 9; Maryland, Baltimore, September 26; Virginia, Alexandria, October 15; South Carolina, Cooper River, November 21; Alaska, Sitka, September 28; Washington, Thurston County, October 25; Montana, Teton County, October 31. Late dates of departure: Alaska, St. Michael, October 8, and St. George Island, October 17.

Casual records.—Accidental in Bermuda (1835 or 1836) and Commander Islands (Bering Island, November 3, 1882). Casual in Mexico (near Colonia Diaz, Chihuahua, January 18, 1904, and at Silao, Guanajuato).

Egg dates.—Arctic Canada: Thirteen records, May 29 to July 5; seven records, June 15 to July 1. Alaska: Ten records, May 17 to July 4; five records, June 4 to 12.

TRUMPETER SWAN

CYGNUS BUCCINATOR Richardson

HABITS

This magnificent bird, the largest of all the North American wild fowl, belongs to a vanishing race; though once common throughout all of the central and northern portions of the continent, it has been gradually receding before the advance of civilization and agriculture; when the great Central West was wild and uncultivated it was known to breed in the uninhabited parts of many of our Central States, even as far south as northern Missouri; but now it probably does not breed anywhere within the limits of the United States, except possibly in some of the wilder portions of Montana or Wyoming; civilization has pushed it farther and farther north until now it is making its last stand in the uninhabited wilds of northern Canada. E. H. Forbush (1912) has summed up the history of its disappearance very well, as follows:

The trumpeter has succumbed to incessant persecution in all parts of its range, and its total extinction is now only a matter of years. Persecution drove it from the northern parts of its winter range to the shores of the Gulf of Mexico; from all the southern portion of its breeding range toward the shores of the Arctic Ocean; and from the Atlantic and Pacific slopes toward the interior. Now it almost has disappeared from the Gulf States. A swan seen at any time of the year in most parts of the United States is the signal for every man with a gun to pursue it. The breeding swans of the United States have been extirpated, and the bird is pursued, even in its farthest northern haunts, by the natives, who capture it in summer, when it has molted its primaries and is unable to fly. The swan lives to a great age. The older birds are about as tough and unfit for food as an old horse. Only the younger are savory, and the gunners might well have spared the adult birds, but it was "sport" to kill them and fashion called for swan's-down. The large size of this bird and its conspicuousness have served, as in the case of the whooping crane, to make it a shining mark, and the trumpetings that were once heard over the breadth of a great continent, as the long converging lines drove on from zone to zone, will soon be heard no more. In the ages to come, like the call of the whooping crane, they will be locked in the silence of the past.

The late E. S. Cameron prepared for me, in 1913 and 1914, some very elaborate notes on the history of this species in Montana, which Mrs. Cameron very kindly sent to me after her husband's death. They are interesting and valuable enough to print in full, but my space will permit only a few quotations and references. Regarding recent records he says:

The trumpeter swan, which 20 years ago was quite common in Montana, has now become exceedingly scarce, and is probably on the verge of extinction everywhere. My investigations during 1912–13 and 1914, show that trumpeters are almost unrepresented among the large numbers of migrant swans which biannually pass over Montana. It seems to be the melancholy fact that thousands of whistling swans are seen to one trumpeter, and at the time of writing I have only two authenticated records of trumpeter swans for the three years above mentioned, the specimens from St. Marys Lake and Cut bank. Mr. J. H. Price informed me that an adult male trumpeter was shot by a boy on the Yellowstone, near Miles City, Custer County, on October 27, 1905, and I have since seen the mounted bird in a saloon keeper's window. At the time of writing the finest specimen of a trumpeter swan, within my knowledge, is the one killed by Mr. Robert Sloane, of Kalispell, when duck shooting on the shore of Flathead Lake. We left Kalispell before daylight on the morning of November 15, 1910, on our way to where the Flathead River debouches from the flat alluvial floor of the valley through a fair-sized delta into Flathead Lake. The morning was chilly, with occasional snow flurries, and we knew that the ducks would be on the move. Leaving the spring wagon behind the strip of brush which fringes the lake shore at this point, we built blinds, set our decoys, and were soon in the midst of a good flight of canvasbacks, redheads, and mallards. The sport was the best in my experience, as the birds, in passing from the lake to the sloughs or bayous inshore, offered fine shots. At this place there is a sandy beach a hundred yards in width, while here, and for some miles on either side, it is possible to wade into the lake for another 200 yards without becoming wet above the knees. About 3 o'clock in the afternoon a violent snow squall arose, and in the thick of it my attention was drawn to some white objects which were rising and falling on the waves about a mile offshore. At times these appeared like small sailing boats, but when they drew nearer I distinguished a flock of eight swans led by a splendid snow-white bird whose every movement was followed by the others. Two more of the swans were white and the remaining five dark colored. I left my blind and, running along a cattle trail through the brush to the wagon, took my 30–30 Winchester and returned to the edge of the beach. I then fired at the big leading swan, and struck it fairly in the neck at the first shot, although the bird was some 200 yards distant. Upon the death of their leader the rest of the flock momentarily bunched up in bewilderment, but, recovering their wits, made a great commotion in their efforts to rise from the water. Having once cleared the lake with their wings, however, they departed at great speed, while I waded through the shallows to retrieve my coveted trophy. The swan was found to weigh a full 31 pounds.

He describes at considerable length the capture of a trumpeter swan by a shepherd employed by G. B. Christian, of Augusta, Montana, in November, 1907. The man fired a rifle at a passing flock and brought down a bird, which he captured unhurt except for one broken pinion. The bird was kept for a year alive and then presented to the Great Falls Park; but, as it was not pinioned, it escaped after the next molt and was shot by a boy and all trace of it was lost. It was a pure white adult of very large size and was supposed to have weighed about 35 pounds.

In 1913 an Indian offered for sale at Kalispell an immense trumpeter swan from St. Marys Lake, Glacier National Park. It was poorly skinned, not properly poisoned, became infested with beetles, and was burned. Dr. Jonathan Dwight has the head and legs of this swan.

A female trumpeter swan, now in the collection of Dr. Jonathan Dwight, was shot by a saloon keeper, Ben Schannberg, at Cutbank, Teton County, Montana, in the first week of November, 1913. This was also an immature bird, supposed to have been about 18 months old, and was in a much emaciated condition, but it weighed 20 pounds.

Henry K. Coale (1915) has published an excellent paper on the present status of this swan, in which he mentions three of the above records, but he does not include them in the list of specimens to which he refers, as follows:

> Of the great multitudes of trumpeter swans which traversed the central and western portion of north America 60 years ago, there are 16 specimens preserved in museums which have authentic data. These were collected between the years 1856 and 1909. There are besides the type, five other Canadian records, Toronto 1863, Fort Resolution 1860, Lake St. Clair 1878, St. Clair Flats 1884, and Manitoba 1887; and one from Wyoming 1856, Idaho 1873, Michigan 1875, Wisconsin 1880, Ohio 1880, Oregon 1881, North Dakota 1891, Minnesota 1893, Montana 1902, and Mexico 1909.

Nesting.—Prof. Wells W. Cooke (1906) says:

> In early times it probably bred south to Indiana, Wisconsin, Iowa, Nebraska, Montana, and Idaho; it nested in Iowa as late as 1871, in Idaho in 1877, in Minnesota in 1886, and in North Dakota probably for a few years later. It is not probable that at the present time the trumpeter nests anywhere in the United States, and even in Alberta no nests seem to have been found later than 1891. The vast wilderness of but a generation ago is now crossed by railroads and thickly dotted with farms. The species is supposed still to breed in the interior of British Columbia at about latitude 53°.

Dr. R. M. Anderson (1907) writes:

> The only definite record of the nesting of the trumpeter swan in Iowa which I have been able to trace was received from the veteran collector, J. W. Preston, in a letter dated March 22, 1904: "A pair of 'trumpeters' reared a brood of young in a slough near Little Twin Lakes, Hancock County, in the season of 1883, not many miles from where some good finds in the way of sets of whooping cranes were made. This was positively *Olor buccinator*. The nest was placed on a large tussock in a marshy slough or creek, and had been used for years by the swans, as I was credibly informed; but the nest mentioned above, so far as I am aware, was the last in that locality."

Roderick MacFarlane (1891) reported:

> Several nests of this species were met with in the Barren Grounds, on islands in Franklin Bay, and one containing 6 eggs was situated near the beach on a sloping knoll. It was composed of a quantity of hay, down, and feathers intermixed, and this was the general mode of structure of the nests of both swans. It usually lays from 4 to 6 eggs, judging from the noted contents of a received total of 24 nests.

Mr. Cameron says that trumpeter swans formerly bred in western Montana, but his diligent investigations have failed to discover any recent nesting sites. " Some birds made great nests of tules, but many more built them on muskrat houses which they flattened out for the purpose." He describes a nest, found in 1871 on the Thompson River, on " a large deserted beaver lodge. On this mound, which measured at least 5 feet across, was a great pile of grass and feathers." The two eggs, which it contained, were concealed under a bunch of down. Of the latest two obtainable records he says:

A Kootenai Indian woman, while hunting with her husband and father-in-law, in the year 1889, saw a pair of swans with two cygnets on a small lake toward the headwaters of the South Fork of the Flathead River. Shortly after this she took two eggs from a swan's nest by the same lake, but can not give the exact year, although it was probably 1890.

The latest information I have comes in a letter from M. P. Skinner, of Yellowstone Park, and is very encouraging; he writes:

Early in the summer of 1919, I noted a swan in the vicinity of Heart Lake. A little later the nest was found on a low island in a lagoon northeast of Lewis Lake, containing five whitish eggs, the nest being made of leaves and grass. On August 14 I returned and then found the tail and flight feathers molted by the adults, but the birds were too far away and too wary to determine the species. On September 6, I again visited this section and found five trumpeter swan (the two parents and three young so nearly grown as to be able to fly well). While I did not feel justified in sacrificing one of these rare birds, there can be no mistake as to identification. I have been familiar since November of 1912 with both of our swans, the whistling swan occuring in comparatively large numbers from October 31 (earliest date ever noted by me) to May 3, the latest date. I saw the trumpeter swans several times, and once within an estimated distance of 50 yards under a pair of 12 X binoculars. Bill and lores of all the birds lacked the yellow spot; they were markedly superior in size to the whistling swans; and their cries were unmistakable. The breeding range of the smaller swan is given as "far northward and probably in British Columbia," whereas the trumpeter has been known to breed as far south as Iowa.

Mr. H. M. Smith, United States Fish Commissioner, reports that on July 16, 1919, he visited a small, unnamed lake lying south of Delusion Lake and found there a pair of swans with six cygnets about the size of teal swimming actively about. Mr. Smith could not identify these as *buccinator*, but in view of my own discovery I believe they were of this variety.

Eggs.—The trumpeter swan has been said to lay from 2 to 10 eggs in a set; the latter number must be very unusual and was probably the product of two birds; probably the usual set consists of from 4 to 6 eggs. Mr. Cameron says that the number of eggs varies from 2 to 8 according to the age of the birds and other circumstances, the smaller sets being laid by the younger birds; at least this is the general opinion among the Indians. The eggs are like those of the whistling swan, but larger. In shape they vary from elliptical ovate or elliptical oval to nearly elliptical. The shell is rough or granulated and more or less pitted. The color is creamy white or dull white, becoming much nest stained. The measurements of 25 eggs, in various collections, average 110 by 71.1 millimeters; the eggs showing the four extremes measure **119.5** by 76, 115 by **76.5**, and **101** by **62.8** millimeters.

Young.—P. M. Silloway (1903) records the following incident which a friend of his witnessed in Montana:

A friend told me of seeing an old swan and a young one upon the "Highland" lakes. The two were in flight between the lakes, and the cygnet flew only a few feet directly above the elder, so that it could drop on the parent's back at frequent intervals. The younger swan would fly 50 or 60 yards alone, then drop lightly upon the parent's back to rest, being carried for 50 to 60 yards in this manner; then it would rise upon its own pinions and flap along above the elder bird until it again became weary of its own exertions.

Mr. Cameron sent me the following statement from Ed. Forbes a rancher of Kalispell:

I punched cows in the Centennial Valley in Beaverhead County from 1883 to 1888. During that time I saw quantities of swans, and killed many young birds which we thought good to eat. We used to paddle after them among the tules (bullrushes) and rope them, as they never seemed to learn to fly until ice formed around the shores, and were fearless, big, and awkward. Even then it took them a long, flapping flight to clear the water. The young birds would dive, and come up at a distance of 600 feet when chased with a boat.

Plumages.—I have never seen a downy young trumpeter swan and can find no description of it in print. We know very little of its molts and plumages. Audubon (1840) describes the first winter plumage as follows:

In winter the young has the bill black, with the middle portion of the ridge, to the length of an inch and a half, light flesh-color, and a large elongated patch of light dull purple on each side; the edge of the lower mandible and the tongue dull yellowish flesh-color. The eye is dark brown. The feet dull yellowish brown, tinged with olive; the claws brownish black; the webs blackish brown. The upper part of the head and the cheeks are light reddish brown, each feather having toward its extremity a small oblong whitish spot, narrowly margined with dusky; the throat nearly white, as well as the edge of the lower eyelid. The general color of the other parts is grayish white, slightly tinged with yellow; the upper part of the neck marked with spots similar to those on the head.

How long it takes for the young bird to reach maturity we do not know, but he speaks of two young birds, seen in captivity, that were about 2 years old and were pure white.

Food.—Audubon (1840) says of the feeding habits of the trumpeter swan:

This swan feeds principally by partially immersing the body and extending the neck under water, in the manner of fresh-water ducks and some species of geese, when the feet often seen working in the air, as if to aid in preserving the balance. Often, however, it resorts to the land, and then picks at the herbage, not sidewise, as geese do, but more in the manner of ducks and poultry. Its food consists of roots of different vegetables, leaves, seeds, various aquatic insects, land snails, small reptiles, and quadrupeds. The flesh of a cygnet is pretty good eating, but that of an old bird is dry and tough.

Behavior.—Referring to the behavior of this species, with which he seems to have been quite familiar, he writes:

The flight of the trumpeter swan is firm, at times greatly elevated and sustained. It passes through the air by regular beats, in the same manner as geese, the neck stretched to its full length, as are the feet, which project beyond the tail. When passing low, I have frequently thought that I heard a rustling sound from the motion of the feathers of their wings. If bound to a distant place, they form themselves in angular lines, and probably the leader of the flock is one of the oldest of the males; but of this I am not at all sure, as I have seen at the head of a line a gray bird, which must have been a young one of that year.

To form a perfect conception of the beauty and elegance of these swans, you must observe them when they are not aware of your proximity, and as they glide over the waters of some secluded inland pond. On such occasions, the neck, which at other times is held stiffly upright, moves in graceful curves, now bent forward, now inclined backward over the body. Now with an extended scooping movement the head becomes immersed for a moment, and with a sudden effort a flood of water is thrown over the back and wings, when it is seen rolling off in sparkling globules, like so many large pearls. The bird then shakes its wings, beats the water, and as if giddy with delight shoots away, gliding over and beneath the surface of the liquid element with surprising agility and grace. Imagine, reader, that a flock of 50 swans are thus sporting before you, as they have more than once been in my sight, and you will feel, as I have felt, more happy and void of care than I can describe.

When swimming unmolested the swan shows the body buoyed up; but when apprehensive of danger, it sinks considerably lower. If resting and basking in the sunshine, it draws one foot expanded curiously toward the back, and in that posture remains often for half an hour at a time. When making off swiftly, the tarsal joint, or knee as it is called, is seen about an inch above the water, which now in wavelets passes over the lower part of the neck and along the sides of the body, as it undulates on the planks of a vessel gliding with a gentle breeze. Unless during the courting season, or while passing by its mate, I never saw a swan with the wings raised and expanded, as it is alleged they do, to profit by the breeze that may blow to assist their progress; and yet I have pursued some in canoes to a considerable distance, and that without overtaking them, or even obliging them to take to wing. You, reader, as well as all the world, have seen swans laboring away on foot, and therefore I will not trouble you with a description of their mode of walking, especially as it is not much to be admired.

The notes of the trumpeter swan are described as loud, resonant trumpetings; differing in tone and volume from those of the whistling swan; the windpipe of the larger species has one more convolution, which enables it to produce a louder and more far-reaching note on a lower key, with the musical resonance of a French horn.

Winter.—For an account of its winter habits, we must again quote Audubon (1840) as follows:

The trumpeter swans make their appearance on the lower portions of the waters of the Ohio about the end of October. They throw themselves at once into the larger ponds or lakes at no great distance from the river, giving a marked preference to those which are closely surrounded by dense and tall canebrakes, and there remain until the water is closed by ice, when they are forced to proceed southward. During mild winters I have seen swans of this species in the ponds about Henderson until the beginning of March, but only a few individuals, which may have stayed there to recover from their wounds. When the cold became intense, most of those which visited the Ohio would remove to the Mississippi, and proceed down that stream as the severity of the weather increased, or return if it diminished; for it has appeared to me, that neither very intense cold nor great heat suit them so well as a medium temperature. I have traced the winter migrations of this species as far southward as Texas, where it is abundant at times.

Whilst encamped in the Tawapatee Bottom, when on a fur-trading voyage, our keel boat was hauled close under the eastern shore of the Mississippi, and our valuables, for I then had a partner in trade, were all disembarked. The great stream was itself so firmly frozen that we were daily in the habit of crossing it from shore to shore. No sooner did the gloom of night become discernible through the gray twilight than the loud-sounding notes of hundreds of trumpeters would burst on the ear; and as I gazed over the ice-bound river, flocks after flocks would be seen coming from afar and in various directions, and alighting about the middle of the stream opposite to our encampment. After pluming themselves awhile they would quietly drop their bodies on the ice, and through the dim light I yet could observe the graceful curve of their necks, as they gently turned them backward, to allow their heads to repose upon the softest and warmest of pillows. Just a dot of black as it were could be observed on the snowy mass, and that dot was about half an inch of the base of the upper mandible, thus exposed, as I think, to enable the bird to breathe with ease. Not a single individual could I ever observe among them to act as a sentinel, and I have since doubted whether their acute sense of hearing was not sufficient to enable them to detect the approach of their enemies. The day quite closed by darkness, no more could be seen until the next dawn; but as often as the howlings of the numerous wolves that prowled through the surrounding woods were heard, the clanging cries of the swans would fill the air. If the morning proved fair, the flocks would rise on their feet, trim their plumage, and as they started with wings extended, as if racing in rivalry, the pattering of their feet would come on the ear like the noise of great muffled drums, accompanied by the loud and clear sounds of their voice. On running 50 yards or so to windward, they would all be on wing. If the weather was thick, drizzly, and cold, or if there were indications of a fall of snow, they would remain on the ice, walking, standing, or lying down, until symptoms of better weather became apparent, when they would all start off.

Mr. Hoyes Lloyd has recently written to me about a flock of wild trumpeter swans that have for several years been spending the winter in a lake in southern British Columbia under the protection of the Canadian National Parks Branch. So there is hope that the species may survive.

DISTRIBUTION

Breeding range.—Probably still breeds sparingly in the wilder portions of Wyoming (Yellowstone Park), western Montana, Alberta, British Columbia (Skeena River), and northwestern Canada. Has bred in the past east to James Bay (Norway House), Manitoba (Shoal Lake, 1893 and 1894), Minnesota (Heron Lake, 1883), and Indiana. South to Iowa (Hancock County, 1883), Nebraska, and Missouri west to British Columbia (Chilcoten) and Alaska (Fort Yukon).

Winter range.—Western United States. South to the Gulf of Mexico and southern California. North to west-central British

Columbia (Skeena River) and the central Mississippi Valley. Now too rare everywhere to outline its range more definitely.

Spring migration.—Average dates of arrival: Nebraska, March 16; South Dakota, April 2; Minnesota, Heron Lake, April 4; Saskatchewan, April 16; British Columbia, April 20. Late date of departure: Arkansas, Helena, April 29, 1891; British Columbia, Osoyoos, April 25.

Fall migration.—Fall dates: Minnesota, Spicer, October 8, 1913; Michigan, St. Clair Flats, November 20, 1875; Washington, Douglas County, November 9, 1912; Colorado, Fort Collins, November 18, 1897, and November 25, 1915.

Egg dates.—Arctic Canada: Five records, June 17 to July 9. Alaska: One record, June 28. Alberta: One record, April 7. Dakota: One record, June 4.

REFERENCES TO BIBLIOGRAPHY

ADAMS, EDWARD.

1878—Notes of the Birds of Michalaski, Norton Sound. The Ibis, 1878, pp. 420–442.

ALFORD, CHARLES E.

1920—Some Notes on Diving Ducks. British Birds, vol. 14, pp. 106–110.

1921—Diving Ducks—Some Notes on their Habits and Courtship. British Birds, vol. 15, pp. 33–38.

ANDERSON, RUDOLPH MARTIN.

1907—The Birds of Iowa. Proceedings of the Davenport Academy of Sciences, vol. 11, pp. 125–417.

ANNANDALE, NELSON.

1905—The Faroes and Iceland.

AUDUBON, JOHN JAMES

1840—The Birds of America. 1840–44.

BAILEY, FLORENCE MERRIAM.

1916—A Populous Shore. The Condor, vol. 18, pp. 100–110.

1919—A Return to the Dakota Lake Region. The Condor, vol. 21, pp. 3–11.

BAILEY, VERNON.

1902—Notes in Handbook of Birds of the Western United States, by Florence Merriam Bailey.

BAIRD, SPENCER FULLERTON; BREWER, THOMAS MAYO; and RIDGWAY, ROBERT.

1884—The Water Birds of North America.

BARNHART, F.

1901—Evolution in the Breeding Habits of the Fulvous Tree-Duck. The Condor, vol. 3, pp. 67–68.

BARNSTON, GEORGE.

1862—The Swans and Geese of Hudson Bay. The Zoologist, 1862, p. 7831.

BARROWS, WALTER BRADFORD.

1912—Michigan Bird Life.

BENDIRE, CHARLES EMIL.

1875—Notes on seventy-nine species of Birds observed in the neighborhood of Camp Harney, Oregon. Proceedings of the Boston Society of Natural History, vol. 18, pp. 153–168.

BEYER, GEORGE EUGENE; ALLISON, ANDREW; KOPMAN, HENRY HAZLITT.

1907—List of the Birds of Louisiana. The Auk, vol. 23, pp. 1–15 and 275–282; vol. 24, pp. 314–321; vol. 25, pp. 173–180 and 439–448.

BISHOP, LOUIS BENNETT.

1921—Notes from Connecticut. The Auk, vol. 38, pp. 582–589.

BLAAUW, FRANS ERNST.

1903—Notes on the Breeding of Ross's Snow-Goose in Captivity. The Ibis, 1903, pp. 245–247.

1905—Letter in The Ibis, 1905, pp. 137, 138.

1916—A Note on the Emperor Goose (*Philacte canagica*) and on the Australian Teal (*Nettion castaneum*). The Ibis, 1916, pp. 252–254.

BLANCHAN, NELTJE. (MRS. F. N. DOUBLEDAY.)

1898—Birds that Hunt and are Hunted.

BOWDISH, BEECHER SCOVILLE.

1909—Ornithological Miscellany from Audubon Wardens. The Auk, vol. 26, pp. 116–128.

BRETHERTON, BERNARD J.

1896—Kadiak Island, A Contribution to the Avifauna of Alaska. The Oregon Naturalist, vol. 3, pp. 45–49; 61–64; 77–79; 100–102.

BREWER, THOMAS MAYO.

1879—The Rocky Mountain Golden-eye (*Bucephala islandica*). Bulletin of the Nuttall Ornithological Club, vol. 4, pp. 148–152.

BREWSTER, WILLIAM.

1900—Notes on the Breeding Habits of the American Golden-eyed Duck or Whistler. The Auk, vol. 17, pp. 207–216.

1909—Snow Geese in Massachusetts. The Auk, vol. 26, pp. 188–189.

1909*a*—Barrows Golden-eye in Massachusetts. The Auk, vol. 26, pp. 153–164.

1911—Courtship of the American Golden-eye or Whistler. The Condor, vol. 13, pp. 22–30.

BROOKS, ALLAN.

1903—Notes on the Birds of the Cariboo District, British Columbia. The Auk, vol. 20, pp. 277–284.

1920—Notes on some American Ducks. The Auk, vol. 37, pp. 353–367.

BROOKS, WINTHROP SPRAGUE.

1912—An Additional Specimen of the Labrador Duck. The Auk, vol. 29, pp. 389–390.

1915—Notes on Birds from East Siberia and Arctic Alaska. Bulletin of the Museum of Comparative Zoology at Harvard College.

BRYANT, EDWIN S.

1899—The White-winged Scoter in North Dakota. The Osprey, vol. 3, pp. 132–133.

BRYANT, HAROLD CHILD.

1914—A Survey of the Breeding Grounds of Ducks in California in 1914. The Condor, vol. 16, pp. 217–239.

BRYANT, WALTER (PIERCE) E.

1890—An Ornithological Retrospect. Zoe, vol. 1, pp. 289–293.

BUTLER, AMOS WILLIAM.

1897—The Birds of Indiana. Department of Geology and Natural Resources, Twenty-second Annual Report.

CAHOON, JOHN CYRUS.

1889—A Quahaug Captures a Tern, and a Sea Clam Drowns a Scoter. Ornithologist and Oologist, vol. 14, p. 36.

CARTWRIGHT, GEORGE.

1792—A Journal of Transactions and Events during a Residence of nearly Sixteen Years on the Coast of Labrador.

CHAPMAN, FRANK MICHLER.

1899—Report of Birds Received through the Peary Expeditions to Greenland. Bulletin of the American Museum of Natural History, vol. 12, pp. 219–244.

CLARK, AUSTIN HOBART.

1910—The Birds collected and observed during the Cruise of the United States Fisheries Steamer "Albatross" in the North Pacific Ocean, and in the Bering, Okhotsk, Japan, and Eastern Seas, from April to December, 1906. Proceedings of the U. S. Nat. Museum, vol. 38, pp. 25–74.

COALE, HENRY KELSO.
1915—The Present Status of the Trumpeter Swan (*Olor buccinator*). The Auk, vol. 32, pp. 82–90.

COLLINS, WILLIAM H.
1881—Those "Brants"—Corrections. Ornithologist and Oologist, vol. 6, p. 55.

COOKE, WELLS WOODBRIDGE.
1906—Distribution and Migration of North American Ducks, Geese, and Swans. U. S. Department of Agriculture, Biological Survey, Bulletin No. 26.

COOPER, JAMES GRAHAM.
1860—The Natural History of Washington Territory and Oregon, by Suckley and Cooper.

CORDEAUX, JOHN.
1898—British Birds with their Nest and Eggs. Order Anseres, vol. 4, pp. 52–203.

COUES, ELLIOTT.
1861—Notes on the Ornithology of Labrador. Proceedings of the Philadelphia Academy of Natural Sciences, 1861, pp. 215–257.
1874—Birds of the North-West.
1897—Branta bernicla glaucogastra. The Auk, vol. 14, pp. 207–208.

CURRIER, EDMONDE SAMUEL.
1902—Winter Water Fowl of the Des Moines Rapids. The Osprey, vol. 6, pp. 71–75.

DAVIS, NATHAN L.
1895—Notes on Whistling Swan *Olor columbianus*. The Museum, vol. 1 pp. 114–116.

DAWSON, WILLIAM LEON.
1909—The Birds of Washington.

DEKAY, JAMES ELLSWORTH.
1844—Zoology of New York, Part 2, Birds.

DICE, LEE RAYMOND.
1920—Notes on some Birds of Interior Alaska. The Condor, vol. 22, pp. 176–185.

DIONNE, CHARLES EUSEBE.
1906—Les Oiseaux de la Province de Quebec.

DIXON, JOSEPH.
1908—Field Notes from Alaska. The Condor, vol. 10, pp. 139–143.

DUTCHER, WILLIAM.
1888—Bird Notes from Long Island, N. Y. The Auk, vol. 5, pp. 169–183.
1891—The Labrador Duck—A Revised List of the Extant Specimens in North America, with some Historical Notes. The Auk, vol. 8, pp. 201–216.
1893—Notes on some Rare Birds in the Collection of the Long Island Historical Society. The Auk, vol. 10, pp. 267–277.
1894—The Labrador Duck—Another Specimen, with Additional Data Respecting Extant Specimens. The Auk, vol. 11, pp. 4–12.

DWIGHT, JONATHAN.
1914—The Moults and Plumages of the Scoters—Genus Oidemia. The Auk, vol. 31, pp. 293–308.

EATON, ELON HOWARD.
1910—Birds of New York.

EIFRIG, CHARLES WILLIAM GUSTAVE.
1905—Ornithological Results of the Canadian "Neptune" Expedition to Hudson Bay and Northward. The Auk, vol. 22, pp. 233–241.

ELLIOT, DANIEL GIRAUD.
1898—The Wild Fowl of North America.

ENGLISH, T. M. SAVAGE.
1916—Notes on some of the Birds of Grand Cayman, West Indies. The Ibis, 1916, pp. 17–35.

FANNIN, JOHN.
1894—The Canada Goose and Osprey laying in the same Nest. The Auk, vol. 11, p. 322.

FIGGINS, JESSE DADE.
1902—Some Food Birds of the Eskimos of Northwestern Greenland. Abstract of the Proceedings of the Linnaean Society of New York for the year ending March 11, 1902, pp. 61–65.
1920—The Status of the Subspecific Races of *Branta canadensis*. The Auk, vol. 37, pp. 94–102.
1922—Additional Notes on the Status of the Subspecific Races of *Branta canadensis*. Proceedings of the Colorado Museum of Natural History, vol. 4, No. 3.

FLEMING, JAMES HENRY.
1908—The Destruction of Whistling Swans (*Olor columbianus*) at Niagara Falls. The Auk, vol. 25, pp. 306–309.
1912—The Niagara Swan Trap. The Auk, vol. 29, pp. 445–448.

FORBUSH, EDWARD HOWE.
1912—A History of the Game Birds, Wild-Fowl and Shore Birds of Massachusetts and Adjacent States.

FORSTER, JOHANN REINHOLD.
1772—An Account of the Birds sent from Hudson's Bay; with Observations relative to their Natural History; and Latin Descriptions of some of the most uncommon. Philosophical Transactions, vol. 62, p. 382.

GÄTKE, HEINRICH.
1895—Heligoland as an Ornithological Observatory.

GIBSON, LANGDON.
1922—Bird Notes from North Greenland. The Auk, vol. 39, pp. 350–362.

GORDON, AUDREY.
1922—Nesting of the Whooper Swan in Scotland. British Birds, vol. 15, p. 170.

GORDON, SETON PAUL.
1920—Periods of Dives made by Long-tailed Ducks. British Birds, vol. 13, pp. 244–245.

GREELY, ADOLPHUS WASHINGTON.
1888—Report on the Proceedings of the United States Expedition to Lady Franklin Bay, Grinnell Land.

GREGG, WILLIAM HENRY.
1879—The American Naturalist, vol. 13, p. 128.

GRINNELL, GEORGE BIRD.
1901—American Duck Shooting.

GRINNELL, JOSEPH.
1909—Birds and Mammals of the 1907 Alexander Expedition to Southeastern Alaska. University of California, Publications in Zoology, vol. 5, pp. 171–264.
1910—Birds of the 1908 Alexander Alaska Expedition. University of California Publications in Zoology, vol. 5, pp. 361–428.

GRINNELL, JOSEPH; BRYANT, HAROLD CHILD; and STORER, TRACY IRWIN.
1918—The Game Birds of California.

GURNEY, JOHN HENRY.
1897—Labrador Duck. The Auk, vol. 14, p. 87.

HAGERUP, ANDREAS THOMSEN.
1891—The Birds of Greenland.

HALKETT, ANDREW.
1905—A Naturalist in the Frozen North. The Ottawa Naturalist, vol. 19, pp. 104–109.

HANNA, G. DALLAS.
1916—Records of Birds New to the Pribilof Islands Including two New to North America. The Auk, vol. 33, pp. 400–403.

HAYES, ISAAC ISRAEL.
1867—The Open Polar Sea.

HAYNES, WILLIAM B.
1901—The Oldsquaw Duck. The Wilson Bulletin, No. 32, pp. 12–13.

HOLLAND, ARTHUR H.
1892—Short Notes on the Birds of Estancia Espartilla, Argentine Republic. The Ibis, 1892, pp. 193–214.

HOWLEY, JAMES P.
1884—The Canada Goose (*Bernicla canadensis*). The Auk, vol. 1, pp. 309–313.

HUBBARD, SAMUEL, Jr.
1893—A Tragedy in Bird Life. Zoe, vol. 3, pp. 361–362.

HUDSON, WILLIAM HENRY.
1920—Birds of La Plata.

HULL, EDWIN D.
1914—Habits of the Oldsquaw (*Harelda hyemalis*) in Jackson Park, Chicago. The Wilson Bulletin, No. 88, pp. 116–123.

JOURDAIN, FRANCIS CHARLES ROBERT.
1922—The Breeding Habits of the Barnacle Goose. The Auk, vol. 9, pp. 166–171.

KING, WILLIAM ROSS.
1866—The Sportsman and Naturalist in Canada.

KNIGHT, ORA WILLIS.
1908—The Birds of Maine.

KUMLIEN, LUDWIG.
1879—Contributions to the Natural History of Arctic America. Bulletin of the United States National Museum, No. 15.

KUMLIEN, LUDWIG, and HOLLISTER, NED.
1903—The Birds of Wisconsin. Bulletin of the Wisconsin Natural History Society, vol. 3, new series, Nos. 1, 2, and 3.

LANGILLE, JAMES HIBBERT.
1884—Our Birds and Their Haunts.

LAWRENCE, GEORGE NEWBOLD.
1846—Description of a New Species of Anser. Annals of the Lyceum of Natural History, New York, vol. 4, pp. 171–172.
1874—Birds of Western and Northwestern Mexico. Memoirs of the Boston Society of Natural History, vol. 2, pp. 265–319.

LEWIS, ELISHA JARRETT.
1885—The American Sportsman.

LEWIS, HARRISON F.
1921—The Greater Snow Goose. The Canadian Naturalist, vol. 25, p. 35.
1922—Notes on some Labrador Birds. The Auk, vol. 39, pp. 507–516.

MACARTNEY, WILLIAM NAPIER.

1918—Golden-eye Duck carrying Young. Bird-Lore, vol. 20, pp. 418–419.

MACFARLANE, RODERICK ROSS.

1891—Notes on and List of Birds and Eggs Collected in Arctic America, 1861–1866. Proceedings of the United States National Museum, vol. 14, pp. 413–446.

MACKAY, GEORGE HENRY.

1890—*Somateria dresseri*, The American Eider. The Auk, vol. 7, pp. 315–319.

1891—The Scoters in New England. The Auk, vol. 8, pp. 279–290.

1892—Habits of the Oldsquaw (*Clangula hyemalis*) in New England. The Auk, vol. 9, pp. 330–337.

1893—Stray Notes from the vicinity of Muskeget Island, Massachusetts. The Auk, vol. 10, pp. 370–371.

MACMILLAN, DONALD BAXTER.

1918—Four Years in the White North.

MACOUN, JOHN.

1909—Catalogue of Canadian Birds. Second Edition.

MANNICHE, A. L. V.

1910—The Terrestrial Mammals and Birds of North-East Greenland. Medelelser om Gronland, vol. 45.

MCATEE, WALDO LEE.

1910—Notes on Chen coerulescens, Chen rossi, and other Waterfowl in Louisiana. The Auk, vol. 27, pp. 337–339.

1911—Winter Ranges of Geese on the Gulf Coast; Notable Bird Records for the same Region. The Auk, vol. 28, pp. 272–274.

MERRIAM, CLINTON HART.

1883—Breeding of the Harlequin Duck. Bulletin of the Nuttall Ornithological Club, vol. 8, p. 220.

MERRILL, JAMES CUSHING.

1878—Notes on the Ornithology of southern Texas, being a List of Birds observed in the Vicinity of Fort Brown, Texas, from February, 1876, to June, 1878. Proceedings of the United States National Museum, vol. 1, pp. 118–173.

1888—Notes on the Birds of Fort Klamath, Oregon. The Auk, vol. 5, pp. 139–146.

MICHAEL, CHARLES W., and ENID.

1922—An Adventure with a Pair of Harlequin Ducks in the Yosemite Valley. The Auk, vol. 39, pp. 14–23.

MILLAIS, JOHN GUILLE.

1913—British Diving Ducks.

MILLER, GERRIT SMITH, Jr.

1891—Further Cape Cod Notes. The Auk, vol. 8, pp. 117–120.

MORRIS, FRANCIS ORPEN.

1903—A History of British Birds. Fifth Edition.

MORTON, THOMAS.

1637—New English Canaan.

MUNRO, JAMES ALEXANDER.

1918—The Barrow Golden-eye in the Okanagan Valley, British Columbia. The Condor, vol. 20, pp. 3–5.

MURDOCH, JOHN.

1885—Report of the International Polar Expedition to Point Barrow, Alaska. Part 4, Natural History.

NELSON, EDWARD WILLIAM.

1881—Habits of the Black Brant in the Vicinity of St. Michaels, Alaska. Bulletin of the Nuttall Ornithological Club, vol. 6, pp. 131–138.

1883—The Birds of Bering Sea and the Arctic Ocean. Cruise of the Revenue-Steamer Corwin in Alaska and the N. W. Arctic Ocean in 1881.

1887—Report upon Natural History Collections made in Alaska.

1913—The Emperor Goose. Educational Leaflet, No. 64. Bird-Lore, vol. 15, pp. 129–132.

NEWTON, ALFRED.

1896—A Dictionary of Birds.

NORTON, ARTHUR HERBERT.

1897—A Noteworthy Plumage observed in the American Eider Drake. The Auk, vol. 14, pp. 303–304.

1909—The Food of Several Maine Water Birds. The Auk, vol. 26, pp. 438–440.

NUTTALL, THOMAS.

1834—A Manual of the Ornithology of the United States and Canada, Water Birds.

PALMER, WILLIAM.

1899—The Avifauna of the Pribilof Islands. The Fur-Seals and Fur-Seal Islands of the North Pacific Ocean, Part 3, p. 355.

PEARSON, THOMAS GILBERT; BRIMLEY, CLEMENT SAMUEL; and BRIMLEY, HERBERT HUTCHINSON.

1919—Birds of North Carolina. North Carolina Geological and Economic Survey, vol. 4.

PHILLIPS, JOHN CHARLES.

1910—Notes on the Autumn Migration of the Canada Goose in Eastern Massachusetts. The Auk, vol. 27, pp. 263–271.

1911—Ten years of Observation on the Migration of Anatidae at Wenham Lake, Massachusetts. The Auk, vol. 28, pp. 188–200.

1911a—Two Unusual Flights of Canada Geese Noted in Massachusetts during the Fall of 1910. The Auk, vol. 28, pp. 319–323.

PREBLE, EDWARD ALEXANDER.

1908—A Biological Investigation of the Athabaska-Mackenzie Region. North American Fauna, No. 27.

PRESTON, JUNIUS WALLACE.

1892—Notes on Bird Flight. Ornithologist and Oologist, vol. 17, pp. 41–42.

RAY, MILTON SMITH.

1912—Nesting of the Canada Goose at Lake Tahoe. The Condor, vol. 14, pp. 67–72.

REEKS, HENRY.

1870—Notes on the Birds of Newfoundland. The Canadian Naturalist, vol. 5, pp. 38–47, 151–159, 289–304, and 406–416.

RHOADS, SAMUEL NICHOLSON.

1895—Contributions to the Zoology of Tennessee, No. 2, Birds. Proceedings of the Academy of Natural Sciences of Philadelphia, 1895, pp. 463–501.

RICH, WALTER HERBERT.

1907—Feathered Game of the Northeast.

ROCKWELL, ROBERT BLANCHARD.

1911—Nesting Notes on the Ducks of the Barr Lake Region, Colorado. The Condor, vol. 13, pp. 121–128 and 186–195.

SANFORD, LEONARD CUTLER; BISHOP, LOUIS BENNETT; and VAN DYKE, THEODORE STRONG.
1903—The Waterfowl Family.

SAUNDERS, WILLIAM ERWIN.
1917—Wild Geese at Moose Factory. The Auk, vol. 34, pp. 334–335.

SCOTT, WILLIAM EARL DODGE.
1891—Observations on the Birds of Jamaica, West Indies. The Auk, vol. 8, pp. 249–256 and 353–365

SEEBOHM, HENRY.
1901—The Birds of Siberia.

SENNETT, GEORGE BURRITT.
1878—Notes on the Ornithology of the Lower Rio Grande of Texas. Bulletin of the United States Geological and Geographical Survey, vol. 4, pp. 1–66.
1879—Further Notes on the Ornithology of the Lower Rio Grande of Texas. Bulletin of the United States Geological and Geographical Survey, vol. 5, pp. 371–440.
1880—An unusual Flight of Whistling Swans in Northwestern Pennsylvania. Bulletin of the Nuttall Ornithological Club, vol. 5, pp. 125–126.

SHIELDS, ALEXANDER MCMILLAN.
1899—Nesting of the Fulvous Tree Duck. Bulletin of the Cooper Ornithological Club, vol. 1, pp. 9–11.

SILLOWAY, PERLEY MILTON.
1903—Birds of Fergus County, Montana. Bulletin No. 1, Fergus County Free High School.

STANSELL, SIDNEY SMITH STOUT.
1909—Birds of Central Alberta. The Auk, vol. 26, pp. 390–400.

STEARNS, WINFRED ALDEN.
1883—Notes on the Natural History of Labrador. Proceedings of the United States National Museum, vol. 6, pp. 111–137.

STONE, WITMER.
1895—List of Birds Collected in North Greenland by the Peary Expedition of 1891–2 and the Relief Expedition of 1892. Proceedings of the Academy of Natural Sciences of Philadelphia, 1895, pp. 502–505.

SWAINSON, WILLIAM; and RICHARDSON, JOHN.
1831—Fauna Boreali-Americana, vol. 2, Birds.

SWARTH, HARRY SCHELWALDT.
1913—A Study of a Collection of Geese of the *Branta canadensis* Group from the San Joaquin Valley, California. University of California Publications in Zoology, vol. 12, pp. 1–21.
1922—Birds and Mammals of the Stikine River Region of Northern British Columbia and Southeastern Alaska. University of California Publications in Zoology, vol. 24, pp. 125–314.

SWARTH, HARRY SCHELWALDT; and BRYANT, HAROLD CHILD.
1917—A Study of the Races of the White-fronted Goose (*Anser albifrons*), occurring in California. University of California Publications in Zoology, vol. 17, pp. 209–222.

TAVERNER, PERCY ALGERNON.
1922—Adventures with the Canada Goose. The Canadian Field Naturalist, vol. 36, pp. 81–83.

THAYER, JOHN ELIOT.

1905—Brant's Nest. The Auk, vol. 22, p. 408.

THAYER, JOHN ELIOT and BANGS, OUTRAM.

1914—Notes on the Birds and Mammals of the Arctic Coast of East Siberia—Birds. Proceedings of the New England Zoological Club, vol. 5, pp. 1–66.

TOWNSEND, CHARLES WENDELL.

1905—The Birds of Essex County, Massachusetts. Memoirs of the Nuttall Ornithological Club. No. 3.

1910—The Courtship of the Golden-eye and Eider Ducks. The Auk, vol. 27, pp. 177–181.

1913—Some More Labrador Notes. The Auk, vol. 30, pp. 1–10.

1914—A plea for the Conservation of the Eider. The Auk, vol. 31, pp. 14–21.

1916—The Courtship of the Merganser, Mallard, Black Duck, Baldpate, Wood Duck, and Bufflehead. The Auk, vol. 33, pp. 9–17.

1916*a*—Notes on the Eider. The Auk, vol. 33, pp. 286–292.

TRUMBULL, GURDON.

1892—Our Scoters. The Auk, vol. 9, pp. 153–160.

1893—Our Scoters. The Auk, vol. 10, pp. 165–176.

TURNER, LUCIEN MCSHAN.

1886—Contributions to the Natural History of Alaska.

WETMORE, ALEXANDER.

1920. Observations on the Habits of Birds at Lake Burford, New Mexico. The Auk, vol. 37, pp. 221–247 and 393–412.

WHITFIELD, ROBERT PARR.

1894—The Food of Wild Ducks. The Auk, vol. 11, p. 323.

WILLIAMS, ROBERT STATHAM.

1886—A Flock of Chen rossii East of the Rocky Mountains. The Auk, vol. 3, p. 274.

WITHERBY, HARRY FORBES.

1913—Barrow's Goldeneye and the Common Goldeneye. British Birds, vol. 6, pp. 272–276.

WITHERBY, HARRY FORBES, and OTHERS.

1920–22.—A Practical Handbook of British Birds.

YARRELL, WILLIAM.

1871—History of British Birds, Fourth Edition, 1871–85. Revised and enlarged by Alfred Newton and Howard Saunders.

YORKE, F. HENRY.

1891—Shooting Canada Geese and Swans. Days with the Waterfowl of America. No. 25. The American Field, vol. 35, No. 26, pp. 641–643.

1899—Our Ducks.

INDEX

PLATES

PLATE 1. SNOW, BLUE, AND CANADA GOOSE. Mixed flock of snow, blue, and Canada geese on a Louisiana Wild Life Refuge, presented by Mr. Stanley Clisby Arthur.

PLATE 2. AMERICAN GOLDENEYE. *Upper:* Nesting site of American goldeye, Devils Lake, North Dakota, June 22, 1898, presented by Mr. Herbert K. Job. *Lower:* Nesting tree of American goldeneye, Nelson County, North Dakota, June 1, 1901, a photograph by the author, referred to on page 3.

PLATE 3. AMERICAN GOLDENEYE. *Upper:* Nesting tree of American goldeneye, near Eskimo Point, Quebec, June 10, 1909, referred to on page 2, presented by Dr. Charles W. Townsend. *Lower:* A pair of adult American goldeneyes feeding, a photograph purchased from Mr. Bonnycastle Dale.

PLATE 4. AMERICAN GOLDENEYE. *Upper:* Nesting tree of American goldeneye, Lake Winnipegosis, Manitoba, June 9, 1913, referred to on page 4. *Lower:* Another nest tree, in the same locality, showing a mass of down clinging to the bark near the opening. Both photographs by the author.

PLATE 5. AMERICAN GOLDENEYE. *Upper:* Female American goldeneye leaning far out of the nesting hole, 25 feet up in a white poplar, near Fort Chipewyan, Alberta. A photograph taken by Mr. Francis Harper, presented by the Biological Survey and Dr. John C. Phillips. *Lower:* Downy young American goldeneyes, Lake Winnipegosis, Manitoba, June, 1913, presented by Herbert K. Job.

PLATE 6. BARROW GOLDENEYE. Nesting stub of Barrow goldeneye, Okanogan, British Columbia, presented by Mr. James A. Munro.

PLATE 7. BARROW GOLDENEYE. Winter resort of Barrow goldeneyes, Yellowstone National Park, Wyoming, presented by Mr. M. P. Skinner.

PLATE 8. BUFFLEHEAD. *Left:* Nesting tree of bufflehead, Buffalo Lake, Alberta, June 8, 1920. The lower hole contained 12 fresh eggs. *Right:* Another nesting tree in the same locality, June 2, 1920. "The photograph shows how far the eggs were below the entrance hole." Both photographs presented by Mr. G. H. Lings.

PLATE 9. BUFFLEHEAD. Nesting stub of bufflehead, Alberta, June 11, 1906, presented by Mr. Walter Raine. The stub also contained the nest and eggs of desert sparrow hawk.

PLATE 10. OLDSQUAW. *Upper:* Nest and eggs of oldsquaw, broken up by gulls, near St. Michael, Alaska, July 5, 1915, from a negative taken by Mr. F. Seymour Hersey for the author, referred to on page 38. *Lower:* Nest and eggs of oldsquaw, Lopp Lagoon, near Cape Prince of Wales, Alaska, July 5, 1922, presented by Mr. Alfred M. Bailey, by courtesy of the Colorado Museum of Natural History.

PLATE 11. OLDSQUAW. *Upper:* Nest and eggs of oldsquaw, Iceland, June, 1911, presented by Mr. C. H. Wells. *Lower:* Pair of adult old-squaws, a photograph purchased from Mr. Bonnycastle Dale.

PLATE 12. OLDSQUAW. *Upper:* Nest of oldsquaw, with eggs covered with down. Cape Prince of Wales, Alaska, July 5, 1922. *Lower:* Same nest, with eggs uncovered. Both photographs presented by Mr. Alfred M. Bailey, by courtesy of the Colorado Museum of Natural History.

PLATE 13. PACIFIC HARLEQUIN DUCK. *Upper:* Male Pacific harlequin duck, Yosemite Valley, California. *Lower:* A pair of the same. Both photographs presented by Mr. Charles W. Michael and referred to on page 59.

PLATE 14. SPECTACLED EIDER. *Upper:* Nesting site of spectacled eider, Point Hope, Alaska, June 15, 1917, from a negative taken by Rev. A. R. Hoare for the author and referred to on page 75. *Lower:* Nest and eggs of spectacled eider, Point Barrow, Alaska, June 26, 1917, from a negative taken by Mr. T. L. Richardson for the author and referred to on page 75.

PLATE 15. NORTHERN EIDER. *Upper:* Two nests of northern eider, near Hopedale, Labrador, July 22, 1912. *Lower:* Another nest, same locality and date. Both photographs by the author, referred to on page 81.

PLATE 16. NORTHERN EIDER. *Upper:* Nest and eggs of northern eider, Sutherland Island, Greenland. *Lower:* Another nest of same, near Etah, Greenland. Both photographs taken by Dr. Donald B. MacMillan and presented by the American Museum of Natural History.

PLATE 17. EIDER. *Upper:* Nest, eggs and young of eider. *Lower:* Female eider on her nest. Both photographs taken in Iceland and presented by Mr. C. H. Wells.

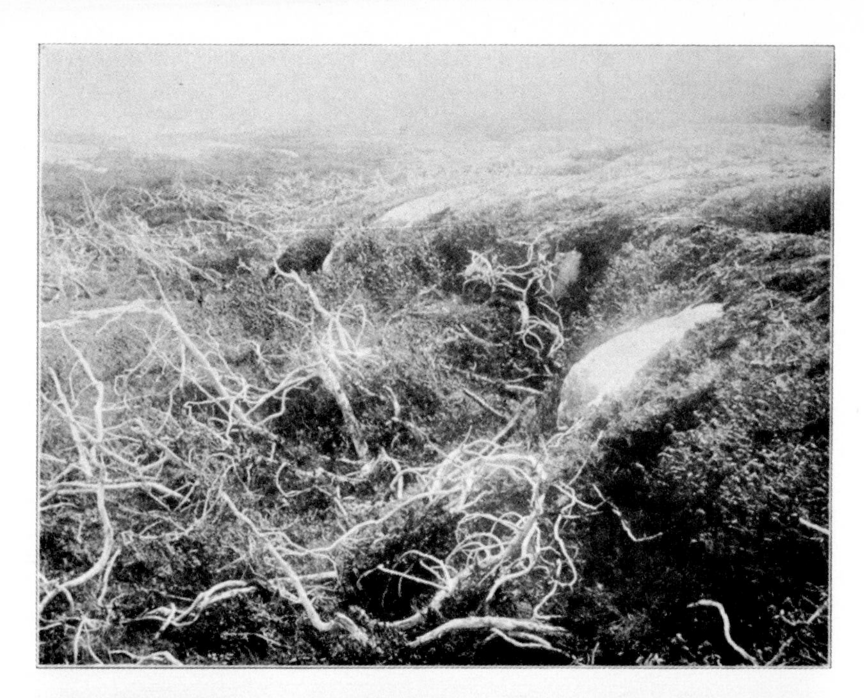

PLATE 18. AMERICAN EIDER. *Upper:* Nesting site of American eider, on an island off the north shore of the Gulf of St. Lawrence, May 26, 1909, referred to on page 97. *Lower:* Closer view of the nest in the same spot. Both photographs by the author.

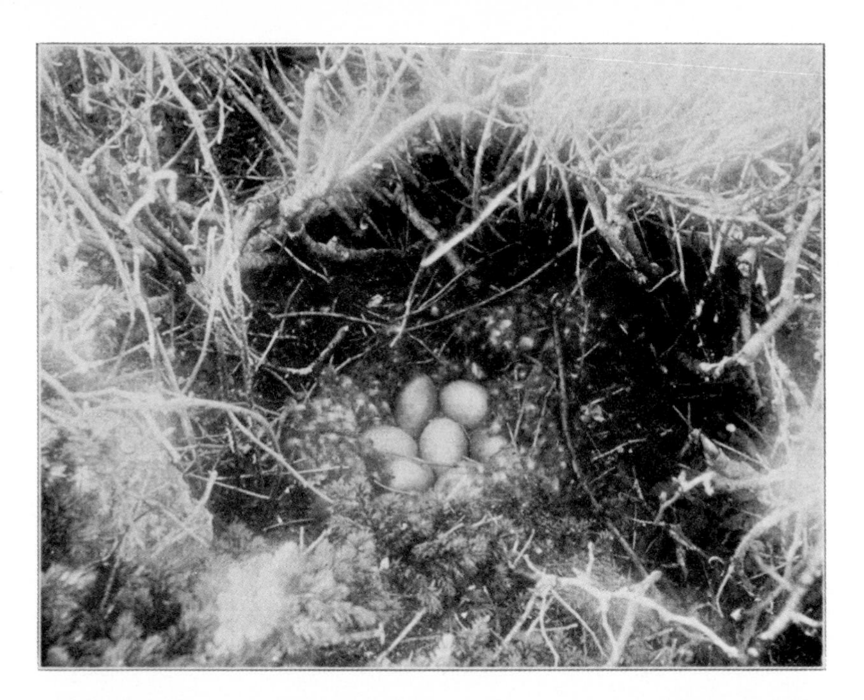

PLATE 19. AMERICAN EIDER. *Upper:* Nest and 7 eggs of American eider on an island off the north shore of the Gulf of St. Lawrence, May 29, 1909, a photograph by the author, referred to on page 98. *Lower:* Young male American eider in first winter plumage beginning to acquire the first white plumage, a photograph purchased from Mr. Bonnycastle Dale.

PLATE 20. PACIFIC EIDER. *Upper:* Nesting site of Pacific eider, Kiska Island, Alaska, June 19, 1911, referred to on page 103. *Lower:* Closer view of the nest in the same spot. Both photographs by the author.

PLATE 21. PACIFIC EIDER. Nest and eggs of Pacific eider, Cape Prince of Wales, Alaska, July 1, 1922, presented by Mr. Alfred M. Bailey, courtesy of the Colorado Museum of Natural History.

PLATE 22. KING EIDER. *Upper:* Nest and eggs of king eider, Spitsbergen, July, 1921. *Lower:* Nesting site of king eider, Spitsbergen, July, 1921. Both photographs taken by Mr. J. S. Huxley, purchased from the Oxford University Expedition to Spitsbergen and referred to on page 110.

PLATE 23. WHITE-WINGED SCOTER. *Upper:* Nest and eggs of white-winged scoter, Stump Lake Reservation, North Dakota, June 27, 1898. *Lower:* Another nest, same locality and date. Both photographs presented by Mr. Herbert K. Job and referred to on page 132.

PLATE 24. WHITE-WINGED SCOTER. *Upper:* Nest and eggs of white-winged scoter, Lake Manitoba, Manitoba, July 4, 1912. *Lower:* Downy young white-winged scoters, St. Marks, Manitoba, July, 1912. Both photographs presented by Mr. Herbert K. Job.

PLATE 25. WHITE-WINGED SCOTER. A flock of migrating white-winged scoters, off Manomet Point, Plymouth, Massachusetts, October 25, 1904, presented by Mr. Herbert K. Job.

PLATE 26. RUDDY DUCK. *Upper:* Pair of adult ruddy ducks, Lake Manitoba, Manitoba, July, 1912. *Lower:* Nest and eggs of ruddy duck, same locality, July 8, 1912. Both photographs presented by Mr. Herbert K. Job.

PLATE 27. RUDDY DUCK. *Upper:* Nest and eggs of ruddy duck, Steele County, North Dakota, June 13, 1901. *Lower:* Another nest, same locality, June 10, 1901. Both photographs by the author, referred to on page 153.

PLATE 28. RUDDY DUCK. *Upper:* Downy young ruddy duck, St. Marks, Manitoba, July, 1912, presented by Mr. Herbert K. Job. *Lower:* Young ruddy ducks, a photograph of dried skins, presented by Mr. Joseph Mailliard.

PLATE 29. SNOW GOOSE. *Upper:* Adult snow goose, Louisiana. *Lower:* Another view of same. Both photographs presented by Mr. Stanley Clisby Arthur.

PLATE 30. BLUE AND SNOW GOOSE. *Upper:* Flock of blue and snow geese, Vermilion Bay, Louisiana, January 3, 1916. *Lower:* Another flock, same locality and date. Both photographs presented by Mr. Herbert K. Job.

PLATE 31. BLUE GOOSE. Adult and young blue geese, same locality and date as preceding plate, presented by Mr. Herbert K. Job.

PLATE 32. BLUE AND SNOW GOOSE. Flock of blue and snow geese, same locality and date as preceding plate, presented by Mr. Herbert K. Job.

PLATE 33. WHITE-FRONTED GOOSE. *Upper:* Nesting site of white-fronted goose, Kolyma Delta, Siberia, July 1, 1917. *Lower:* Closer view of nest in same spot. Both from negatives taken by Mr. Johan Koren for the author.

PLATE 34. PINK-FOOTED GOOSE. *Left:* Nest and eggs of pink-footed goose, Spitsbergen, July, 1922, taken by Mr. W. M. Congreve. *Right:* Another nest of same, Prince Charles Foreland, 1921, taken by Mr. T. G. Longstaff. Both photographs purchased from the Oxford University Expedition to Spitsbergen and referred to on page 201.

PLATE 35. PINK-FOOTED GOOSE. Nest and eggs of pink-footed goose, Bruce City, Spitsbergen, July, 1921, a photograph taken by Mr. J. D. Brown and purchased from the Oxford University Expedition to Spitsbergen.

PLATE 36. CANADA GOOSE. *Upper:* Pair of Canada geese nesting under semi-wild conditions, Taunton, Massachusetts, April 27, 1918. *Lower:* Deserted nest of Canada goose in a slough, Steele County, North Dakota, June 10, 1901, referred to on page 206. Both photographs by the author.

PLATE 37. CANADA GOOSE. *Upper:* Nest and eggs of Canada goose, on an island in Crane Lake, Saskatchewan, June 2, 1905. *Lower:* Another nest on the same island. Both photographs by the author, referred to on page 206.

PLATE 38. CANADA GOOSE. *Upper:* Immature Canada geese. *Lower:* Downy young Canada goose. Both photographs taken on Klamath River, Oregon, and presented by Messrs. H. T. Bohlman and William L. Finley.

PLATE 39. CANADA GOOSE. Nest of Canada goose in an old red-tailed hawk's nest, Alberta, April 25, 1896, presented by Mr. Walter Raine.

PLATE 40. HUTCHINS GOOSE. *Upper:* Nest and eggs of Hutchins goose, Point Barrow, Alaska, June 20, 1917, from a negative taken by Mr. T. L. Richardson for the author. *Lower:* Adult Hutchins goose, Louisiana, presented by Mr. Stanley Clisby Arthur.

PLATE 41. BRANT. *Upper:* Nest and eggs of brant, Eider Duck Island, Greenland. *Lower:* Another nest, same locality. Both photographs taken by Dr. Donald B. MacMillan, presented by the American Museum of Natural History and referred to on page 242.

PLATE 42. BRANT. *Upper:* Nest and eggs of brant, Moffen Island, Spitsbergen, July 8, 1921, a photograph taken by Mr. J. D. Brown. *Lower:* Another nest, same locality and date, a photograph taken by Mr. Seton P. Gordon. Both photographs purchased from the Oxford University Expedition to Spitsbergen and referred to on page 241.

PLATE 43. BRANT. Nesting site of brant, Moffen Island, 80° N., Spitsbergen, July 8, 1921, a photograph taken by Mr. J. D. Brown and purchased from the Oxford University Expedition to Spitsbergen.

PLATE 44. BLACK BRANT. *Upper:* Nesting site of black brant, Point Hope, Alaska, June 15, 1917. *Lower:* Closer view of the nest in the same spot. Both photographs from negatives taken by Rev. A. R. Hoare for the author and referred to on page 252.

PLATE 45. BARNACLE GOOSE. *Upper:* Barnacle goose on its nest, 1,200 feet above the bed of the valley, Spitsbergen, June, 1922. *Lower:* Same nest as above. Both photographs taken by Mr. W. M. Congreve, purchased from the Oxford University Expedition to Spitsbergen and referred to on page 260.

PLATE 46. BARNACLE GOOSE. Nest and eggs of barnacle goose, Advent Bay, Spitsbergen, June 28, 1921, a photograph purchased from the Oxford University Expedition to Spitsbergen and referred to on page 260.

PLATE 47. EMPEROR GOOSE. *Upper:* Nest of emperor goose, with eggs concealed, Cape Prince of Wales, Alaska, July 5, 1922. *Lower:* Another nest of same. Mint River, Cape Prince of Wales, Alaska, July 12, 1922. Both photographs presented by Mr. Alfred M. Bailey, by courtesy of the Colorado Museum of Natural History.

PLATE 48. EMPEROR GOOSE. Nesting site of emperor goose, nest on small islet to right of stump, Lopp Lagoon, near Cape Prince of Wales, Alaska, July 5, 1922, presented by Mr. Alfred M. Bailey, by courtesy of the Colorado Museum of Natural History.

PLATE 49. FULVOUS TREE-DUCK. Nest of fulvous tree-duck, concealed in thick tules, Los Banos, California, June 4, 1914, presented by Mr. W. Leon Dawson, by courtesy of the South Moulton Company.

PLATE 50. FULVOUS TREE-DUCK. Closer view of same nest shown on previous plate, presented by Mr. W. Leon Dawson, by courtesy of the South Moulton Company.

PLATE 51. FULVOUS TREE-DUCK. Fulvous tree-ducks flying, near Santa Barbara, California, May 2, 1912, presented by Mr. W. Leon Dawson, by courtesy of the South Moulton Company.

PLATE 52. FULVOUS TREE-DUCK. Nesting site of fulvous tree-duck, Kern County, California, presented by Mr. Donald R. Dickey.

PLATE 53. FULVOUS TREE-DUCK. Nest and eggs of fulvous tree-duck, Kern County, California, presented by Mr. Donald R. Dickey.

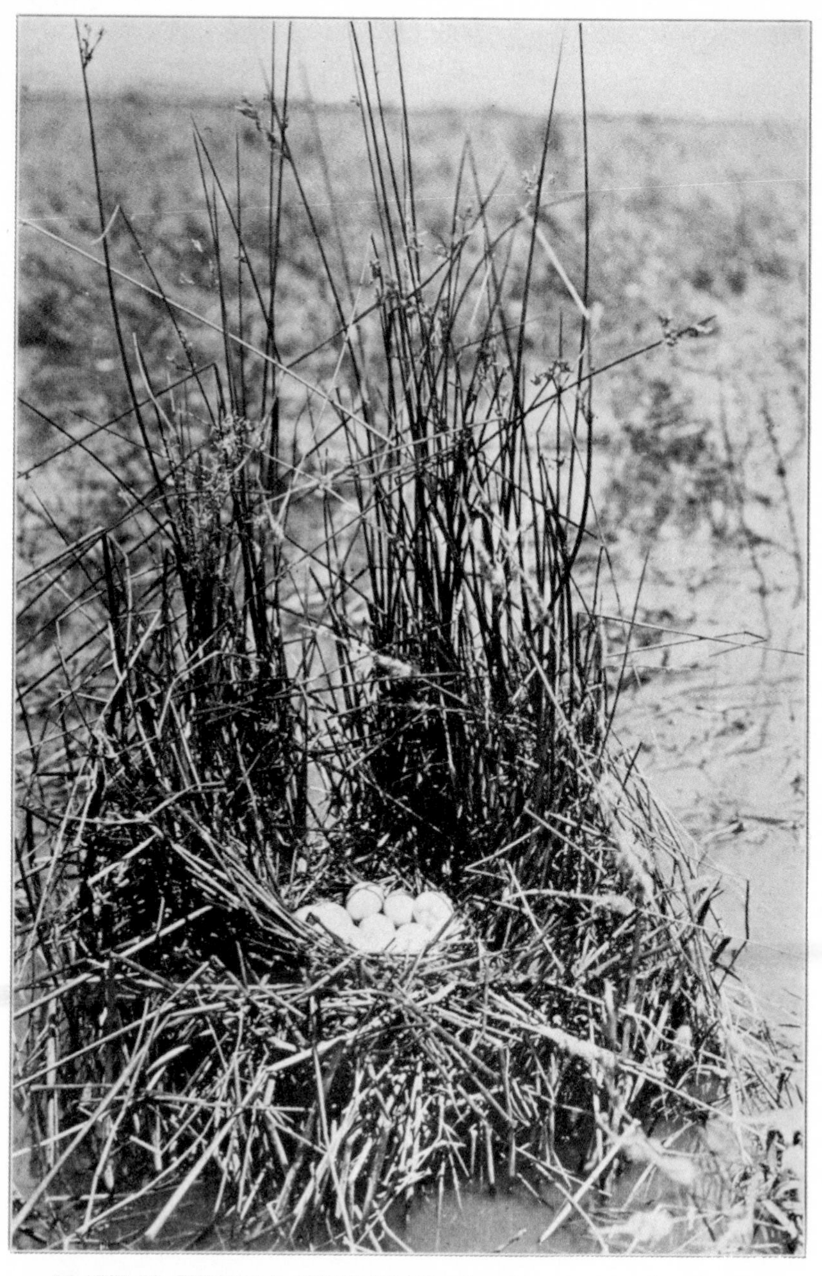

PLATE 54. FULVOUS TREE-DUCK. Another nest and eggs of fulvous tree-duck, Kern County, California, presented by Mr. Donald R. Dickey.

PLATE 55. WHOOPING SWAN. *Right:* Nest, eggs and young of whooping swan, Iceland, June, 1911, presented by Mr. C. H. Wells. *Left:* Adult whooping swan standing over its nest, West Highlands, Scotland, presented by Mrs. Audrey Gordon, and referred to on page 279.

PLATE 56. WHISTLING SWAN. Nesting site of whistling swan, Mint River, Cape Prince of Wales, Alaska, July 13, 1922, presented by Mr. Alfred M. Bailey, by courtesy of the Colorado Museum of Natural History.

PLATE 57. WHISTLING SWAN. Nest and downy young of whistling swan, Mint River, Cape Prince of Wales, Alaska, July 13, 1922, presented by Mr. Alfred M. Bailey, by courtesy of the Colorado Museum of Natural History.

PLATE 58. WHISTLING SWAN. Pair of whistling swans, Fergus County, Montana, presented by Mr. E. S. Cameron.

PLATE 59. WHISTLING SWAN. Flock of wild whistling swans, Point Pelee, Ontario, April 13, 1924, presented by Mr. Walter E. Hastings.

PLATE 60. TRUMPETER SWAN. Adult trumpeter swan, Louisiana State Wild Life Reserve, February, 1915, presented by Mr. Stanley Clisby Arthur.

Catalog of DOVER BOOKS

BOOKS EXPLAINING SCIENCE

(Note: The books listed under this category are general introductions, surveys, reviews, and non-technical expositions of science for the interested layman or scientist who wishes to brush up. Dover also publishes the largest list of inexpensive reprints of books on intermediate and higher mathematics, mathematical physics, engineering, chemistry, astronomy, etc., for the professional mathematician or scientist. For our complete Science Catalog, write Dept. catrr., Dover Publications, Inc., 180 Varick Street, New York 14, N. Y.)

CONCERNING THE NATURE OF THINGS, Sir William Bragg. Royal Institute Christmas Lectures by Nobel Laureate. Excellent plain-language introduction to gases, molecules, crystal structure, etc. explains "building blocks" of universe, basic properties of matter, with simplest, clearest examples, demonstrations. 32pp. of photos; 57 figures. 244pp. 5⅜ x 8.
T31 Paperbound **$1.35**

MATTER AND LIGHT, THE NEW PHYSICS, Louis de Broglie. Non-technical explanations by a Nobel Laureate of electro-magnetic theory, relativity, wave mechanics, quantum physics, philosophies of science, etc. Simple, yet accurate introduction to work of Planck, Bohr, Einstein, other modern physicists. Only 2 of 12 chapters require mathematics. 300pp. 5⅜ x 8.
T35 Paperbound **$1.60**

THE COMMON SENSE OF THE EXACT SCIENCES, W. K. Clifford. For 70 years, Clifford's work has been acclaimed as one of the clearest, yet most precise introductions to mathematical symbolism, measurement, surface boundaries, position, space, motion, mass and force, etc. Prefaces by Bertrand Russell and Karl Pearson. Introduction by James Newman. 130 figures. 249pp. 5⅜ x 8.
T61 Paperbound **$1.60**

THE NATURE OF LIGHT AND COLOUR IN THE OPEN AIR, M. Minnaert. What causes mirages? haloes? "multiple" suns and moons? Professor Minnaert explains these and hundreds of other fascinating natural optical phenomena in simple terms, tells how to observe them, suggests hundreds of experiments. 200 illus; 42 photos. xvi + 362pp.
T196 Paperbound **$1.95**

SPINNING TOPS AND GYROSCOPIC MOTION, John Perry. Classic elementary text on dynamics of rotation treats gyroscopes, tops, how quasi-rigidity is induced in paper disks, smoke rings, chains, etc, by rapid motion, precession, earth's motion, etc. Contains many easy-to-perform experiments. Appendix on practical uses of gyroscopes. 62 figures. 128pp.
T416 Paperbound **$1.00**

A CONCISE HISTORY OF MATHEMATICS, D. Struik. This lucid, easily followed history of mathematics from the Ancient Near East to modern times requires no mathematical background itself, yet introduces both mathematicians and laymen to basic concepts and discoveries and the men who made them. Contains a collection of 31 portraits of eminent mathematicians. Bibliography. xix + 299pp. 5⅜ x 8.
T255 Paperbound **$1.75**

THE RESTLESS UNIVERSE, Max Born. A remarkably clear, thorough exposition of gases, electrons, ions, waves and particles, electronic structure of the atom, nuclear physics, written for the layman by a Nobel Laureate. "Much more thorough and deep than most attempts . . . easy and delightful," CHEMICAL AND ENGINEERING NEWS. Includes 7 animated sequences showing motion of molecules, alpha particles, etc. 11 full-page plates of photographs. Total of nearly 600 illus. 315pp. 6⅛ x 9¼.
T412 Paperbound **$2.00**

WHAT IS SCIENCE?, N. Campbell. The role of experiment, the function of mathematics, the nature of scientific laws, the limitations of science, and many other provocative topics are explored without technicalities by an eminent scientist. "Still an excellent introduction to scientific philosophy," H. Margenau in PHYSICS TODAY. 192pp. 5⅜ x 8.
S43 Paperbound **$1.25**

TRADE

FADS AND FALLACIES IN THE NAME OF SCIENCE, Martin Gardner. The standard account of the various cults, quack systems and delusions which have recently masqueraded as science: hollow earth theory, Atlantis, dianetics, Reich's orgone theory, flying saucers, Bridey Murphy, psionics, irridiagnosis, many other fascinating fallacies that deluded tens of thousands. "Should be read by everyone, scientist and non-scientist alike," R. T. Birge, Prof. Emeritus, Univ. of California; Former President, American Physical Society. Formerly titled, "In the Name of Science." Revised and enlarged edition. x + 365pp. 5⅜ x 8.
T394 Paperbound **$1.50**

THE STUDY OF THE HISTORY OF MATHEMATICS, THE STUDY OF THE HISTORY OF SCIENCE, G. Sarton. Two books bound as one. Both volumes are standard introductions to their fields by an eminent science historian. They discuss problems of historical research, teaching, pitfalls, other matters of interest to the historically oriented writer, teacher, or student. Both have extensive bibliographies. 10 illustrations. 188pp. 5⅜ x 8. T240 Paperbound **$1.25**

THE PRINCIPLES OF SCIENCE, W. S. Jevons. Unabridged reprinting of a milestone in the development of symbolic logic and other subjects concerning scientific methodology, probability, inferential validity, etc. Also describes Jevons' "logic machine," an early precursor of modern electronic calculators. Preface by E. Nagel. 839pp. 5⅜ x 8. S446 Paperbound **$2.98**

SCIENCE THEORY AND MAN, Erwin Schroedinger. Complete, unabridged reprinting of "Science and the Human Temperament" plus an additional essay "What is an Elementary Particle?" Nobel Laureate Schroedinger discusses many aspects of modern physics from novel points of view which provide unusual insights for both laymen and physicists. 192 pp. 5⅜ x 8.
T428 Paperbound **$1.35**

BRIDGES AND THEIR BUILDERS, D. B. Steinman & S. R. Watson. Information about ancient, medieval, modern bridges; how they were built; who built them; the structural principles employed; the materials they are built of; etc. Written by one of the world's leading authorities on bridge design and construction. New, revised, expanded edition. 23 photos; 26 line drawings, xvii + 401pp. 5⅜ x 8. T431 Paperbound **$1.95**

HISTORY OF MATHEMATICS, D. E. Smith. Most comprehensive non-technical history of math in English. In two volumes. Vol. I: A chronological examination of the growth of mathematics from primitive concepts up to 1900. Vol. II: The development of ideas in specific fields and areas, up through elementary calculus. The lives and works of over a thousand mathematicians are covered; thousands of specific historical problems and their solutions are clearly explained. Total of 510 illustrations, 1355pp. 5⅜ x 8. Set boxed in attractive container. T429, T430 Paperbound, the set **$5.00**

PHILOSOPHY AND THE PHYSICISTS, L. S. Stebbing. A philosopher examines the philosophical implications of modern science by posing a lively critical attack on the popular science expositions of Sir James Jeans and Arthur Eddington. xvi + 295pp. 5⅜ x 8.
T480 Paperbound **$1.65**

ON MATHEMATICS AND MATHEMATICIANS, R. E. Moritz. The first collection of quotations by and about mathematicians in English. 1140 anecdotes, aphorisms, definitions, speculations, etc. give both mathematicians and layman stimulating new insights into what mathematics is, and into the personalities of the great mathematicians from Archimedes to Euler, Gauss, Klein, Weierstrass. Invaluable to teachers, writers. Extensive cross index. 410pp. 5⅜ x 8.
T489 Paperbound **$1.95**

NATURAL SCIENCE, BIOLOGY, GEOLOGY, TRAVEL

A SHORT HISTORY OF ANATOMY AND PHYSIOLOGY FROM THE GREEKS TO HARVEY, C. Singer. A great medical historian's fascinating intermediate account of the slow advance of anatomical and physiological knowledge from pre-scientific times to Vesalius, Harvey. 139 unusually interesting illustrations. 221pp. 5⅜ x 8. T389 Paperbound **$1.75**

THE BEHAVIOUR AND SOCIAL LIFE OF HONEYBEES, Ronald Ribbands. The most comprehensive, lucid and authoritative book on bee habits, communication, duties, cell life, motivations, etc. "A MUST for every scientist, experimenter, and educator, and a happy and valuable selection for all interested in the honeybee," AMERICAN BEE JOURNAL. 690-item bibliography. 127 illus.; 11 photographic plates. 352pp. 5⅜ x 8⅜. S410 Clothbound **$4.50**

TRAVELS OF WILLIAM BARTRAM, edited by **Mark Van Doren.** One of the 18th century's most delightful books, and one of the few first-hand sources of information about American geography, natural history, and anthropology of American Indian tribes of the time. "The mind of a scientist with the soul of a poet," John Livingston Lowes. 13 original illustrations, maps. Introduction by Mark Van Doren. 448pp. 5⅜ x 8. T326 Paperbound **$2.00**

☆ **STUDIES ON THE STRUCTURE AND DEVELOPMENT OF VERTEBRATES, Edwin Goodrich.** The definitive study of the skeleton, fins and limbs, head region, divisions of the body cavity, vascular, respiratory, excretory systems, etc., of vertebrates from fish to higher mammals, by the greatest comparative anatomist of recent times. "The standard textbook," JOURNAL OF ANATOMY. 754 illus. 69-page biographical study. 1186-item bibliography. 2 vols. Total of 906pp. 5⅜ x 8.
Vol. I: S449 Paperbound **$2.50**
Vol. II: S450 Paperbound **$2.50**

THE BIRTH AND DEVELOPMENT OF THE GEOLOGICAL SCIENCES, F. D. Adams. The most complete and thorough history of the earth sciences in print. Covers over 300 geological thinkers and systems; treats fossils, theories of stone growth, paleontology, earthquakes, vulcanists vs. neptunists, odd theories, etc. 91 illustrations, including medieval, Renaissance wood cuts, etc. 632 footnotes and bibliographic notes. 511pp. 308pp. 5⅜ x 8. T5 Paperbound **$2.00**

FROM MAGIC TO SCIENCE, Charles Singer. A close study of aspects of medical science from the Roman Empire through the Renaissance. The sections on early herbals, and "The Visions of Hildegarde of Bingen," are probably the best studies of these subjects available. 158 unusual classic and medieval illustrations. xxvii + 365pp. 5⅜ x 8. T390 Paperbound **$2.00**

SAILING ALONE AROUND THE WORLD, Captain Joshua Slocum. Captain Slocum's personal account of his single-handed voyage around the world in a 34-foot boat he rebuilt himself. A classic of both seamanship and descriptive writing. "A nautical equivalent of Thoreau's account," Van Wyck Brooks. 67 illus. 308pp. 5⅜ x 8. T326 Paperbound **$1.00**

TREES OF THE EASTERN AND CENTRAL UNITED STATES AND CANADA, W. M. Harlow. Standard middle-level guide designed to help you know the characteristics of Eastern trees and identify them at sight by means of an 8-page synoptic key. More than 600 drawings and photographs of twigs, leaves, fruit, other features. xiii + 288pp. 4⅝ x 6½. T395 Paperbound **$1.35**

FRUIT KEY AND TWIG KEY ("Fruit Key to Northeastern Trees," "Twig Key to Deciduous Woody Plants of Eastern North America"), **W. M. Harlow.** Identify trees in fall, winter, spring. Easy-to-use, synoptic keys, with photographs of every twig and fruit identified. Covers 120 different fruits, 160 different twigs. Over 350 photos. Bibliographies. Glossaries. Total of 143pp. 5⅝ x 8⅜. T511 Paperbound **$1.25**

INTRODUCTION TO THE STUDY OF EXPERIMENTAL MEDICINE, Claude Bernard. This classic records Bernard's far-reaching efforts to transform physiology into an exact science. It covers problems of vivisection, the limits of physiological experiment, hypotheses in medical experimentation, hundreds of others. Many of his own famous experiments on the liver, the pancreas, etc., are used as examples. Foreword by I. B. Cohen. xxv + 266pp. 5⅜ x 8. T400 Paperbound **$1.50**

THE ORIGIN OF LIFE, A. I. Oparin. The first modern statement that life evolved from complex nitro-carbon compounds, carefully presented according to modern biochemical knowledge of primary colloids, organic molecules, etc. Begins with historical introduction to the problem of the origin of life. Bibliography. xxv + 270pp. 5⅜ x 8. S213 Paperbound **$1.75**

A HISTORY OF ASTRONOMY FROM THALES TO KEPLER, J. L. E. Dreyer. The only work in English which provides a detailed picture of man's cosmological views from Egypt, Babylonia, Greece, and Alexandria to Copernicus, Tycho Brahe and Kepler. "Standard reference on Greek astronomy and the Copernican revolution," SKY AND TELESCOPE. Formerly called "A History of Planetary Systems From Thales to Kepler." Bibliography. 21 diagrams. xvii + 430pp. 5⅜ x 8. S79 Paperbound **$1.98**

URANIUM PROSPECTING, H. L. Barnes. A professional geologist tells you what you need to know. Hundreds of facts about minerals, tests, detectors, sampling, assays, claiming, developing, government regulations, etc. Glossary of technical terms. Annotated bibliography. x + 117pp. 5⅜ x 8. T309 Paperbound **$1.00**

DE RE METALLICA, Georgius Agricola. All 12 books of this 400 year old classic on metals and metal production, fully annotated, and containing all 289 of the 16th century woodcuts which made the original an artistic masterpiece. A superb gift for geologists, engineers, libraries, artists, historians. Translated by Herbert Hoover & L. H. Hoover. Bibliography, survey of ancient authors. 289 illustrations of the excavating, assaying, smelting, refining, and countless other metal production operations described in the text. 672pp. 6¾ x 10¾. Deluxe library edition. S6 Clothbound **$10.00**

DE MAGNETE, William Gilbert. A landmark of science by the man who first used the word "electricity," distinguished between static electricity and magnetism, and founded a new science. P. F. Mottelay translation. 90 figures. lix + 368pp. 5⅜ x 8. S470 Paperbound **$2.00**

THE AUTOBIOGRAPHY OF CHARLES DARWIN AND SELECTED LETTERS, Francis Darwin, ed. Fascinating documents on Darwin's early life, the voyage of the "Beagle," the discovery of evolution, Darwin's thought on mimicry, plant development, vivisection, evolution, many other subjects Letters to Henslow, Lyell, Hooker, Wallace, Kingsley, etc. Appendix. 365pp. 5⅜ x 8. T479 Paperbound **$1.65**

A WAY OF LIFE AND OTHER SELECTED WRITINGS OF SIR WILLIAM OSLER. 16 of the great physician, teacher and humanist's most inspiring writings on a practical philosophy of life, science and the humanities, and the history of medicine. 5 photographs. Introduction by G. L. Keynes, M.D., F.R.C.S. xx + 278pp. 5⅜ x 8. T488 Paperbound **$1.50**

LITERATURE

WORLD DRAMA, B. H. Clark. 46 plays from Ancient Greece, Rome, to India, China, Japan. Plays by Aeschylus, Sophocles, Euripides, Aristophanes, Plautus, Marlowe, Jonson, Farquhar, Goldsmith, Cervantes, Molière, Dumas, Goethe, Schiller, Ibsen, many others. One of the most comprehensive collections of important plays from all literature available in English. Over 1/3 of this material is unavailable in any other current edition. Reading lists. 2 volumes. Total of 1364pp. 5⅜ x 8. Vol. I, T57 Paperbound **$2.00**
Vol. II, T59 Paperbound **$2.00**

MASTERS OF THE DRAMA, John Gassner. The most comprehensive history of the drama in print. Covers more than 800 dramatists and over 2000 plays from the Greeks to modern Western, Near Eastern, Oriental drama. Plot summaries, theatre history, etc. "Best of its kind in English," NEW REPUBLIC. 35 pages of bibliography. 77 photos and drawings. Deluxe edition. xxii + 890pp. 5⅜ x 8. T100 Clothbound **$5.95**

THE DRAMA OF LUIGI PIRANDELLO, D. Vittorini. All 38 of Pirandello's plays (to 1935) summarized and analyzed in terms of symbolic techniques, plot structure, etc. The only authorized work. Foreword by Pirandello. Biography. Bibliography. xiii + 350pp. 5⅜ x 8.
T435 Paperbound **$1.98**

ARISTOTLE'S THEORY OF POETRY AND THE FINE ARTS, S. H. Butcher, ed. The celebrated "Butcher translation" faced page by page with the Greek text; Butcher's 300-page introduction to Greek poetic, dramatic thought. Modern Aristotelian criticism discussed by John Gassner. lxxvi + 421pp. 5⅜ x 8.
T42 Paperbound **$2.00**

EUGENE O'NEILL: THE MAN AND HIS PLAYS, B. H. Clark. The first published source-book on O'Neill's life and work. Analyzes each play from the early THE WEB up to THE ICEMAN COMETH. Supplies much information about environmental and dramatic influences. ix + 182pp. 5⅜ x 8. T379 Paperbound **$1.25**

INTRODUCTION TO ENGLISH LITERATURE, B. Dobrée, ed. Most compendious literary aid in its price range. Extensive, categorized bibliography (with entries up to 1949) of more than 5,000 poets, dramatists, novelists, as well as historians, philosophers, economists, religious writers, travellers, and scientists of literary stature. Information about manuscripts, important biographical data. Critical, historical, background works not simply listed, but evaluated. Each volume also contains a long introduction to the period it covers.

Vol. I: **THE BEGINNINGS OF ENGLISH LITERATURE TO SKELTON, 1509, W. L. Renwick. H. Orton.** 450pp. 5⅛ x 7⅛. T75 Clothbound **$3.50**

Vol. II: **THE ENGLISH RENAISSANCE, 1510-1688, V. de Sola Pinto.** 381pp. 5⅛ x 7⅛.
T76 Clothbound **$3.50**

Vol. III: **THE AUGUSTANS AND ROMANTICS, 1689-1830, H. Dyson, J. Butt.** 320pp. 5⅛ x 7⅛.
T77 Clothbound **$3.50**

Vol. IV: **THE VICTORIANS AND AFTER, 1830-1914, E. Batho, B. Dobrée.** 360pp. 5⅛ x 7⅛.
T78 Clothbound **$3.50**

EPIC AND ROMANCE, W. P. Ker. The standard survey of Medieval epic and romance by a foremost authority on Medieval literature. Covers historical background, plot, literary analysis, significance of Teutonic epics, Icelandic sagas, Beowulf, French chansons de geste, the Niebelungenlied, Arthurian romances, much more. 422pp. 5⅜ x 8. T355 Paperbound **$1.95**

THE HEART OF EMERSON'S JOURNALS, Bliss Perry, ed. Emerson's most intimate thoughts, impressions, records of conversations with Channing, Hawthorne, Thoreau, etc., carefully chosen from the 10 volumes of The Journals. "The essays do not reveal the power of Emerson's mind . . .as do these hasty and informal writings," N. Y. TIMES. Preface by B. Perry. 370pp. 5⅜ x 8. T447 Paperbound **$1.85**

A SOURCE BOOK IN THEATRICAL HISTORY, A. M. Nagler. (Formerly, "Sources of Theatrical History.") Over 300 selected passages by contemporary observers tell about styles of acting, direction, make-up, scene designing, etc., in the theatre's great periods from ancient Greece to the Théâtre Libre. "Indispensable complement to the study of drama," EDUCATIONAL THEATRE JOURNAL. Prof. Nagler, Yale Univ. School of Drama, also supplies notes, references. 85 illustrations. 611pp. 5⅜ x 8. T515 Paperbound **$2.75**

THE ART OF THE STORY-TELLER, M. L. Shedlock. Regarded as the finest, most helpful book on telling stories to children, by a great story-teller. How to catch, hold, recapture attention; how to choose material; many other aspects. Also includes: a 99-page selection of Miss Shedlock's most successful stories; extensive bibliography of other stories. xxi + 320pp. 5⅜ x 8. T245 Clothbound **$3.50**

THE DEVIL'S DICTIONARY, Ambrose Bierce. Over 1000 short, ironic definitions in alphabetical order, by America's greatest satirist in the classical tradition. "Some of the most gorgeous witticisms in the English language," H. L. Mencken. 144pp. 5⅜ x 8. T487 Paperbound **$1.00**

MUSIC

A DICTIONARY OF HYMNOLOGY, John Julian. More than 30,000 entries on individual hymns, their authorship, textual variations, location of texts, dates and circumstances of composition, denominational and ritual usages, the biographies of more than 9,000 hymn writers, essays on important topics such as children's hymns and Christmas carols, and hundreds of thousands of other important facts about hymns which are virtually impossible to find anywhere else. Convenient alphabetical listing, and a 200-page double-columned index of first lines enable you to track down virtually any hymn ever written. Total of 1786pp. 6¼ x 9¼. 2 volumes. T133. The Set, Clothbound **$15.00**

STRUCTURAL HEARING, TONAL COHERENCE IN MUSIC, Felix Salzer. Extends the well-known Schenker approach to include modern music, music of the middle ages, and Renaissance music. Explores the phenomenon of tonal organization by discussing more than 500 compositions, and offers unusual new insights into the theory of composition and musical relationships. "The foundation on which all teaching in music theory has been based at this college," Leopold Mannes, President, The Mannes College of Music. Total of 658pp. 6½ x 9¼. 2 volumes. S418 The set, Clothbound **$8.00**

A GENERAL HISTORY OF MUSIC, Charles Burney. The complete history of music from the Greeks up to 1789 by the 18th century musical historian who personally knew the great Baroque composers. Covers sacred and secular, vocal and instrumental, operatic and symphonic music; treats theory, notation, forms, instruments; discusses composers, performers, important works. Invaluable as a source of information on the period for students, historians, musicians. "Surprisingly few of Burney's statements have been invalidated by modern research . . . still of great value," NEW YORK TIMES. Edited and corrected by Frank Mercer. 35 figures. 1915pp. 5½ x 8½. 2 volumes. T36 The set, Clothbound **$12.50**

JOHANN SEBASTIAN BACH, Phillip Spitta. Recognized as one of the greatest accomplishments of musical scholarship and far and away the definitive coverage of Bach's works. Hundreds of individual pieces are analyzed. Major works, such as the B Minor Mass and the St. Matthew Passion are examined in minute detail. Spitta also deals with the works of Buxtehude, Pachelbel, and others of the period. Can be read with profit even by those without a knowledge of the technicalities of musical composition. "Unchallenged as the last word on one of the supreme geniuses of music," John Barkham, SATURDAY REVIEW SYNDICATE. Total of 1819pp. 5⅜ x 8. 2 volumes. T252 The set, Clothbound **$10.00**

HISTORY

THE IDEA OF PROGRESS, J. B. Bury. Prof. Bury traces the evolution of a central concept of Western civilization in Greek, Roman, Medieval, and Renaissance thought to its flowering in the 17th and 18th centuries. Introduction by Charles Beard. xl + 357pp. 5⅜ x 8.
T39 Clothbound **$3.95**
T40 Paperbound **$1.95**

THE ANCIENT GREEK HISTORIANS, J. B. Bury. Greek historians such as Herodotus, Thucydides, Xenophon; Roman historians such as Tacitus, Caesar, Livy; scores of others fully analyzed in terms of sources, concepts, influences, etc., by a great scholar and historian. 291pp. 5⅜ x 8. T397 Paperbound **$1.50**

HISTORY OF THE LATER ROMAN EMPIRE, J. B. Bury. The standard work on the Byzantine Empire from 395 A.D. to the death of Justinian in 565 A.D., by the leading Byzantine scholar of our time. Covers political, social, cultural, theological, military history. Quotes contemporary documents extensively. "Most unlikely that it will ever be superseded," Glanville Downey, Dumbarton Oaks Research Library. Genealogical tables. 5 maps. Bibliography. 2 vols. Total of 965pp. 5⅜ x 8. T398, T399 Paperbound, the set **$4.00**

GARDNER'S PHOTOGRAPHIC SKETCH BOOK OF THE CIVIL WAR, Alexander Gardner. One of the rarest and most valuable Civil War photographic collections exactly reproduced for the first time since 1866. Scenes of Manassas, Bull Run, Harper's Ferry, Appomattox, Mechanicsville, Fredericksburg, Gettysburg, etc.; battle ruins, prisons, arsenals, a slave pen, fortifications; Lincoln on the field, officers, men, corpses. By one of the most famous pioneers in documentary photography. Original copies of the "Sketch Book" sold for $425 in 1952. Introduction by E. Bleiler. 100 full-page 7 x 10 photographs (original size). 244pp. 10¾ x 8½
T476 Clothbound **$6.00**

THE WORLD'S GREAT SPEECHES, L. Copeland and L. Lamm, eds. 255 speeches from Pericles to Churchill, Dylan Thomas. Invaluable as a guide to speakers; fascinating as history past and present; a source of much difficult-to-find material. Includes an extensive section of informal and humorous speeches. 3 indices: Topic, Author, Nation. xx + 745pp. 5⅜ x 8.
T468 Paperbound **$2.49**

FOUNDERS OF THE MIDDLE AGES, E. K. Rand. The best non-technical discussion of the transformation of Latin paganism into medieval civilization. Tertullian, Gregory, Jerome, Boethius, Augustine, the Neoplatonists, other crucial figures, philosophies examined. Excellent for the intelligent non-specialist. "Extraordinarily accurate," Richard McKeon, THE NATION. ix + 365pp. 5⅜ x 8. T369 Paperbound **$1.85**

THE POLITICAL THOUGHT OF PLATO AND ARISTOTLE, Ernest Barker. The standard, comprehensive exposition of Greek political thought. Covers every aspect of the "Republic" and the "Politics" as well as minor writings, other philosophers, theorists of the period, and the later history of Greek political thought. Unabridged edition. 584pp. 5⅜ x 8.
T521 Paperbound **$1.85**

PHILOSOPHY

THE GIFT OF LANGUAGE, M. Schlauch. (Formerly, "The Gift of Tongues.") A sound, middle-level treatment of linguistic families, word histories, grammatical processes, semantics, language taboos, word-coining of Joyce, Cummings, Stein, etc. 232 bibliographical notes. 350pp. 5⅜ x 8.
T243 Paperbound **$1.85**

THE PHILOSOPHY OF HEGEL, W. T. Stace. The first work in English to give a complete and connected view of Hegel's entire system. Especially valuable to those who do not have time to study the highly complicated original texts, yet want an accurate presentation by a most reputable scholar of one of the most influential 19th century thinkers. Includes a 14 x 20 fold-out chart of Hegelian system. 536pp. 5⅜ x 8.
T254 Paperbound **$2.00**

ARISTOTLE, A. E. Taylor. A lucid, non-technical account of Aristotle written by a foremost Platonist. Covers life and works; thought on matter, form, causes, logic, God, physics, metaphysics, etc. Bibliography. New index compiled for this edition. 128pp. 5⅜ x 8.
T280 Paperbound **$1.00**

GUIDE TO PHILOSOPHY, C. E. M. Joad. This basic work describes the major philosophic problems and evaluates the answers propounded by great philosophers from the Greeks to Whitehead, Russell. "The finest introduction," BOSTON TRANSCRIPT. Bibliography, 592pp. 5⅜ x 8.
T297 Paperbound **$2.00**

LANGUAGE AND MYTH, E. Cassirer. Cassirer's brilliant demonstration that beneath both language and myth lies an unconscious "grammar" of experience whose categories and canons are not those of logical thought. Introduction and translation by Susanne Langer. Index. x + 103pp. 5⅜ x 8.
T51 Paperbound **$1.25**

SUBSTANCE AND FUNCTION, EINSTEIN'S THEORY OF RELATIVITY, E. Cassirer. This double volume contains the German philosopher's profound philosophical formulation of the differences between traditional logic and the new logic of science. Number, space, energy, relativity, many other topics are treated in detail. Authorized translation by W. C. and M. C. Swabey. xii + 465pp. 5⅜ x 8.
T50 Paperbound **$2.00**

THE PHILOSOPHICAL WORKS OF DESCARTES. The definitive English edition, in two volumes, of all major philosophical works and letters of René Descartes, father of modern philosophy of knowledge and science. Translated by E. S. Haldane and G. Ross. Introductory notes. Total of 842pp. 5⅜ x 8.
T71 Vol. 1, Paperbound **$2.00**
T72 Vol. 2, Paperbound **$2.00**

ESSAYS IN EXPERIMENTAL LOGIC, J. Dewey. Based upon Dewey's theory that knowledge implies a judgment which in turn implies an inquiry, these papers consider such topics as the thought of Bertrand Russell, pragmatism, the logic of values, antecedents of thought, data and meanings. 452pp. 5⅜ x 8.
T73 Paperbound **$1.95**

THE PHILOSOPHY OF HISTORY, G. W. F. Hegel. This classic of Western thought is Hegel's detailed formulation of the thesis that history is not chance but a rational process, the realization of the Spirit of Freedom. Translated and introduced by J. Sibree. Introduction by C. Hegel. Special introduction for this edition by Prof. Carl Friedrich, Harvard University. xxxix + 447pp. 5⅜ x 8.
T112 Paperbound **$1.85**

THE WILL TO BELIEVE and **HUMAN IMMORTALITY, W. James.** Two of James's most profound investigations of human belief in God and immortality, bound as one volume. Both are powerful expressions of James's views on chance vs. determinism, pluralism vs. monism, will and intellect, arguments for survival after death, etc. Two prefaces. 429pp. 5⅜ x 8.
T294 Clothbound **$3.75**
T291 Paperbound **$1.65**

INTRODUCTION TO SYMBOLIC LOGIC, S. Langer. A lucid, general introduction to modern logic, covering forms, classes, the use of symbols, the calculus of propositions, the Boole-Schroeder and the Russell-Whitehead systems, etc. "One of the clearest and simplest introductions," MATHEMATICS GAZETTE. Second, enlarged, revised edition. 368pp. 5⅜ x 8.
S164 Paperbound **$1.75**

MIND AND THE WORLD-ORDER, C. I. Lewis. Building upon the work of Peirce, James, and Dewey, Professor Lewis outlines a theory of knowledge in terms of "conceptual pragmatism," and demonstrates why the traditional understanding of the a priori must be abandoned. Appendices. xiv + 446pp. 5⅜ x 8.
T359 Paperbound **$1.95**

THE GUIDE FOR THE PERPLEXED, M. Maimonides One of the great philosophical works of all time, Maimonides' formulation of the meeting-ground between Old Testament and Aristotelian thought is essential to anyone interested in Jewish, Christian, and Moslem thought in the Middle Ages. 2nd revised edition of the Friedlander translation. Extensive introduction. lix + 414pp. 5⅜ x 8.
T351 Paperbound **$1.85**

ORIENTALIA AND RELIGION

THE MYSTERIES OF MITHRA, F. Cumont. The great Belgian scholar's definitive study of the Persian mystery religion that almost vanquished Christianity in the ideological struggle for the Roman Empire. A masterpiece of scholarly detection that reconstructs secret doctrines, organization, rites. Mithraic art is discussed and analyzed. 70 illus. 239pp. 5⅜ x 8.
T323 Paperbound **$1.85**

CHRISTIAN AND ORIENTAL PHILOSOPHY OF ART. A. K. Coomaraswamy. The late art historian and orientalist discusses artistic symbolism, the role of traditional culture in enriching art, medieval art, folklore, philosophy of art, other similar topics. Bibliography. 148pp. 5⅜ x 8.
T378 Paperbound **$1.25**

TRANSFORMATION OF NATURE IN ART, A. K. Coomaraswamy. A basic work on Asiatic religious art. Includes discussions of religious art in Asia and Medieval Europe (exemplified by Meister Eckhart), the origin and use of images in Indian art, Indian Medieval aesthetic manuals, and other fascinating, little known topics. Glossaries of Sanskrit and Chinese terms. Bibliography. 41pp. of notes. 245pp. 5⅜ x 8. T368 Paperbound **$1.75**

ORIENTAL RELIGIONS IN ROMAN PAGANISM, F. Cumont. This well-known study treats the ecstatic cults of Syria and Phrygia (Cybele, Attis, Adonis, their orgies and mutilatory rites); the mysteries of Egypt (Serapis, Isis, Osiris); Persian dualism; Mithraic cults; Hermes Trismegistus, Ishtar, Astarte, etc. and their influence on the religious thought of the Roman Empire. Introduction. 55pp. of notes; extensive bibliography. xxiv + 298pp. 5⅜ x 8.
T321 Paperbound **$1.75**

ANTHROPOLOGY, SOCIOLOGY, AND PSYCHOLOGY

PRIMITIVE MAN AS PHILOSOPHER, P. Radin. A standard anthropological work based on Radin's investigations of the Winnebago, Maori, Batak, Zuni, other primitive tribes. Describes primitive thought on the purpose of life, marital relations, death, personality, gods, etc. Extensive selections of original primitive documents. Bibliography. xviii + 420pp. 5⅜ x 8.
T392 Paperbound **$2.00**

PRIMITIVE RELIGION, P. Radin. Radin's thoroughgoing treatment of supernatural beliefs, shamanism, initiations, religious expression, etc. in primitive societies. Arunta, Ashanti, Aztec, Bushman, Crow, Fijian, many other tribes examined. "Excellent," NATURE. New preface by the author. Bibliographic notes. x + 322pp. 5⅜ x 8. T393 Paperbound **$1.85**

SEX IN PSYCHO-ANALYSIS, S. Ferenczi. (Formerly, "Contributions to Psycho-analysis.") 14 selected papers on impotence, transference, analysis and children, dreams, obscene words, homosexuality, paranoia, etc. by an associate of Freud. Also included: THE DEVELOPMENT OF PSYCHO-ANALYSIS, by Ferenczi and Otto Rank. Two books bound as one. Total of 406pp. 5⅜ x 8. T324 Paperbound **$1.85**

THE PRINCIPLES OF PSYCHOLOGY, William James. The complete text of the famous "long course," one of the great books of Western thought. An almost incredible amount of information about psychological processes, the stream of consciousness, habit, time perception, memory, emotions, reason, consciousness of self, abnormal phenomena, and similar topics. Based on James's own discoveries integrated with the work of Descartes, Locke, Hume, Royce, Wundt, Berkeley, Lotse, Herbart, scores of others. "A classic of interpretation," PSYCHIATRIC QUARTERLY. 94 illus. 1408pp. 2 volumes. 5⅜ x 8.
T381 Vol. 1, Paperbound **$2.50**
T382 Vol. 2, Paperbound **$2.50**

THE POLISH PEASANT IN EUROPE AND AMERICA, W. I. Thomas, F. Znaniecki. Monumental sociological study of peasant primary groups (family and community) and the disruptions produced by a new industrial system and emigration to America, by two of the foremost sociologists of recent times. One of the most important works in sociological thought. Includes hundreds of pages of primary documentation; point by point analysis of causes of social decay, breakdown of morality, crime, drunkenness, prostitution, etc. 2nd revised edition. 2 volumes. Total of 2250pp. 6 x 9. T478 2 volume set, Clothbound **$12.50**

FOLKWAYS, W. G. Sumner. The great Yale sociologist's detailed exposition of thousands of social, sexual, and religious customs in hundreds of cultures from ancient Greece to Modern Western societies. Preface by A. G. Keller. Introduction by William Lyon Phelps. 705pp. 5⅜ x 8. S508 Paperbound **$2.49**

BEYOND PSYCHOLOGY, Otto Rank. The author, an early associate of Freud, uses psychoanalytic techniques of myth-analysis to explore ultimates of human existence. Treats love, immortality, the soul, sexual identity, kingship, sources of state power, many other topics which illuminate the irrational basis of human existence. 291pp. 5⅜ x 8. T485 Paperbound **$1.75**

ILLUSIONS AND DELUSIONS OF THE SUPERNATURAL AND THE OCCULT, D. H. Rawcliffe. A rational, scientific examination of crystal gazing, automatic writing, table turning, stigmata, the Indian rope trick, dowsing, telepathy, clairvoyance, ghosts, ESP, PK, thousands of other supposedly occult phenomena. Originally titled "The Psychology of the Occult." 14 illustrations. 551pp. 5⅜ x 8. T503 Paperbound **$2.00**

YOGA: A SCIENTIFIC EVALUATION, Kovoor T. Behanan. A scientific study of the physiological and psychological effects of Yoga discipline, written under the auspices of the Yale University Institute of Human Relations. Foreword by W. A. Miles, Yale Univ. 17 photographs. 290pp. 5⅜ x 8. T505 Paperbound **$1.65**

HOAXES, C. D. MacDougall. Delightful, entertaining, yet scholarly exposition of how hoaxes start, why they succeed, documented with stories of hundreds of the most famous hoaxes. "A stupendous collection . . . and shrewd analysis, "NEW YORKER. New, revised edition. 54 photographs. 320pp. 5⅜ x 8. T465 Paperbound **$1.75**

CREATIVE POWER: THE EDUCATION OF YOUTH IN THE CREATIVE ARTS, Hughes Mearns. Named by the National Education Association as one of the 20 foremost books on education in recent times. Tells how to help children express themselves in drama, poetry, music, art, develop latent creative power. Should be read by every parent, teacher. New, enlarged, revised edition. Introduction. 272pp. 5⅜ x 8. T490 Paperbound **$1.50**

LANGUAGES

NEW RUSSIAN-ENGLISH, ENGLISH-RUSSIAN DICTIONARY, M. A. O'Brien. Over 70,000 entries in new orthography! Idiomatic usages, colloquialisms. One of the few dictionaries that indicate accent changes in conjugation and declension. "One of the best," Prof. E. J. Simmons, Cornell. First names, geographical terms, bibliography, many other features. 738pp. 4½ x 6¼. T208 Paperbound **$2.00**

MONEY CONVERTER AND TIPPING GUIDE FOR EUROPEAN TRAVEL, C. Vomacka. Invaluable, handy source of currency regulations, conversion tables, tipping rules, postal rates, much other travel information for every European country plus Israel, Egypt and Turkey. 128pp. 3½ x 5¼. T260 Paperbound **60¢**

MONEY CONVERTER AND TIPPING GUIDE FOR TRAVEL IN THE AMERICAS (including the United States and Canada), **C. Vomacka.** The information you need for informed and confident travel in the Americas: money conversion tables, tipping guide, postal, telephone rates, etc. 128pp. 3½ x 5¼. T261 Paperbound **65¢**

DUTCH-ENGLISH, ENGLISH-DUTCH DICTIONARY, F. G. Renier. The most convenient, practical Dutch-English dictionary on the market. New orthography. More than 60,000 entries: idioms, compounds, technical terms, etc. Gender of nouns indicated. xviii + 571pp. 5½ x 6¼. T224 Clothbound **$2.50**

LEARN DUTCH!, F. G. Renier. The most satisfactory and easily-used grammar of modern Dutch. Used and recommended by the Fulbright Committee in the Netherlands. Over 1200 simple exercises lead to mastery of spoken and written Dutch. Dutch-English, English-Dutch vocabularies. 181pp. 4¼ x 7¼. T441 Clothbound **$1.75**

PHRASE AND SENTENCE DICTIONARY OF SPOKEN RUSSIAN, English-Russian, Russian-English. Based on phrases and complete sentences, rather than isolated words; recognized as one of the best methods of learning the idiomatic speech of a country. Over 11,500 entries, indexed by single words, with more than 32,000 English and Russian sentences and phrases, in immediately usable form. Probably the largest list ever published. Shows accent changes in conjugation and declension; irregular forms listed in both alphabetical place and under main form of word. 15,000 word introduction covering Russian sounds, writing, grammar, syntax. 15-page appendix of geographical names, money, important signs, given names, foods, special Soviet terms, etc. Travellers, businessmen, students, government employees have found this their best source for Russian expressions. Originally published as U.S. Government Technical Manual TM 30-944. iv + 573pp. 5⅝ x 8⅜. T496 Paperbound **$2.75**

PHRASE AND SENTENCE DICTIONARY OF SPOKEN SPANISH, Spanish-English, English-Spanish. Compiled from spoken Spanish, emphasizing idiom and colloquial usage in both Castilian and Latin-American. More than 16,000 entries containing over 25,000 idioms—the largest list of idiomatic constructions ever published. Complete sentences given, indexed under single words —language in immediately usable form, for travellers, businessmen, students, etc. 25-page introduction provides rapid survey of sounds, grammar, syntax, with full consideration of irregular verbs. Especially apt in modern treatment of phrases and structure. 17-page glossary gives translations of geographical names, money values, numbers, national holidays, important street signs, useful expressions of high frequency, plus unique 7-page glossary of Spanish and Spanish-American foods and dishes. Originally published as U.S. Government Technical Manual TM 30-900. iv + 513pp. 5⅝ x 8⅜. T495 Paperbound **$1.75**

SAY IT language phrase books

"SAY IT" in the foreign language of your choice! We have sold over ½ million copies of these popular, useful language books. They will not make you an expert linguist overnight, but they do cover most practical matters of everyday life abroad.

Over 1000 useful phrases, expressions, with additional variants, substitutions.

Modern! Useful! Hundreds of phrases not available in other texts: "Nylon," "air-conditioned," etc.

The ONLY inexpensive phrase book **completely indexed.** Everything is available at a flip of your finger, ready for use.

Prepared by native linguists, travel experts.

Based on years of travel experience abroad.

This handy phrase book may be used by itself, or it may supplement any other text or course; it provides a living element. Used by many colleges and institutions: Hunter College; Barnard College; Army Ordnance School, Aberdeen; and many others.

Available, 1 book per language:

Danish (T818) 75¢
Dutch T(817) 75¢
English (for German-speaking people) (T801) 60¢
English (for Italian-speaking people) (T816) 60¢
English (for Spanish-speaking people) (T802) 60¢
Esperanto (T820) 75¢
French (T803) 60¢
German (T804) 60¢
Modern Greek (T813) 75¢
Hebrew (T805) 60¢
Italian (T806) 60¢
Japanese (T807) 60¢
Norwegian (T814) 75¢
Russian (T810) 75¢
Spanish (T811) 60¢
Turkish (T821) 75¢
Yiddish (T815) 75¢
Swedish (T812) 75¢
Polish (T808) 75¢
Portuguese (T809) 75¢

LISTEN & LEARN language record sets

LISTEN & LEARN is the only language record course designed especially to meet your travel needs, or help you learn essential foreign language quickly by yourself, or in conjunction with any school course, by means of the automatic association method. Each set contains three 33⅓ rpm long- playing records — 1½ hours of recorded speech by eminent native speakers who are professors at Columbia, N.Y.U., Queens College and other leading universities. The sets are priced far below other sets of similar quality, yet they contain many special features not found in other record sets:

* Over 800 selected phrases and sentences, a basic vocabulary of over 3200 words.
* Both English and foreign language recorded; with a pause for your repetition.
* Designed for persons with limited time; no time wasted on material you cannot use immediately.
* Living, modern expressions that answer modern needs: drugstore items, "air-conditioned," etc.
* 128-196 page manuals contain everything on the records, plus simple pronunciation guides.
* Manual is fully indexed; find the phrase you want instantly.
* High fidelity recording—equal to any records costing up to $6 each.

The phrases on these records cover 41 different categories useful to the traveller or student interested in learning the living, spoken language: greetings, introductions, making yourself understood, passing customs, planes, trains, boats, buses, taxis, nightclubs, restaurants, menu items, sports, concerts, cameras, automobile travel, repairs, drugstores, doctors, dentists, medicines, barber shops, beauty parlors, laundries, many, many more.

"Excellent . . . among the very best on the market," Prof. Mario Pei, Dept. of Romance Languages, Columbia University. "Inexpensive and well-done . . . an ideal present," CHICAGO SUNDAY TRIBUNE. "More genuinely helpful than anything of its kind which I have previously encountered," Sidney Clark, well-known author of "ALL THE BEST" travel books. Each set contains 3 33⅓ rpm pure vinyl records, 128- 196 page with full record text, and album. One language per set. LISTEN & LEARN record sets are now available in—

FRENCH the set $4.95
ITALIAN the set $4.95
RUSSIAN the set $5.95
GERMAN the set $4.95
SPANISH the set $4.95
JAPANESE * the set $5.95

* Available Sept. 1, 1959

UNCONDITIONAL GUARANTEE: Dover Publications stands behind every Listen and Learn record set. If you are dissatisfied with these sets for any reason whatever, return them within 10 days and your money will be refunded in full.

ART HISTORY

STICKS AND STONES, Lewis Mumford. An examination of forces influencing American architecture: the medieval tradition in early New England, the classical influence in Jefferson's time, the Brown Decades, the imperial facade, the machine age, etc. "A truly remarkable book," SAT. REV. OF LITERATURE. 2nd revised edition. 21 illus. xvii + 228pp. 5⅜ x 8.
T202 Paperbound **$1.60**

THE AUTOBIOGRAPHY OF AN IDEA, Louis Sullivan. The architect whom Frank Lloyd Wright called "the master," records the development of the theories that revolutionized America's skyline. 34 full-page plates of Sullivan's finest work. New introduction by R. M. Line. xiv + 335pp. 5⅜ x 8. T281 Paperbound **$1.85**

THE MATERIALS AND TECHNIQUES OF MEDIEVAL PAINTING, D. V. Thompson. An invaluable study of carriers and grounds, binding media, pigments, metals used in painting, al fresco and al secco techniques, burnishing, etc. used by the medieval masters. Preface by Bernard Berenson. 239pp. 5⅜ x 8. T327 Paperbound **$1.85**

PRINCIPLES OF ART HISTORY, H. Wölfflin. This remarkably instructive work demonstrates the tremendous change in artistic conception from the 14th to the 18th centuries, by analyzing 164 works by Botticelli, Dürer, Hobbema, Holbein, Hals, Titian, Rembrandt, Vermeer, etc., and pointing out exactly what is meant by "baroque," "classic," "primitive," "picturesque," and other basic terms of art history and criticism. "A remarkable lesson in the art of seeing," SAT. REV. OF LITERATURE. Translated from the 7th German edition. 150 illus. 254pp. 6⅛ x 9¼. T276 Paperbound **$2.00**

FOUNDATIONS OF MODERN ART, A. Ozenfant. Stimulating discussion of human creativity from paleolithic cave painting to modern painting, architecture, decorative arts. Fully illustrated with works of Gris, Lipchitz, Léger, Picasso, primitive, modern artifacts, architecture, industrial art, much more. 226 illustrations. 368pp. 6⅛ x 9¼. T215 Paperbound **$1.95**

HANDICRAFTS, APPLIED ART, ART SOURCES, ETC.

WILD FOWL DECOYS, J. Barber. The standard work on this fascinating branch of folk art, ranging from Indian mud and grass devices to realistic wooden decoys. Discusses styles, types, periods; gives full information on how to make decoys. 140 illustrations (including 14 new plates) show decoys and provide full sets of plans for handicrafters, artists, hunters, and students of folk art. 281pp. 7⅞ x 10¾. Deluxe edition. T11 Clothbound **$8.50**

METALWORK AND ENAMELLING, H. Maryon. Probably the best book ever written on the subject. Tells everything necessary for the home manufacture of jewelry, rings, ear pendants, bowls, etc. Covers materials, tools, soldering, filigree, setting stones, raising patterns, repoussé work, damascening, niello, cloisonné, polishing, assaying, casting, and dozens of other techniques. The best substitute for apprenticeship to a master metalworker. 363 photos and figures. 374pp. 5½ x 8½. T183 Clothbound **$7.50**

SHAKER FURNITURE, E. D. and F. Andrews. The most illuminating study of Shaker furniture ever written. Covers chronology, craftsmanship, houses, shops, etc. Includes over 200 photographs of chairs, tables, clocks, beds, benches, etc. "Mr. & Mrs. Andrews know all there is to know about Shaker furniture," Mark Van Doren, NATION. 48 full-page plates. 192pp. Deluxe cloth binding. 7⅞ x 10¾. T7 Clothbound **$6.00**

PRIMITIVE ART, Franz Boas. A great American anthropologist covers theory, technical virtuosity, styles, symbolism, patterns, etc. of primitive art. The more than 900 illustrations will interest artists, designers, craftworkers. Over 900 illustrations. 376pp. 5⅜ x 8.
T25 Paperbound **$1.95**

ON THE LAWS OF JAPANESE PAINTING, H. Bowie. The best possible substitute for lessons from an oriental master. Treats both spirit and technique; exercises for control of the brush; inks, brushes, colors; use of dots, lines to express whole moods, etc. 220 illus. 132pp. 6⅛ x 9¼. T30 Paperbound **$1.95**

HANDBOOK OF ORNAMENT, F. S. Meyer. One of the largest collections of copyright-free traditional art: over 3300 line cuts of Greek, Roman, Medieval, Renaissance, Baroque, 18th and 19th century art motifs (tracery, geometric elements, flower and animal motifs, etc.) and decorated objects (chairs, thrones, weapons, vases, jewelry, armor, etc.). Full text. 3300 illustrations. 562pp. 5⅜ x 8. T302 Paperbound **$2.00**

THREE CLASSICS OF ITALIAN CALLIGRAPHY. Oscar Ogg, ed. Exact reproductions of three famous Renaissance calligraphic works: Arrighi's OPERINA and IL MODO, Tagliente's LO PRESENTE LIBRO, and Palatino's LIBRO NUOVO. More than 200 complete alphabets, thousands of lettered specimens, in Papal Chancery and other beautiful, ornate handwriting. Introduction. 245 plates. 282pp. 6⅛ x 9¼. T212 Paperbound **$1.95**

THE HISTORY AND TECHNIQUES OF LETTERING, A. Nesbitt. A thorough history of lettering from the ancient Egyptians to the present, and a 65-page course in lettering for artists. Every major development in lettering history is illustrated by a complete alphabet. Fully analyzes such masters as Caslon, Koch, Garamont, Jenson, and many more. 89 alphabets, 165 other specimens. 317pp. 5⅜ x 8. T427 Paperbound **$2.00**

LETTERING AND ALPHABETS, J. A. Cavanagh. An unabridged reissue of "Lettering," containing the full discussion, analysis, illustration of 89 basic hand lettering tyles based on Caslon, Bodoni, Gothic, many other types. Hundreds of technical hints on construction, strokes, pens, brushes, etc. 89 alphabets, 72 lettered specimens, which may be reproduced permission-free. 121pp. 9¾ x 8. T53 Paperbound **$1.25**

THE HUMAN FIGURE IN MOTION, Eadweard Muybridge. The largest collection in print of Muybridge's famous high-speed action photos. 4789 photographs in more than 500 action-strip-sequences (at shutter speeds up to 1/6000th of a second) illustrate men, women, children—mostly undraped—performing such actions as walking, running, getting up, lying down, carrying objects, throwing, etc. "An unparalleled dictionary of action for all artists," AMERICAN ARTIST. 390 full-page plates, with 4789 photographs. Heavy glossy stock, reinforced binding with headbands. 7⅞ x 10¾. T204 Clothbound **$10.00**

ANIMALS IN MOTION, Eadweard Muybridge. The largest collection of animal action photos in print. 34 different animals (horses, mules, oxen, goats, camels, pigs, cats, lions, gnus, deer, monkeys, eagles—and 22 others) in 132 characteristic actions. All 3919 photographs are taken in series at speeds up to 1/1600th of a second, offering artists, biologists, cartoonists a remarkable opportunity to see exactly how an ostrich's head bobs when running, how a lion puts his foot down, how an elephant's knee bends, how a bird flaps his wings, thousands of other hard-to-catch details. "A really marvelous series of plates," NATURE. 380 full-pages of plates. Heavy glossy stock, reinforced binding with headbands. 7⅞ x10¾. T203 Clothbound **$10.00**

THE BOOK OF SIGNS, R. Koch. 493 symbols—crosses, monograms, astrological, biological symbols, runes, etc.—from ancient manuscripts, cathedrals, coins, catacombs, pottery. May be reproduced permission-free. 493 illustrations by Fritz Kredel. 104pp. 6⅛ x 9¼. T162 Paperbound **$1.00**

A HANDBOOK OF EARLY ADVERTISING ART, C. P. Hornung. The largest collection of copyright-free early advertising art ever compiled. Vol. I: 2,000 illustrations of animals, old automobiles, buildings, allegorical figures, fire engines, Indians, ships, trains, more than 33 other categories! Vol II: Over 4,000 typographical specimens; 600 Roman, Gothic, Barnum, Old English faces; 630 ornamental type faces; hundreds of scrolls, initials, flourishes, etc. "A remarkable collection," PRINTERS' INK.

Vol. I: Pictorial Volume. Over 2000 illustrations. 256pp. 9 x 12. T122 Clothbound **$10.00**

Vol. II: Typographical Volume. Over 4000 speciments. 319pp. 9 x 12. T123 Clothbound **$10.00**

Two volume set, Clothbound, only **$18.50**

DESIGN FOR ARTISTS AND CRAFTSMEN, L. Wolchonok. The most thorough course on the creation of art motifs and designs. Shows you step-by-step, with hundreds of examples and 113 detailed exercises, how to create original designs from geometric patterns, plants, birds, animals, humans, and man-made objects. "A great contribution to the field of design and crafts," N. Y. SOCIETY OF CRAFTSMEN. More than 1300 entirely new illustrations. xv + 207pp. 7⅞ x 10¾. T274 Clothbound **$4.95**

HANDBOOK OF DESIGNS AND DEVICES, C. P. Hornung. A remarkable working collection of 1836 basic designs and variations, all copyright-free. Variations of circle, line, cross, diamond, swastika, star, scroll, shield, many more. Notes on symbolism. "A necessity to every designer who would be original without having to labor heavily," ARTIST and ADVERTISER. 204 plates. 240pp. 5⅜ x 8. T125 Paperbound **$1.90**

THE UNIVERSAL PENMAN, George Bickham. Exact reproduction of beautiful 18th century book of handwriting. 22 complete alphabets in finest English roundhand, other scripts, over 2000 elaborate flourishes, 122 calligraphic illustrations, etc. Material is copyright-free. "An essential part of any art library, and a book of permanent value," AMERICAN ARTIST. 212 plates. 224pp. 9 x 13¾. T20 Clothbound **$10.00**

AN ATLAS OF ANATOMY FOR ARTISTS, F. Schider. This standard work contains 189 full-page plates, more than 647 illustrations of all aspects of the human skeleton, musculature, cutaway portions of the body, each part of the anatomy, hand forms, eyelids, breasts, location of muscles under the flesh, etc. 59 plates illustrate how Michelangelo, da Vinci, Goya, 15 others, drew human anatomy. New 3rd edition enlarged by 52 new illustrations by Cloquet, Barcsay. "The standard reference tool," AMERICAN LIBRARY ASSOCIATION. "Excellent," AMERICAN ARTIST. 189 plates, 647 illustrations. xxvi + 192pp. 7⅞ x 10⅝. T241 Clothbound **$6.00**

AN ATLAS OF ANIMAL ANATOMY FOR ARTISTS, W. Ellenberger, H. Baum, H. Dittrich. The largest, richest animal anatomy for artists in English. Form, musculature, tendons, bone structure, expression, detailed cross sections of head, other features, of the horse, lion, dog, cat, deer, seal, kangaroo, cow, bull, goat, monkey, hare, many other animals. "Highly recommended," DESIGN. Second, revised, enlarged edition with new plates from Cuvier, Stubbs, etc. 288 illustrations. 153pp. 11⅜ x 9. T82 Clothbound **$6.00**

ANIMAL DRAWING: ANATOMY AND ACTION FOR ARTISTS, C. R. Knight. 158 studies, with full accompanying text, of such animals as the gorilla, bear, bison, dromedary, camel, vulture, pelican, iguana, shark, etc., by one of the greatest modern masters of animal drawing. Innumerable tips on how to get life expression into your work. "An excellent reference work,' SAN FRANCISCO CHRONICLE. 158 illustrations. 156pp. 10½ x 8½. T426 Paperbound **$2.00**

CATALOG OF
PUZZLES, GAMES, AND ENTERTAINMENTS

MATHEMATICS, MAGIC AND MYSTERY, Martin Gardner. Astonishing feats of mind reading, mystifying "magic" tricks, are often based on mathematical principles anyone can learn. This book shows you how to perform scores of tricks with cards, dice, coins, knots, numbers, etc., by using simple principles from set theory, theory of numbers, topology, other areas of mathematics, fascinating in themselves. No special knowledge required. 135 illus. 186pp. 5⅜ x 8.
T335 Paperbound **$1.00**

MATHEMATICAL PUZZLES FOR BEGINNERS AND ENTHUSIASTS, G. Mott-Smith. Test your problem-solving techniques and powers of inference on 188 challenging, amusing puzzles based on algebra, dissection of plane figures, permutations, probabilities, etc. Appendix of primes, square roots, etc. 135 illus. 2nd revised edition. 248pp. 5⅜ x 8.
T198 Paperbound **$1.00**

LEARN CHESS FROM THE MASTERS, F. Reinfeld. Play 10 games against Marshall, Bronstein, Najdorf, other masters, and grade yourself on each move. Detailed annotations reveal principles of play, strategy, etc. as you proceed. An excellent way to get a real insight into the game. Formerly titled, "Chess by Yourself." 91 diagrams. vii + 144pp. 5⅜ x 8.
T362 Paperbound **$1.00**

REINFELD ON THE END GAME IN CHESS, F. Reinfeld. 62 end games of Alekhine, Tarrasch, Morphy, other masters, are carefully analyzed with emphasis on transition from middle game to end play. Tempo moves, queen endings, weak squares, other basic principles clearly illustrated. Excellent for understanding why some moves are weak or incorrect, how to avoid errors. Formerly titled, "Practical End-game Play." 62 diagrams. vi + 177pp. 5⅜ x 8.
T417 Paperbound **$1.25**

101 PUZZLES IN THOUGHT AND LOGIC, C. R. Wylie, Jr. Brand new puzzles you need no special knowledge to solve! Each one is a gem of ingenuity that will really challenge your problem-solving technique. Introduction with simplified explanation of scientic puzzle solving. 128pp. 5⅜ x 8.
T167 Paperbound **$1.00**

THE COMPLETE NONSENSE OF EDWARD LEAR. The only complete edition of this master of gentle madness at a popular price. The Dong with the Luminous Nose, The Jumblies, The Owl and the Pussycat, hundreds of other bits of wonderful nonsense. 214 limericks, 3 sets of Nonsense Botany, 5 Nonsense Alphabets, 546 fantastic drawings, much more. 320pp. 5⅜ x 8.
T167 Paperbound **$1.00**

28 SCIENCE FICTION STORIES OF H. G. WELLS. Two complete novels, "Men Like Gods" and "Star Begotten," plus 26 short stories by the master science-fiction writer of all time. Stories of space, time, future adventure that are among the all-time classics of science fiction. 928pp. 5⅜ x 8.
T265 Clothbound **$3.95**

SEVEN SCIENCE FICTION NOVELS, H. G. Wells. Unabridged texts of "The Time Machine," "The Island of Dr. Moreau," "First Men in the Moon," "The Invisible Man," "The War of the Worlds," "The Food of the Gods," "In the Days of the Comet." "One will have to go far to match this for entertainment, excitement, and sheer pleasure," N. Y. TIMES. 1015pp. 5⅜ x 8.
T264 Clothbound **$3.95**

MATHEMAGIC, MAGIC PUZZLES, AND GAMES WITH NUMBERS, R. V. Heath. More than 60 new puzzles and stunts based on number properties: multiplying large numbers mentally, finding the date of any day in the year, etc. Edited by J. S. Meyer. 76 illus. 129pp. 5⅜ x 8.
T110 Paperbound **$1.00**

FIVE ADVENTURE NOVELS OF H. RIDER HAGGARD. The master story-teller's five best tales of mystery and adventure set against authentic African backgrounds: "She," "King Solomon's Mines," "Allan Quatermain," "Allan's Wife," "Maiwa's Revenge." 821pp. 5⅜ x 8.
T108 Clothbound **$3.95**

WIN AT CHECKERS, M. Hopper. (Formerly "Checkers.") The former World's Unrestricted Checker Champion gives you valuable lessons in openings, traps, end games, ways to draw when you are behind, etc. More than 100 questions and answers anticipate your problems. Appendix. 75 problems diagrammed, solved. 79 figures. xi + 107pp. 5⅜ x 8.
T363 Paperbound **$1.00**

CRYPTOGRAPHY, L. D. Smith. Excellent introductory work on ciphers and their solution, history of secret writing, techniques, etc. Appendices on Japanese methods, the Baconian cipher, frequency tables. Bibliography. Over 150 problems, solutions. 160pp. 5⅜ x 8.
T247 Paperbound **$1.00**

CRYPTANALYSIS, H. F. Gaines. (Formerly, "Elementary Cryptanalysis.") The best book available on cryptograms and how to solve them. Contains all major techniques: substitution, transposition, mixed alphabets, multafid, Kasiski and Vignere methods, etc. Word frequency appendix. 167 problems, solutions. 173 figures. 236pp. 5⅜ x 8.
T97 Paperbound **$1.95**

FLATLAND, E. A. Abbot. The science-fiction classic of life in a 2-dimensional world that is considered a first-rate introduction to relativity and hyperspace, as well as a scathing satire on society, politics and religion. 7th edition. 16 illus. 128pp. 5⅜ x 8.
T1 Paperbound **$1.00**